TRANSLATIONS OF MATHEMATICAL MONOGRAPHS

VOLUME 68

Introduction to Analytic Number Theory

A. G. POSTNIKOV

American Mathematical Society · Providence · Rhode Island

АЛЕКСЕЙ ГЕОРГИЕВИЧ ПОСТНИКОВ

ВВЕДЕНИЕ В АНАЛИТИЧЕСКУЮ
ТЕОРИЮ ЧИСЕЛ

«НАУКА», МОСКВА, 1971

Translated from the Russian by G. A. Kandall
Translation edited by Ben Silver

1980 *Mathematics Subject Classification* (1985 *Revision*). Primary 11-02, 11M45, 11N60, 11Q05.

ABSTRACT. This book is intended as a mid-level view of analytic number theory, somewhere between a textbook and the latest monographs. The author has attempted to give as broad a picture as possible of the problems of analytic number theory, while avoiding specialization and those topics already sufficiently well treated in the literature. This is the meaning of the title, *Introduction to analytic number theory.* The deep results in analytic number theory involve, of course, the use of well-developed machinery. However, it is advisable for a young scholar to have, in addition to a mastery of powerful tools, a supply of problems to which he can apply these strong techniques. In this the author has tried to be of assistance to his young colleagues. This book is directed to scholars, teachers, and graduate students interested in number theory and its connections with other branches of science.

Library of Congress Cataloging-in-Publication Data

Postnikov, A. G. (Alekseĭ Georgievich)
[Vvedenie v analiticheskuiu teoriiu chisel. English]
Introduction to analytic number theory / A.G. Postnikov ; translated from the Russian by G.A. Kandall ; translation edited by Ben Silver].
p. cm. -- (Translations of mathematical monographs ; v. 68)
Translation of: Vvedenie v analiticheskuiu teoriiu chisel.
ISBN 0-8218-4521-7
1. Numbers, Theory of. I. Title. II. Series.
QA241.P6713 1987
512'.73--dc19 87-33428
CIP

Contents

This book is dedicated
to the memory of
Vyacheslav Vasil'evich Stepanov

Preface

The literature on analytic number theory is very rich and naturally reflects the historical development of this science. As an introduction to analytic number theory one can look to the supplementary material (problems and solutions) in Vinogradov's textbook, *Elements of number theory* [148]. The monograph *Number theory* [13] by Borevich and Shafarevich discusses the foundations of the method of Dirichlet generating series (applied to algebraic number theory). The book *Elementary methods in analytic number theory* [60] by Gel′fond and Linnik can be regarded as an elementary reader on the subject. One can find books in Russian dealing with Diophantine approximation (Cassels [17]), the approximation of algebraic and transcendental numbers (Gel′fond [59]), the method of trigonometric sums (Vinogradov [149]), the distribution of primes and the theory of Dirichlet L-functions (Ingham [74], Chudakov [20], Titchmarsh [139], and Prachar [118]), probabilistic number theory (Kubilius [87]), and so on. Nevertheless, there are still some fundamental questions in analytic number theory that have not been adequately discussed in the literature in any systematic way—these are questions directly or indirectly connected with the concept of a numerical semigroup, in other words, with general additive number theory. It is to this aspect of number theory that the present book is devoted.

It is assumed that the reader is familiar with the fundamentals of analysis, number theory, and probability theory. The necessary analysis is covered by Fikhtengol′ts [48], [49], [50], the number theory by Vinogradov [148] or Bukhshtab [16], and the probability theory by the first eight chapters of Gnedenko [61].

The author thanks G. A. Freĭman for his simple proof of the Hardy-Ramanujan asymptotic formula. He is also grateful to K. Yu. Bulota for taking on the difficult task of editing this book; his constructive criticism was most helpful. Finally, the author expresses his sincere gratitude to all who helped in the preparation of this volume.

Standard Notation

To avoid being repetitious let us agree on the meaning of the following notation (unless stated otherwise):

γ	Euler's constant,
$\chi_4(n)$	the nonprincipal character modulo 4,
$\varphi(n)$	Euler's function,
$\tau(n)$	the number of natural divisors of n,
$\sigma(n)$	the sum of the natural divisors of n,
$\nu(n)$	the number of distinct prime divisors of n,
$r(n)$	the number of representations of n as a sum of two squares,
$\alpha_p(n)$	the exponent of the prime p in the canonical factorization of n,
$\mu(n)$	the Möbius function,
$\Lambda(n)$	von Mangoldt's function, $\Lambda(n) = \begin{cases} \ln p, & n = p^\alpha, \\ 0, & n \neq p^\alpha, \end{cases}$
(x)	the distance from the real number x to the nearest integer,
[x]	the integral part of the real number x,
$\{x\}$	the fractional part of x ($\{x\} = x - [x]$),
$\Gamma(s)$	the gamma-function,
$\varsigma(s)$	the zeta-function,
$\pi(x)$	the number of primes not exceeding x.

Introduction

We will give here an overview of the content of this book and reflect on these thoughts in the course of our exposition.

Modern life and, in particular, our mathematical life are inconceivable without a continual "separation" of machinery from problems. Consequently, before entering into a discussion of the problems of analytic number theory it is convenient to present a chapter of an auxiliary nature.

The first chapter is devoted to an exposition of some facts from analysis. Here we will be concerned mainly with Tauberian theorems.

In analytic number theory the method of generating functions has broad applicability (power series are mainly used in additive number theory, Dirichlet series in the theory of multiplicative functions, and characteristic functions in questions concerning the distribution of values of arithmetical functions). An indispensable feature of the method of generating functions is the use of inversion (which leads in various problems to explicit formulas). The generating series of the classical problems of number theory admit of investigation in the whole complex plane; hence the most convenient tools for inversion are Cauchy's formula for the coefficients of power series and Perron's formula for the summation function of a Dirichlet series. In general problems, information about the Dirichlet generating series is much more meager. This is the reason for using Tauberian theory in general number theory problems.

In the first chapter we also discuss Esseen's theorem pertaining to the method of characteristic functions.

Many problems of analytic number theory are of an additive nature. To clarify this we formulate several problems.

We first consider direct additive problems.

Suppose we are given two finite or infinite sequences of real numbers, arranged in increasing order:

$$\lambda_1, \lambda_2, \lambda_3, \dots, \tag{1}$$

$$\mu_1, \mu_2, \mu_3, \dots \tag{2}$$

(we may assume that each of these two sequences has at most one limit point at infinity). We form the set of pairwise sums

$$\tau = \lambda_i + \mu_j. \tag{3}$$

To a point τ of this set we attach a multiplicity equal to the number of representations of τ in the form (3). One problem is to obtain properties of the set of pairwise sums from properties of sequences (1) and (2).

This general formulation includes many problems, among which we distinguish two types.

a) *Problems on the distribution of multiplicities.* These problems will be called *local.* Here we want to study the quantity

$$p(\tau) = \sum_{\tau=\lambda_i+\mu_j} 1. \tag{4}$$

A more general problem is the following. Suppose we are given two functions, $f_1(x)$ and $f_2(x)$, of a real variable. We want to study the quantity

$$\tilde{p}(\tau) = \sum_{\tau=\lambda_i+\mu_j} f_1(\lambda_i) f_2(\mu_j). \tag{4'}$$

We will adopt this terminology: a function $f(x)$ defined on some sequence will be called a *law*, and the quantity $\tilde{p}(\tau)$ defined by (4′) will be called the *convolution* of f_1 and f_2 and will be denoted by $\tilde{p} = f_1 * f_2$.

b) We denote by $n_1(u)$ and $n_2(u)$ the number of elements of the sequences (1) and (2), respectively, that do not exceed u (u is an asymptotically increasing parameter); we denote by $q(u)$ the number of solutions of the inequality $\lambda_i + \mu_j \le u$. The problem is to study $q(u)$, given $n_1(u)$ and $n_2(u)$. Such problems will be called *integral.* Obviously,

$$q(u) = \sum_{\lambda_i \le u} n_2(u - \lambda_i) = \sum_{\mu_j \le n} n_1(u - \mu_j).$$

Bachmann's book *Die Analytische Zahlentheorie* [6], which appeared at the end of the last century, played a large role in the development of analytic number theory. On pp. 447–449 of that book the following theorem is proved.

THEOREM. *Suppose A, B, α, and β are given positive numbers. Let $q(N)$ denote the number of pairs (x, y) of natural numbers such that $Ax^\alpha + By^\beta \le N$. Then*

$$q(N) = \frac{N^{1/\alpha+1/\beta}}{A^{1/\alpha}B^{1/\beta}} \frac{1}{\alpha+\beta} \frac{\Gamma(1/\alpha)\Gamma(1/\beta)}{\Gamma(1/\alpha+1/\beta)} - \theta\left(\left(\frac{N}{A}\right)^{1/\alpha} + \left(\frac{N}{B}\right)^{1/\beta}\right),$$
$$0 < \theta < 1.$$

For the cases $A = 1$ and $B = 1$ this theorem can be viewed as a result pertaining to a direct additive problem.

Let us turn to formulating additive problems with an unbounded number of summands. Suppose we are given a nondecreasing sequence

$$\lambda_1, \lambda_2, \lambda_3, \ldots \tag{1}$$

of positive numbers. We add to this sequence a zero, getting $0, \lambda_1, \lambda_2, \lambda_3, \ldots$. By definition, the sequence (1) added to itself ad infinitum consists of the numbers of the form

$$n_1\lambda_1 + n_2\lambda_2 + \cdots + n_s\lambda_s + \cdots, \tag{5}$$

where $n_1, n_2, \ldots$ are nonnegative integers; of course, in each expression (5) there occur only finitely many nonzero n_i. An example of such a problem is the one considered by Freĭman [55] concerning an asymptotic formula for the number of solutions in nonnegative integers of the inequality

$$2^{\alpha_2} 3^{\alpha_3} \cdots r^{\alpha_r} \cdots \leq n$$

as $n \to \infty$.

On the subject of additive problems with an unbounded number of summands we mention in this book the direct additive theorem of Bredikhin [15]. Suppose G is a multiplicatively written free semigroup with a countable system P of generators $\omega_1, \omega_2, \ldots$, and suppose N is a homomorphism of G onto some numerical multiplicative semigroup $\overline{G}$ such that G contains only finitely many elements α with $N(\alpha) \leq x$, where $N(\alpha)$ denotes the image (or norm) of the element $\alpha \in G$ under the homomorphism N. Let

$$\pi_G(x) = \sum_{\substack{\omega \in P \\ N(\omega) \leq x}} 1 \quad \text{and} \quad \nu_G(x) = \sum_{\substack{N(\alpha) \leq x \\ \alpha \in G}} 1.$$

THEOREM. *If for some $\varepsilon > 0$*

$$\pi_G(x) = \tau \frac{x}{\ln x} + O\left(\frac{x}{\ln^{1+\varepsilon} x}\right),$$

then

$$\nu_G(x) = C_G x \ln^{\tau-1} x + O\left(\frac{x(\ln x)^{\tau-1}}{(\ln\ln x)^{\varepsilon_1}}\right).$$

where $\varepsilon_1 = \min(1, \varepsilon)$.

We will show that Bredikhin's theorem can be included among the direct additive problems. First of all, $N(\omega) > 1$ for each $\omega \in P$, because if $N(\omega_1) \leq 1$, then we would have infinitely many elements α ($\alpha = \omega_1^n$, $n = 1, 2, \ldots$) such that $N(\alpha) \le 1$. Put $\ln N(\omega_j) = \lambda_j$. Let us arrange these λ in nondecreasing order: $\lambda_1 \le \lambda_2 \le \lambda_3 \le \cdots$. Obviously,

$$n(u) = \sum_{\lambda_j \le u} 1 = \pi_G(e^u).$$

We add the sequence $0, \lambda_1, \lambda_2, \lambda_3, \ldots$ to itself ad infinitum. Then the quantity $q(u)$ expressing the number of solutions in integers $n_1, n_2, \ldots$ of the inequality $n_1\lambda_1 + n_2\lambda_2 + \cdots \le u$ is equal to $\nu_G(e^u)$. Thus in Bredikhin's theorem $q(u)$ can be reconstructed from $n(u)$—a typical direct additive problem.

Suppose G is a free semigroup consisting of natural numbers. For convenience we will always include the number 1 in such a semigroup. Let $f(n)$ denote the indicator of G:

$$f(n) = \begin{cases} 1, & n \in G, \\ 0, & n \notin G. \end{cases}$$

Obviously,

$$\nu_G(x) = \sum_{\substack{n \in G \\ n \le x}} 1 = \sum_{n \le x} f(n).$$

Bredikhin's direct additive theorem can be applied to the problem of summing the indicator $f(n)$. We give two examples.

1. Given a modulus k along with r residues $l_1, \ldots, l_r$ modulo k that are relatively prime to k. Let P denote the set of all primes congruent modulo k to at least one of $l_1, \ldots, l_r$. Let G denote the set of natural numbers formed from the primes belonging to P. It is obvious that G is a semigroup. We have

$$\pi_G(x) = \sum_{\substack{p \equiv l_1, \ldots, l_r \pmod{k} \\ p \le x}} 1 = \frac{r}{\varphi(k)} \frac{x}{\ln x} + O\left(\frac{x}{\ln^2 x}\right).$$

Therefore, by Bredikhin's theorem,

$$\nu_G(x) = \frac{C_G x}{(\ln x)^{1-r/\varphi(k)}} + O\left(\frac{x}{(\ln x)^{1-r/\varphi(k)} \ln\ln x}\right).$$

This result is due to Landau ([91], pp. 641–669).

2. As is well known, the set of numbers representable as a sum of two squares of integers is a free semigroup, a basis of which is formed by the number 2, the primes of the form $4k+1$, and the squares of the primes of the form $4k+3$. For this semigroup G we have

$$\pi_G(x) = 1 + \sum_{\substack{p \equiv 1 \pmod{4} \\ p \le x}} 1 + \sum_{\substack{p \equiv 3 \pmod{4} \\ p \le \sqrt{x}}} 1 = \frac{1}{2} \frac{x}{\ln x} + O\left(\frac{x}{\ln^2 x}\right).$$

Therefore, by Bredikhin's theorem,

$$\nu_G(x) = \sum_{\substack{n\in G\\ n\le x}} 1 = C\frac{x}{\sqrt{\ln x}} + O\left(\frac{x}{\sqrt{\ln x}\ln\ln x}\right).$$

This result was established by Landau [90].

An example of a local problem with an infinite number of summands is the classical Hardy-Ramanujan problem. It involves the function $p(n)$ of a natural argument defined as the number of representations of n as a sum of natural summands. For example, $p(4) = 5$:

$$4 = 1+1+1+1, \qquad 4 = 2+1+1,$$
$$4 = 2+2, \qquad 4 = 3+1, \qquad 4 = 4.$$

In this problem the original sequence is the natural one, which is added to itself ad infinitum.

The function $p(n)$ increases rapidly. For example,

$$p(200) = 3\,972\,999\,029\,388.$$

It can be shown that as $n \to \infty$

$$\ln p(n) \sim \pi\sqrt{2/3}\cdot\sqrt{n}.$$

Asymptotic formulas for $p(n)$ were obtained by Hardy and Ramanujan (see Chapter 8 of Hardy [65]) and Uspensky [143]; the results of Hardy and Ramanujan were the more precise. As $n \to \infty$

$$p(n) \sim \exp(\pi\sqrt{2n/3})/4\sqrt{3}n.$$

An exposition of the Hardy-Ramanujan problem can be found in Ayoub [4], which provides fairly complete information on the subject.

The function $p(n)$ in the Hardy-Ramanujan problem can be defined as the number of solutions of the equation

$$n = 1\cdot x_1 + 2x_2 + \cdots + nx_n + \cdots$$

in nonnegative integers x_i. Indeed, this equality means that in the partition of n into a sum of natural summands the number 1 occurs x_1 times, the number 2 occurs x_2 times, and so on. This formulation of the problem can be generalized in a natural way. Suppose we are given a monotone increasing sequence of natural numbers $a_1, a_2, \ldots$; for a given natural number n we denote by $p_{\text{gen}}(n)$ the number of solutions in nonnegative integers x_i of the equation $n = a_1x_1 + a_2x_2 + \cdots$. The asymptotic behavior of $p_{\text{gen}}(n)$ has been investigated by three methods. One of these is elementary (see Freĭman [75]). Another method, based on Tauberian theorems for the Laplace transform of rapidly increasing functions, was developed by Ingham [75] (Ingham's memoir is recalled in this book). Finally, an analytic method based on inversion formulas for generating functions was developed by Haselgrove and Temperley [70] and Roth and Szekeres [128]. All these methods yield formulas of the same type for $p_{\mathrm{gen}}(n)$.

We now consider inverse additive problems. In direct additive problems the investigator proceeds from $n(u)$ to $q(u)$, but in inverse problems in the opposite direction. An example of a result pertaining to an inverse additive problem is the following theorem of Tashbaev [138].

THEOREM. *Assume that sequence* (1) *consists of nonnegative integers and is strictly increasing. Suppose* $C > 0$ *is a constant,*

$$\alpha_1 < \frac{\alpha(1+\alpha)}{2} - 1, \qquad \alpha_1 < \alpha,$$

and

$$q(u) = Cu^{\alpha} + O(u^{\alpha_1}).$$

Then

$$n(u) = Bu^{\alpha/2} + O(u^{\omega}\ln u) + O(u^{\alpha/2-(1-\omega)}\ln u),$$

where

$$\omega = \frac{\alpha_1+1}{\alpha+1}, \qquad B = \frac{\sqrt{C\Gamma(\alpha+1)}}{\Gamma(\alpha/2+1)}.$$

A systematic study of inverse additive problems pertaining to duplication of sequences was carried out by Freĭman [57].

Erdős [36] and Freĭman [52], [53] considered inverse additive problems on partitioning numbers into an unbounded number of summands. For a given sequence (1) of real numbers we denote by $q(u)$ the number of solutions in nonnegative integers of the inequality

$$\lambda_i + \lambda_j \le u. \tag{6}$$

How can the order of growth of $n(u)$ be determined from that of $q(u)$?

Suppose that sequence (1) consists of the logarithms of the prime numbers:

$$\ln 2,\ \ln 3,\ \ln 5, \dots . \tag{7}$$

Since each integer $n \ge 2$ can be uniquely represented as a product of powers of primes, it follows that sequence (7) added to itself ad infinitum is the sequence of logarithms of integers greater than 1, and each term of this sequence has multiplicity 1. Therefore the number of solutions of (6) is equal to the number of natural numbers not exceeding e^u:

$$q(u) = [e^u] = e^u + O(1).$$

This quantity has a simple structure. The quantity $n(u) = \pi(e^u)$, however, differs from the main function of the theory of distribution of primes by only a change of argument. Thus the problem of the distribution of primes can be viewed as an inverse additive problem. We find such a point of view in Beurling [12], Nyman [112], and Bredikhin [14]; see also Chapter III of Wintner [153].

The treatment of problems on the distribution of primes as additive problems can be illustrated by the following proof that the set of primes is infinite (the proof is taken from Shnirel′man (Schnirelmann [132], pp. 44–45).

Suppose there are only finitely many primes $p_1, \dots, p_r$. By the fundamental theorem of arithmetic, any natural number n can be uniquely written in the form $n = p_1^{\alpha_1} \cdots p_r^{\alpha_r}$, where $p_1 < p_2 < \cdots < p_r$ and each α_i is a nonnegative integer. Consider the numbers $1, \dots, N$ $(N \geq 2)$. The number of natural numbers n not exceeding N is exactly N. On the other hand, there is a one-to-one correspondence between the natural numbers n and the r-tuples $(\alpha_1, \dots, \alpha_r)$ of nonnegative integers. Consequently, the number of solutions $(\alpha_1, \dots, \alpha_r)$ of the inequality $p_1^{\alpha_1} \cdots p_r^{\alpha_r} \leq N$ in nonnegative integers is exactly N. But if $p_1^{\alpha_1} \cdots p_r^{\alpha_r} \leq N$, then

$$\alpha_i \leq \frac{\ln N}{\ln p_i} + 1 \leq \frac{\ln N}{\ln 2} + 1 \leq 2\frac{\ln N}{\ln 2}.$$

It follows that $N \leq 2^r(\ln N/\ln 2)^r$. This inequality is impossible in view of the slow rate of growth of the logarithm function for sufficiently large N.

The fundamental law of analytic number theory—the asymptotic law of the distribution of primes—says that $\pi(x) \sim x/\ln x$ as $x \to \infty$. A natural generalization of the question of the distribution of primes is that of the distribution of the prime ideals of a given algebraic number field K. Suppose K is an algebraic number field of degree n over the field of rational numbers, and $\pi_K(x)$ is the number of prime ideals whose norm does not exceed x. The asymptotic law is then $\pi_K(x) \sim x/\ln x$.

We have mentioned a theorem of Bredikhin pertaining to a direct additive problem. In [14] Bredikhin solved the inverse problem of which we were speaking; he established the following

THEOREM. *Suppose* $\nu_G(x) = C_G x^\theta + O(x^{\theta_1})$, *where* $0 \leq \theta_1 < \theta$. *Then*

$$\pi_G(x) \sim \frac{1}{\theta}\frac{x^\theta}{\ln x}.$$

This theorem generalizes not only the asymptotic law of the distribution of primes, but also the asymptotic law of the distribution of prime ideals of an algebraic number field. Indeed, consider a finite extension K of the rationals of degree n. The set G of nonzero integral ideals of K is a free semigroup with a basis consisting of the prime ideals $\mathfrak{P}_1, \mathfrak{P}_2, \dots$ of this field. Let $N(\mathfrak{A})$ denote the norm of the ideal $\mathfrak{A}$, and let

$$\nu_K(x) = \sum_{\substack{\mathfrak{A} \in G \\ N(\mathfrak{A}) \le x}} 1 \quad \text{and} \quad \pi_K(x) = \sum_{\substack{\mathfrak{P} \in G \\ N(\mathfrak{P}) \le x}} 1.$$

THEOREM. *As $x \to \infty$,*

$$\pi_K(x) \sim x / \ln x.$$

Indeed, by Dirichlet's theorem with remainder term (see, for example, Landau [92]),

$$\nu_K(x) = C_0 x + O(x^{(n-1)/(n+1)}).$$

We can now apply Bredikhin's theorem with $\alpha = \mathfrak{A}$, $\omega_i = \mathfrak{P}_i$, $N(\alpha) = N(\mathfrak{A})$, and $\theta = 1$.

Let us recall, by the way, some facts about the remainder term in the asymptotic law of the distribution of primes. As is well known, if we take as the principal term in the asymptotic formula for $\pi(x)$ the expression

$$\operatorname{li} x = \int_2^x \frac{dt}{\ln t},$$

then we can obtain an asymptotic expansion for $\pi(x)$ in powers of $\ln x$.

The most precise estimates of the difference $\pi(x) - \operatorname{li} x$ have been attained by the analytic method. In these investigations, reduction in the remainder term depends on the extent to which we can guarantee absence of zeros of the Riemann zeta-function $\zeta(s)$ in the critical strip (to the left of the straight line $\operatorname{Re} s = 1$ and to the right of $\operatorname{Re} s = 1/2$). Let $s = \sigma + it$. Hadamard and de la Vallée Poussin showed that the zeta-function has no zeros in the region

$$\sigma \ge 1 - \frac{c}{\ln(|t| + 2)},$$

where $c > 0$ is some absolute constant. This result leads to the estimate

$$\pi(x) = \operatorname{li} x + O(xe^{-c\sqrt{\ln x}}),$$

where $c > 0$ is a constant. Further improvements depend on refining estimates of the growth of the zeta-function near the straight line $\sigma = 1$. This is related to estimates of trigonometric sums of the form

$$\sum_{N_1 \le n < N_2} e^{it \ln n}.$$

It was shown by means of estimates obtained by Vinogradov [147] and Korobov [86] in 1958 (a detailed exposition can be found in Walfisz [151]) that $\zeta(s)$ has no zeros in the region

$$\sigma \ge 1 - c(\ln |t| + 2)^{-2/3}.$$

This result leads to the estimate

$$\pi(x) = \operatorname{li} x + O\left(x \exp\left(-c\frac{(\ln x)^{3/5}}{(\ln\ln x)^{1/5}}\right)\right)$$

(see Ingham [74], Prachar [118], Titchmarsh [139], and Walfisz [151]).

Suppose $d \geq 2$, $1 \leq l < d$, and $(l, d) = 1$. Let $\pi(x, d, l)$ denote the number of primes not exceeding x belonging to the progression $l + du$, $u = 0, 1, \ldots$. The following formula for $\pi(x, d, l)$ can be established:

$$\pi(x, d, l) = \frac{1}{\varphi(d)} \int_2^x \frac{du}{\ln u} + O\left(x \exp\left(-c\frac{(\ln x)^{3/5}}{(\ln\ln x)^{1/5}}\right)\right),$$

where the constant $c > 0$ in the symbol O depends on d.

Additive problems with an increasing number of summands are usually formulated in the language of probability theory, and one speaks of increasing sums of random variables. To the "uninitiated" such a formulation often causes confusion: it seems that the problem is firmly connected with some "concretely proceeding" stochastic process, but the results obtained by probabilistic arguments are appraised as unreliable. The fact is that in a number of situations the formalized stochastic process problem turns out to be identical to the additive problem.

We will be concerned with two types of problems: integral and local.

We will now describe the formulation of integral problems. Suppose we are given an infinite sequence of sets of integers $\mathfrak{M}_1, \mathfrak{M}_2, \ldots$; the numbers occurring in the sets $\mathfrak{M}_i$, $i = 1, 2, \ldots$, may be assumed distinct. Let t be a real parameter. The problem is to find an asymptotic formula as $t \to \infty$ for the number of solutions of the inequality $x_1 + \cdots + x_n \leq t$, where $x_i \in \mathfrak{M}_i$, $i = 1, \ldots, n$.

In this book we give an example of an application of an integral limit theorem illustrating how arithmetical results can be obtained by means of a probabilistic interpretation. One of the main methods in the theory of approximation of algebraic numbers is that of Thue, which was developed by Siegel and Roth. Schneider [131] (Hilfsatz 1) pointed out that in this method it is essential that in the hypercube of large dimension an overwhelming number of lattice points lie near the diagonal hyperplane of the cube. In the section called "Schneider's lemma" (§2.2) we will give a precise formulation of this assertion and show how a probabilistic interpretation turns out to be useful in this problem.

We will now describe the formulation of local problems with an increasing number of summands. Suppose we are given an infinite sequence $\mathfrak{M}_1, \mathfrak{M}_2, \ldots$ of finite sets of integers. Let N be an integer. The problem is to find an asymptotic formula for the number of solutions of the Diophantine equation

$x_1 + \cdots + x_n = N$, $x_i \in \mathfrak{M}_i$, $i = 1, \ldots, n$, where the increasing parameter is n. We only consider problems in which all of the sets $\mathfrak{M}_i$, $i = 1, 2, \ldots$, are the same. The tool for solving these problems is the local limit theorem of probability theory for identically distributed lattice random variables (see Gnedenko [61], §43). Gnedenko proved this theorem by the method of characteristic functions. He made use of a representation of the solution of an additive problem as an integral of a trigonometric sum. The situation is the same as in analytic number theory. However a direct approach to the local limit theorem does not require the subtle investigation of the behavior of a trigonometric sum carried out in classical analytic number theory problems. We will give, in the appropriate section of this book, examples of the use of local limit theorems in arithmetic.

Suppose $f(n)$ is an integer-valued function of an integral argument n, and suppose P is a fixed natural number. Consider the problem of representing an integer N in the form $N = f(x_1) + \cdots + f(x_n)$, where $0 \le x_1, \ldots, x_n \le P - 1$; here the parameter n increases without bound. It is easy to see that this problem can be reduced to the local limit theorem for identically distributed independent random variables. The restriction of a fixed P is not typical for number theory. To solve the problem with an increasing P we need local theorems that are uniform with respect to the initial distributions. Although such theorems exist in the integral formulation, it is necessary to restrict the examples in local limit theorems. We will consider the problem of Usol′tsev [142] on representing a number N in the form $N = g^{x_1} + \cdots + g^{x_n}$, where $g \ge 2$ is fixed, $0 \le x_1, \ldots, x_n \le P - 1$, $n \to \infty$, and $P \to \infty$.

The third chapter of this book is entitled "Theory of functions of a natural argument". The problem of determining average values of functions of a natural argument was handed down to us by the great number theorists of the past. Gauss established the asymptotic formula

$$\sum_{n \le N} r(n) = \pi N + O(\sqrt{N}),$$

and Dirichlet proved that

$$\sum_{n \le N} \tau(n) = N \ln N + (2\gamma - 1)N + O(\sqrt{N}).$$

Simplifying the situation, we can say that the development of problems on the summation of values of arithmetical functions has proceeded in two directions.

1. In one direction the problem was to obtain summation formulas for the classical functions of number theory with as precise remainder terms as possible. Here an important role was played by the memoir of Voronoĭ [150],

in which the above-mentioned formula for the sum of the values $\tau(n)$ was improved. Of central importance in this direction is a method of estimating trigonometric sums (due to Weyl, Vinogradov, and others). Introductory material on this subject can be found in the exercises of Vinogradov's textbook [148].

2. The other direction, in which our main interest lies, involves the construction of the most general possible theorems pertaining to the summation of functions of a natural argument.

An example of such a theorem is the following result of Wintner (see [153], p. 9).

THEOREM. *Suppose a function $f(n)$ of a natural argument is represented in the form $f(n) = \sum_{d|n} \Phi(d)$ and the series $\sum_{d=1}^{\infty} \Phi(d)/d$ is absolutely convergent. Then we have the limit relation*

$$\lim_{N\to\infty} \frac{1}{N} \sum_{n=1}^{N} f(n) = \sum_{d=1}^{\infty} \frac{\Phi(d)}{d}.$$

There are many generalizations of Wintner's theorem. In this book we present Axer's theorem as an example of one of them. The whole first chapter of Wintner's book [153] is devoted to the study of functions $f(n)$ and $\Phi(n)$ connected by the relation

$$f(n) = \sum_{d|n} \Phi(d).$$

Questions on the summation of functions of a natural argument are grounds for us to dwell on the analogy between number theory and analysis. We list analogous concepts:

Number theory	**Analysis**
The natural sequence of integers.	The interval $[0, 1]$ of the real axis.
An arithmetic progression with modulus D.	A half-open interval $[a/D, (a+1)/D]$.
Asymptotic density of a set $\mathfrak{M}$ of natural numbers: $D(\mathfrak{M}) = \lim_{N\to\infty} \frac{1}{N} \sum_{n\le N,\ n\in\mathfrak{M}} 1$.	Lebesgue measure of a set of numbers in $[0, 1]$.

In constructing the analogy between asymptotic density and Lebesgue measure we encounter the following difficulty: Lebesgue measure is countably additive, whereas asymptotic density is not. Indeed, the asymptotic density of a set consisting of a single natural number is zero, but a countable union of such sets yields the natural sequence, whose asymptotic density is 1. The fact that asymptotic density is not countably additive prevents us from directly applying the machinery of probability theory to problems on the distribution of values of arithmetical functions. Kubilius (see [87], Chapter II) suggested several ways of getting around this difficulty. These suggestions constitute the methodological basis for applying probability theory to the study of the distribution of values of arithmetical functions.

Number theory	**Analysis**
The sum $N^{-1}\sum_{n=1}^{N} f(n)$.	The integral sum of a function $f(x)$ defined on $[0,1]$.
A periodic function of a natural argument with integral period D.	A function piecewise constant on half-open intervals $[a/D,(a+1)/D]$.

A function $f(n)$ of a natural argument will be said to be *well approximated by periodic functions* if for any $\varepsilon > 0$ there exists an A-periodic function $f(n, A)$, where A is a natural number such that

$$\overline{\lim_{N\to\infty}}\, \frac{1}{N}\sum_{n\le N} |f(n) - f(n,A)| \le \varepsilon.$$

It can be proved that any function of a natural argument that satisfies the conditions of Wintner's theorem is well approximated by periodic functions. Such functions inherit many properties of periodic functions; in particular,

$$\lim_{N\to\infty} \frac{1}{N}\sum_{n\le N} f(n) = M(f).$$

The analogy of which we are speaking acquires a sharper character if we consider the natural sequence of integers as a subset of a certain ring. We have in mind the ring of polyadic numbers first introduced, it seems, by Prüfer [119]. Let us briefly recall its construction. We denote the ring of integers by Z. We define a norm on Z by putting

$$N(x) = \sum_{k=1}^{\infty} \frac{1}{2^k}\left(\frac{x}{k}\right), \qquad x \in Z,$$

where (t) is the distance from t to the nearest integer. It is not difficult to show that the function

$$\rho(x, y) = N(x - y)$$

of a pair of numbers has all of the properties of a metric relative to which the operations of addition and multiplication are continuous. The completion of Z relative to the metric $\rho(x, y)$ is a compact ring $\mathfrak{S}$ containing Z as a dense subset. Novoselov [107] interpreted several propositions of elementary number theory as topologico-algebraic theorems about the ring $\mathfrak{S}$.

There exists on the additive group of $\mathfrak{S}$ a normalized Haar measure. This measure is not complete. Completion of the Haar measure leads to the construction of a probability measure on $\mathfrak{S}$. The measure of an arithmetic progression $\alpha + \beta D$, where $\alpha, \beta \in \mathfrak{S}$ and D is a natural number, is $1/D$. Therefore, embedding Z in $\mathfrak{S}$ eliminates the difficulty associated with the fact that asymptotic density is not completely additive. Novoselov [110] treated problems concerning the calculation of average values of functions of specific classes as problems of the integral calculus on $\mathfrak{S}$. Embedding the ring of integers in the ring of polyadic numbers enabled Novoselov to interpret, from a new point of view, the theory of functions well approximated by periodic functions and a number of propositions (of probabilistic number theory) pertaining to the distribution of values of arithmetical functions.

Chapter 4 deals with the theory of multiplicative functions. The theory of multiplicative functions is so multifaceted and rich in results that one feels a pressing need for a special monograph on this subject. We will discuss only selected topics of this theory.

In the first section of this chapter we consider upper bounds of multiplicative functions. Wigert [152] proved that for each $\varepsilon > 0$ and for all sufficiently large natural numbers N we have

$$\tau(N) \le 2^{(\ln N/\ln\ln N)(1+\varepsilon)}$$

and there exists an infinite sequence of natural numbers N_i such that

$$\tau(N_i) > 2^{(\ln N_i/\ln\ln N_i)(1+\varepsilon)}.$$

Wigert's result was strengthened somewhat by Ramanujan (see [121], pp. 78–128). Generalizing these results, Drozdova and Freĭman [33] proved the following theorem.

THEOREM. *Suppose the canonical decomposition of N is $N = p_1^{\alpha_1} \cdots p_s^{\alpha_s}$. Let there be given a function $l(\alpha)$, defined for natural numbers and taking positive values, such that*

$$\ln \frac{l(r+1)}{l(r)} = O\left(\frac{1}{r}\right).$$

Let $f(N) = l(\alpha_1) \cdots l(\alpha_s)$. Then

$$f(N) \le \exp\left[\sup_{m\ge 1} \frac{\ln l(m)}{m} \frac{\ln N}{\ln\ln N} + O\left(\frac{\ln N \ln\ln\ln N}{(\ln\ln N)^2}\right)\right].$$

We can suggest a more general formulation of this question. Suppose G is a free multiplicative semigroup consisting of natural numbers. Let $\omega_1, \omega_2, \ldots$

be a basis of G. Given a positive function $l(\alpha)$ of a natural argument satisfying

$$\ln \frac{l(r+1)}{l(r)} = O\left(\frac{1}{r}\right),$$

we define a function $f(N)$ on G as follows: if $N = \omega_1^{\alpha_1} \cdots \omega_s^{\alpha_s}$, then $f(N) = l(\alpha_1) \cdots l(\alpha_s)$.

THEOREM. *Arrange the basis elements of the semigroup G in increasing order: $\omega_1 < \omega_2 < \cdots$. Assume that as $s \to \infty$*

$$\ln \omega_s = \ln s + O(\ln \ln s).$$

Then

$$f(N) \leq \exp\left[\sup_{m \geq 1} \frac{\ln l(m)}{m} \frac{\ln N}{\ln \ln N} + O\left(\frac{\ln N}{(\ln \ln N)^2} \ln \ln \ln N\right)\right].$$

The theorem of Drozdova and Freĭman is a special case of this one, if we take as G the multiplicative semigroup of all natural numbers. Indeed, the set of all primes is a basis of G. By a theorem of Tchebycheff, $\pi(x) \asymp x/\ln x$; hence $p_s \asymp s \ln s$, and therefore

$$\ln p_s = \ln s + O(\ln \ln s),$$

i.e., the condition on the growth of the basis is satisfied.

Central to the fourth chapter are questions on the asymptotic behavior (as $N \to \infty$) of sums of the form $\sum_{n \leq N} f(n)$, where $f(n)$ is a multiplicative function of a natural argument—i.e., questions concerning the summation of values of multiplicative functions.

Summation of multiplicative functions is encountered in almost half the problems of analytic number theory. In particular, this formulation includes broadly applicable questions on the sums

$$\sum_{A \leq n \leq B} e^{it \ln n},$$

and on sums involving Hecke characters. We will be able to touch upon only a small part of the immense theory of summation of multiplicative functions. In particular, we have to almost totally ignore questions on sums of multiplicative functions of polynomials.

We can apply to the summation of values of multiplicative functions the methods suggested in Chapter 3, "Theory of functions of a natural argument", but they do not take account of the specifics of multiplicative functions and so their application is not always advantageous.

In Chapter 4, on multiplicative functions, we carry out the program of which we spoke in connection with additive problems with an infinite number of summands.

Suppose we are given a semigroup G of natural numbers, a basis of which is some set of primes. The indicator of this semigroup, which we denote by $f(n)$, is a totally multiplicative function. Bredikhin's direct additive theorem on the behavior of $\sum_{p\le N} f(p)$ establishes the asymptotic behavior of $\sum_{n\le N} f(n)$. This idea can be extensively developed.

A multiplicative function $f(n)$ is determined as soon as we know its values on the set of prime powers. Since the set of primes raised to powers higher than the first is sparse, it follows that under certain restrictions on $f(p^\alpha)$ with $\alpha \ge 2$ the asymptotic behavior of the sum

$$\sum_{n\le N} f(n) \tag{8}$$

is determined by the behavior of

$$\sum_{p\le N} f(p). \tag{9}$$

The calculation of the latter sum can be carried out by using asymptotic laws of the distribution of primes. Thus the asymptotic laws of the distribution of primes can be applied to the summation of multiplicative functions.

Here we should first mention a theorem of Wirsing [154].

THEOREM. *Suppose $f(n)$ is a multiplicative function satisfying the following conditions:*

1) $f(n) \ge 0$, $n = 1, 2, \dots$.
2) $f(p^r) \le C_1 C_2^r$, $C_2 < 2$.
3) *As* $x \to \infty$

$$\sum_{p\le x} f(p)\ln p = (\tau + o(1))x,$$

where $\tau \ge 0$ is a constant.

Then as $x \to \infty$

$$\sum_{n\le x} f(n) = \left(\frac{e^{-\gamma\tau}}{\Gamma(\tau)} + o(1)\right)\frac{x}{\ln x}\prod_{p\le x}\left(1 + \frac{f(p)}{p} + \frac{f(p^2)}{p^2} + \cdots\right).$$

The proof of Wirsing's theorem is elementary. It consists of two parts. In the first it is shown that, under the conditions of the theorem,

$$\sum_{n\le x} f(n) = (\tau + o(1))\frac{x}{\ln x}\sum_{n\le x}\frac{f(n)}{n}.$$

In the second, the asymptotic relation

$$\sum_{n\le x}\frac{f(n)}{n} = \left(\frac{e^{-\gamma\tau}}{\Gamma(1+\tau)} + o(1)\right)\prod_{p\le x}\left(1 + \frac{f(p)}{p} + \frac{f(p^2)}{p^2} + \cdots\right)$$

is established.

Another example of a theorem proceeding from properties of the sequence $f(p)$ to properties of the sequence $f(n)$ is the following result of Delange (see [28]).

THEOREM. *Suppose* $f(n)$ *is a multiplicative function satisfying the following conditions*:

1) $|f(n)| \leq 1$, $n = 1, 2, \ldots$.
2) $\sum_p (f(p) - 1)/p$ *converges.*

Then the limit

$$M(f) = \lim_{N\to\infty} \frac{1}{N} \sum_{n\leq N} f(n)$$

exists, and

$$M(f) = \prod_p \left(1 - \frac{1}{p}\right)\left(1 + \frac{f(p)}{p} + \frac{f(p^2)}{p^2} + \cdots\right).$$

By an *additive* arithmetical function we mean a sequence $g(n)$, $n = 1, 2, \ldots$, such that, for any two relatively prime numbers n_1 and n_2, $g(n_1 n_2) = g(n_1) + g(n_2)$. Examples of additive functions include $\ln(\varphi(n)/n)$ and $\nu(n)$.

We mention in passing, even though it is not important for our present purposes, that both $\ln(\varphi(n)/n)$ and $\nu(n)$ are strongly additive functions, i.e., they satisfy the additional condition $g(p^\alpha) = g(p)$ for any prime p and natural number α.

Let

$$x^* = \begin{cases} x, & \text{if } |x| \leq 1, \\ 1, & \text{if } |x| > 1. \end{cases} \tag{10}$$

Additive functions $g(n)$ can be classified according to the convergence or divergence of the series

$$\sum_p \frac{(g(p)^*)^2}{p}. \tag{11}$$

If series (11) converges, we will say that the function $g(n)$ is *stable*; if (11) diverges, we will say that $g(n)$ is *unstable.* Since for any prime p we have

$$\left(\ln \frac{\varphi(p)}{p}\right)^* = \ln\left(1 - \frac{1}{p}\right) \sim -\frac{1}{p},$$

it follows that the series

$$\sum_p \frac{1}{p}\left(\left(\ln \frac{\varphi(p)}{p}\right)^*\right)^2$$

converges, i.e., the function $\ln(\varphi(n)/n)$ is stable. The function $\nu(n)$ is obviously unstable.

A significant part of this book is devoted to questions pertaining to the distribution of values of additive functions. Consider, for example, the question of the distribution of values of $\ln(\varphi(n)/n)$. For any real number u we put

$$F_P(u) = \frac{1}{P} \sum_{\substack{n \le P \\ \ln(\varphi(n)/n) < u}} 1.$$

The function $F_P(u)$ is obviously a distribution function. We are interested in the behavior of $F_P(u)$ for large P. It can be proved that the sequence of distribution functions

$$F_1(u),\ F_2(u), \ldots,\ F_P(u), \ldots \tag{12}$$

is weakly convergent (or essentially convergent); by this we mean convergence to some distribution function at each point of continuity of the latter. Three questions arise:

1. What conditions on the function $g(n)$ guarantee the existence of a limit function $F(u)$? This will be called the question on integral limit theorems.
2. What are the properties of the limit function $F(u)$?
3. How well does $F_P(u)$ approximate $F(u)$?

Jessen and Wintner [76] suggested using the method of characteristic functions to study the distribution of values of additive functions. The characteristic function of the distribution function $F_P(u)$ (sequence (12)) has the form

$$f_P(t) = \int_{-\infty}^{\infty} e^{itu}\, dF_P(u) = \frac{1}{P} \sum_{n=1}^{P} e^{it \ln(\varphi(n)/n)}.$$

The function $e^{it\ln(\varphi(n)/n)} = (\varphi(n)/n)^{it}$ is obviously multiplicative. The characteristic function of an empirical distribution function of an additive function turns out to be a sum of values of multiplicative functions. This general property opens up the possibility of applying summation methods for multiplicative functions to the question of the distribution of values of additive functions. In 1938 Erdős [34] proved that, for a given real additive function $g(n)$, convergence of the series

$$\sum_p \frac{g(p)^*}{p}, \tag{13}$$

$$\sum_p \frac{(g(p)^*)^2}{p} \tag{11}$$

is a sufficient condition for the sequence of empirical distribution functions

$$F_P(x) = \frac{1}{P} \sum_{n \le P,\ g(n) < x} 1$$

to weakly converge to a limit function. In this book we will give a proof of Erdős's theorem based on a theorem of Delange. In 1939 Erdős and Wintner [41] showed that this condition is also necessary.

Let us now consider properties of the limit function. Erdős and Wintner [41] proved the following theorem.

THEOREM. *A distribution function that is the asymptotic distribution function of an additive function has no discontinuities (i.e., is continuous) if and only if* $\sum_{g(p) \ne 0} 1/p$ *diverges.*

This theorem follows from the fact that the asymptotic distribution function of an additive function $g(n)$ can be represented as an infinite convolution of certain purely discontinuous distribution functions and from the criterion (due to Lévy [98]) for the continuity of an infinite convolution. Van Kampen [79] gave a different proof of Lévy's criterion.

In their previously cited paper [76] Jessen and Wintner proved (p. 86) the following general theorem.

THEOREM. *If* $F = F_1 * F_2 * \cdots$ *is a convergent infinite convolution of distribution functions, each of which is purely discontinuous, then* F *is either purely discontinuous, or singular, or absolutely continuous.*

Thus a continuous distribution function that is the asymptotic distribution function on an additive function $g(n)$ can be either singular or absolutely continuous. Erdős [35] showed that both cases can occur.

The question of how fast is the approach to the limit function was investigated by Faĭnleĭb [45].

Uždavinys [144] obtained results extending Erdős's theorem to sequences of values of stable additive functions whose argument is a polynomial.

The case of real-valued additive arithmetical functions $g(n)$ for which the series

$$\sum_p \frac{g(p)^*}{p}, \tag{13}$$

diverges is handled by the following theorem of Delange [31].

THEOREM. *Suppose* $g(n)$ *is a real additive function for which the series* (11) *converges, but the series* (13) *diverges. Then the sequence of distribution functions*

$$\frac{1}{P} \sum_{n \le P,\ g(n) < x + S_P} 1,$$

where $S_P = \sum_{p \le P} g(p)^*/p$, *is weakly convergent.*

The rest of the survey of the theory of distribution of values of additive functions is given in §4.10, "Distribution of values of unstable additive functions".

The original problem in the investigation of the distribution of values of multiplicative functions concerned the function $\varphi(n)/n$. Schoenberg [133] used the method of moments to study the distribution of values of this function.

If a multiplicative function $f(n)$ is positive (as is, for example, $f(n) = \varphi(n)/n$), then the problem of the distribution of values of this function can be immediately reduced to that of the distribution of values of an additive function by putting $g(n) = \ln f(n)$.

Let us mention some results pertaining to the distribution of values of $\varphi(n)/n$. For any x such that $0 \leq x \leq 1$ we put

$$V_P(x) = \frac{1}{P} \sum_{n \leq P,\ \varphi(n) \leq nx} 1.$$

This is a distribution function whose variation is concentrated on $[0, 1]$. If we consider the distribution functions $F_P(u)$ of sequence (12), then obviously

$$F_P(u) = V_P(e^u), \qquad V_P(x) = F_P(\ln x).$$

We have already observed that the sequence of distribution functions $V_P(x)$ weakly converges to some distribution function $V(x)$. Schoenberg [133] proved that the limit function $V(x)$ is continuous. An explicit expression for $V(x)$ was obtained by Davenport [23]; this expression enabled Venkov [146] to establish a number of properties of $V(x)$. The rate of convergence of $V_P(x)$ to $V(x)$ was first estimated by Tyan (Tian) [141], who obtained a rather crude result. More precise estimates were obtained by Faĭnleĭb [44], [45]. He showed that

$$\frac{1}{P} \sum_{\substack{n \leq P \\ \varphi(n) \leq nx}} 1 = V(x) + O\left(\frac{1}{\ln P}\left(\frac{\ln\ln P}{\ln\ln\ln P}\right)^2\right)$$

uniformly in x, $0 \leq x \leq 1$.

There naturally arises the question of studying the distribution of values of real-valued functions that do not assume only positive values. The first result in this direction was due to Erdős [37], [38].

THEOREM. *Suppose $f(n) \geq 0$ is a multiplicative function for which the series*

$$\sum_p \frac{(f(p) - 1)^*}{p} \quad \text{and} \quad \sum_p \frac{((f(p) - 1)^*)^2}{p} \tag{14}$$

converge. Then the sequence of distribution functions

$$F_N(x) = \frac{1}{N} \sum_{n \le N,\ f(n) \le x} 1$$

converges weakly as $N \to \infty$ *to some limit distribution function. Conversely, if for a multiplicative function* $f(n)$, *where* $f(n) \ge 0$, *the sequence of functions* $F_N(x)$ *converges weakly to some distribution function* $F(x)$ *and*

$$F(x) \ne \begin{cases} 0, & x \le 0, \\ 1, & x > 0, \end{cases}$$

then the two series (14) *converge.*

Bakštys [7] obtained an analogue of the Erdős-Wintner theorem and an extension of Erdős's theorem to real-valued multiplicative functions of arbitrary sign, and in [8] and [9] he obtained results not requiring convergence of the series (14). The main investigatory tool in these papers is Zolotarev's theory (see [156]) of multiplication of independent random variables. In this book we will give a summary of Bakštys's paper [7].

We have been speaking of the distribution of values of stable additive functions. A systematic construction of probabilistic number theory, with a view towards applications to the distribution of values of unstable additive functions, was effected by Kubilius [87]. In our book we present only selected topics from the probabilistic theory of unstable additive functions.

In the theorems on summation formulas for multiplicative functions mentioned so far (Wirsing's theorem, Delange's theorem) broad conditions were imposed on the function being summed, and the results had the character of asymptotic equivalence. For certain arithmetical quantities we have completely natural asymptotic expansions. Recall, for example, the asymptotic expansion of $\pi(x)$ in powers of $\ln x$.

We give an example pertaining to the summation of multiplicative functions. Suppose k is some modulus and $l_1, \dots, l_r$ are r residues modulo k. We define a function $f(n)$ as follows: $f(n) = 1$ if the canonical decomposition of n contains only primes congruent to $l_1, \dots, l_r$ modulo k; $f(n) = 0$ for the other natural numbers n. It is clear that $f(n)$ is multiplicative. The sum

$$N(x, k, l_1, l_2, \dots, l_r) = \sum_{n \le x} f(n)$$

is the number of natural numbers not exceeding x, each prime divisor of which satisfies a congruence $p \equiv l_\mu \pmod{k}$, $\mu = 1, \dots, r$. We have already mentioned Landau's asymptotic formula for this quantity. The following result is due to Levin and Faĭnleĭb [94].

THEOREM. *As $x \to \infty$ the following asymptotic expansion holds*:

$$N(x, k, l_1, l_2, \ldots, l_r) \sim x \sum_{\nu=1}^{\infty} b_\nu (\ln x)^{r/\varphi(m)-\nu}.$$

Landau's result is a generalization of a weak form of the asymptotic law $\pi(x) \sim x/\ln x$ of the distribution of primes, while that of Levin and Faĭnleĭb is a generalization of the approximation of $\pi(x)$ by an integral logarithm.

We have also spoken of the asymptotic formula for the sum

$$\nu_G(x) = \sum_{\substack{n=u^2+v^2 \\ n \le x}} 1.$$

Ramanujan conjectured that there exists an asymptotic expansion for $\nu_G(x)$ in powers of $\ln x$. This conjecture turned out to be true; for the history of this problem and a proof of the theorem see Luthar [99].

Similar expansions are within the classical sphere of influence of the method of Dirichlet generating series. Levin and Faĭnleĭb [14] proved a general theorem on sums of values of multiplicative functions (Theorem 2.4.1), which, under certain restrictions on the summable function $f(n)$, can be given concrete expression in an asymptotic expansion (in powers of $\ln x$) of the sum $\sum_{n \le x} f(n)$. In determining the asymptotic behavior of this sum with a specific function $f(n)$ one uses either the asymptotic law of the distribution of primes or the asymptotic law of the distribution of prime ideals of an algebraic number field. Thus the method of Levin and Faĭnleĭb is one of transfer of information.

Following [94], we will illustrate the method by means of the problem on the number of natural numbers representable as a sum of two squares, i.e., we will obtain an asymptotic expansion for the quantity

$$\nu_G(x) = \sum_{\substack{n=u^2+v^2 \\ n \le x}} 1.$$

Some classical problems of number theory (e.g., cyclotomy problems and divisor problems) involve the summation of multiplicative functions such that in the asymptotic formula for $\sum_{n \le x} f(n)$ there is power reduction. There are theorems of a general nature providing an orientation in such formulations of these problems (see Kuz′min [89], Chandrasekharan and Narasimhan [19], and Faĭnleĭb [46]). To obtain the most precise results in these problems it is usually necessary to use estimates of trigonometric sums.

In this book we consider only one problem on the summation of a multiplicative function in which the remainder term of the asymptotic formula has power reduction—the problem of the behavior as $x \to \infty$ of the sum $\sum_{n\le x}\tau(n^2)$. This is a simplified variant of a question considered by Stronina [62]. The exposition is elementary, but is based on two deep theorems, the proofs of which involve Dirichlet series and the use of modern tools of analytic number theory. One of these theorems is an estimate pertaining to the divisor problem:

$$\sum_{n\le x}\tau_3(n) = xP(\ln x) + O(x^{37/75+\varepsilon}),$$

where $\tau_3(n)$ is the number of representations of n as a product of three natural factors and $P(x)$ is a certain second-degree polynomial (see [139], p. 270). The second is an estimate for the sum of values of the Möbius function:

$$\sum_{n\le x}\mu(n) = O\left(x\exp\left(-c\frac{(\ln x)^{3/5}}{(\ln\ln x)^{1/5}}\right)\right),$$

where $c > 0$ is a constant (see [151], p. 191).

We conclude this introduction with a diagram exhibiting the logical interdependence of the chapters of this book.

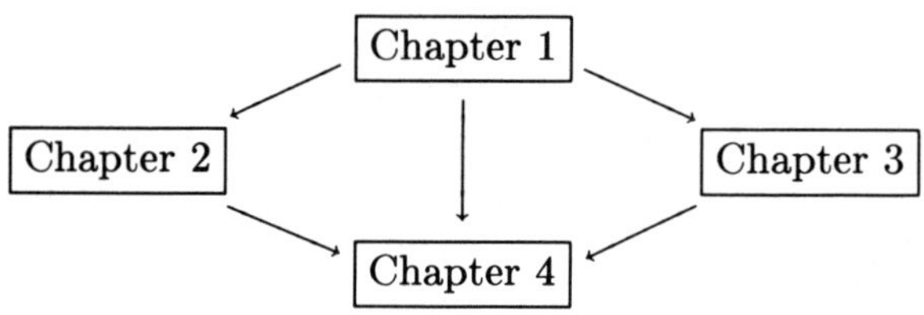

CHAPTER 1

Facts from Analysis

§1.1. Tauberian theorems for power series

In analytic number theory, as in several other branches of science, the method of generating functions is found to be useful. The generating functions of analytic number theory are power series

$$\sum_{n=0}^{\infty} a_n z^n \tag{1}$$

and ordinary Dirichlet series

$$\sum_{n=0}^{\infty} \frac{a_n}{n^s}. \tag{2}$$

Power series and ordinary Dirichlet series are special forms of a more general concept of Dirichlet series, i.e., expressions of the form

$$\sum_{n=0}^{\infty} a_n e^{-\lambda_n s}, \tag{3}$$

where $\lambda_1 < \lambda_2 < \cdots$ and $\lambda_n \to \infty$. Indeed, after the substitution $z = e^{-s}$ the series (1) assumes the form (3), and in (2) $\lambda_n = \ln n$. A still more general object is the Laplace-Stieltjes transform

$$f(s) = \int_0^{\infty} e^{-ts}\, d\alpha(t), \tag{4}$$

where $\alpha(t)$ is a function defined on $[0, \infty)$ having bounded variation on each finite t-interval.

A classical result in the theory of power series is the following theorem of Abel.

THEOREM. *Suppose a power series $f(z) = \sum_0^{\infty} a_n z^n$ with radius of convergence 1 converges at the point $z = 1$ to the sum A. Then as $z \to 1$ along the real axis*

$$\lim_{z \to 1-0} f(z) = A.$$

In Abel's theorem a certain property of the coefficients of the power series is postulated (convergence of $\sum_0^\infty a_n$) and a property of the function $f(z)$ is deduced. Theorems proceeding from properties of the coefficients of a power series to properties of the function represented by the power series will be called *Abelian theorems* for the power series.

The following inversion of Abel's theorem is due to Littlewood.

THEOREM. *Suppose the coefficients of a power series* (1) *converging in* $|z| < 1$ *are real numbers and* $na_n \leq C$. *Suppose also that*

$$\lim_{z \to 1-0} \sum_{n=0}^{\infty} a_n z^n = A.$$

Then $\sum_0^\infty a_n$ *converges, and its sum is equal to* A. *The same result holds under the condition* $na_n \geq -C$.

In Littlewood's theorem, a property of the coefficients of a power series is deduced from properties of the function represented by the power series. Theorems of this type will be called *Tauberian theorems.*

We will use the terms "Abelian theorem" and "Tauberian theorem" not only for power series, but also for the Laplace-Stieltjes transform (4); thus, theorems proceeding from properties of the function $\alpha(t)$ to properties of $f(s)$ will be called Abelian, while theorems proceeding in the reverse direction will be called Tauberian. Note that the terms "Abelian theorem" and "Tauberian theorem" are also used in a broader sense (see Hardy [66]).

Our immediate goal is to review some important facts from Tauberian theory.

We first consider the question of generalizing Littlewood's theorem to the Laplace-Stieltjes transform (4). A function $\alpha(t)$ defined on $[0, \infty)$ is called *slowly oscillating* if $\alpha(t') - \alpha(t) \to 0$ as $t \to \infty$, $t' > t$, $t'/t \to 1$. The following result will be called the *generalized Littlewood theorem.*

THEOREM. *Suppose the integral* $f(s) = \int_0^\infty e^{-st}\, d\alpha(t)$ *converges for* $s > 0$, *and* $f(s) \to A$ *as* $s \to 0$ *from the right. Suppose also that* $\alpha(t)$ *is a slowly oscillating function. Then* $\alpha(t) \to A$ *as* $t \to \infty$.

For a proof of this theorem we refer the reader to Hardy's *Divergent series* [66], p. 164, where this result is formulated as Theorem 105. Korevaar [84] investigated the question of the remainder term in Littlewood's theorem and significantly generalized this theorem. Korevaar's theorem (not stated in full generality) is as follows:

THEOREM. *Suppose* α *is a real number and* $K_1 > 0$ *is a constant. Let* $\varphi(n) = K_1 n^\alpha$ *(here* $\varphi(n)$ *does not denote Euler's function). Suppose* $\omega(u)$

is a positive function defined for $u > 0$ that tends monotonically to zero as $u \to 0$ and satisfies the condition

$$\varliminf_{u \to 0} u \ln \omega(u) > -\infty.$$

Assume the power series $f(z) = \sum_0^\infty a_n e^{-nz}$ is such that $a_n > -\varphi(n)$, $n = 1, 2, \dots$, and $|f(z) - A| < \omega(z)$ for $z > 0$. Then for $n = 1, 2, \dots$

$$\left| \sum_{k \le n} a_k - A \right| \le \min_{p \ge K_0} \left(K_2 \frac{n\varphi(n)}{p} + K_3^p \omega\left(\frac{p}{n}\right)\right),$$

where K_0, K_2, and K_3 are constants depending only on the functions φ and ω, and where $K_3 > 1$.

We mention two special cases of this general theorem.

1. If $a_n \ge -K_1/n$ and as $z \to 0$

$$f(z) = A + O\left(\frac{1}{(\ln(1/z))^\lambda}\right),$$

where $\lambda > 0$, then

$$\sum_{k \le n} a_k = A + O\left(\frac{1}{\ln \ln n}\right).$$

2. If $a_n \ge -K_1 n^\alpha$ and as $z \to 0$

$$f(z) = A + O(z^\lambda),$$

where $\lambda > -\alpha - 1$, then

$$\sum_{k \le n} a_k = A + O(n^{1+\alpha} \ln^{-1} n).$$

Korevaar [84] gave examples showing that these estimates are unimprovable for a wide class of functions $\omega(u)$.

The following Abelian theorem for a power series is easy to prove.

THEOREM. *For an infinite numerical sequence $a_1, a_2, \dots$ assume that*

$$\lim_{n \to \infty} \frac{1}{n} \sum_{k \le n} a_k = 1.$$

Then

$$\lim_{z \to 1-0} (1 - z) \sum_{n=0}^{\infty} a_n z^n = 1.$$

Hardy and Littlewood [67] discovered that this theorem can be inverted; namely, they proved the following fact.

THEOREM. *Assume that the power series $\sum_0^\infty a_n z^n$ has nonnegative coefficients and satisfies the limit relation*

$$\lim_{z\to 1-0}(1-z)\sum_{n=0}^{\infty} a_n z^n = 1.$$

Then

$$\lim_{n\to\infty}\frac{1}{n}\sum_{k\le n} a_k = 1.$$

There are examples showing that the restriction $a_n \ge 0$ cannot be removed.

The Hardy-Littlewood Tauberian theorem admits a generalization from power series to Dirichlet series. We state the variant for ordinary Dirichlet series.

THEOREM. *Suppose the ordinary Dirichlet series $\sum_{n=1}^\infty a_n/n^\sigma$ has nonnegative coefficients and is equivalent to $1/\sigma$ as $\sigma \to 0$ from the right along the real axis, i.e.,*

$$\lim_{\sigma\to 0}\sigma\sum_{n=1}^{\infty}\frac{a_n}{n^\sigma} = 1.$$

Then as $n \to \infty$

$$\sum_{k\le n} a_k \sim \ln n.$$

There have been many investigations of the remainder term in the Hardy-Littlewood Tauberian theorem. The definitive solution was obtained by Freud [58]; the word definitive means that Freud's theorem cannot, in general, be strengthened (see Korevaar's example [83]). In particular, Freud's theorem says that if a power series has nonnegative coefficients a_n and as $z \to 1-0$

$$\sum_{n=0}^{\infty} a_n z^n = \frac{1}{(1-z)^\alpha}(1 + O((1-z)^\varepsilon)),$$

where $\alpha > 0$ and $\varepsilon > 0$, then

$$\sum_{n\le x} a_n = \frac{1}{\Gamma(1+\alpha)}x^\alpha + O\left(\frac{x^\alpha}{\ln x}\right).$$

We will devote one section of this book to a detailed account of Freud's theorem.

We now consider complex Tauberian theorems, i.e., theorems in which a condition is imposed on the behavior of the Laplace-Stieltjes integral in the complex domain. Complex Tauberian theorems are, first of all, inversion formulas of the theory of functions of a complex variable.

Given a power series $f(z) = \sum_0^\infty a_n z^n$ with radius of convergence 1, let C be a circle with center at the origin and radius less than 1. By Cauchy's formula for the coefficients of a power series,

$$a_n = \frac{1}{2\pi i}\int_C \frac{f(z)}{z^{n+1}}\,dz.$$

Then (assuming N is an integer)

$$\begin{aligned}\sum_{n=0}^{N} a_n &= \frac{1}{2\pi i}\int_C f(z)\left(\frac{1}{z}+\frac{1}{z^2}+\cdots+\frac{1}{z^{N+1}}\right)dz \\ &= \frac{1}{2\pi i}\int_C \frac{f(z)}{(1-z)z^{N+1}}\,dz - \frac{1}{2\pi i}\int_C \frac{f(z)}{1-z}\,dz \\ &= \frac{1}{2\pi i}\int_C \frac{f(z)}{1-z}\frac{1}{z^{N+1}}\,dz. \qquad (5)\end{aligned}$$

The analogue of (5) for Dirichlet series is Perron's formula. We will not give it here, because in analytic number theory it is used in a somewhat modified form (see Titchmarsh [139], §3.12).

THEOREM. *Suppose a function $f(s)$, $s = \sigma + it$, is represented by an ordinary Dirichlet series $\sum_{n=1}^\infty a_n/n^s$ that converges for $s > 1$, where $a_n = O(\psi(n))$, $\psi(n)$ is nondecreasing, and*

$$\sum_{n=1}^{\infty}\frac{|a_n|}{n^\sigma} = O\left(\frac{1}{(\sigma-1)^\alpha}\right)$$

as $\sigma \to 1$. If $C > 0$, $\sigma + C > 1$, x is not an integer, and N is the integer closest to x, then

$$\sum_{n<x}\frac{a_n}{n^s} = \frac{1}{2\pi i}\int_{C-iT}^{C+iT} f(s+w)\frac{x^w}{w}\,dw + O\left(\frac{x^C}{T(\sigma+C-1)^\alpha}\right) + O\left(\frac{\psi(2x)x^{1-\sigma}\ln x}{T}\right) + O\left(\frac{\psi(N)x^{1-\sigma}}{T|x-N|}\right).$$

If x is an integer, the corresponding result is

$$\sum_{n=1}^{x-1}\frac{a_n}{n^s} + \frac{a_x}{2x^s} = \frac{1}{2\pi i}\int_{C-iT}^{C+iT} f(s+w)\frac{x^w}{w}\,dw + O\left(\frac{x^C}{T(\sigma+C-1)^\alpha}\right) + O\left(\frac{\psi(2x)x^{1-\sigma}\ln x}{T}\right).$$

We can invert a power series by examining its behavior in a neighborhood of the point $z = 1$. In this connection there appear analogues of real Tauberian theorems, namely Littlewood's theorem and the Hardy-Littlewood theorem.

The analogue of Littlewood's theorem is the well-known theorem of Fatou [47].

THEOREM. *Suppose a function $f(z)$ is defined by a power series $\sum_0^\infty a_n z^n$ that converges for $|z| < 1$. Suppose also that $f(z)$ is analytic at the point $z = 1$. If $a_n \to 0$, then $\sum_0^\infty a_n$ converges.*

It can be shown that if the coefficients a_n are real, then the condition $\lim_{n\to\infty} a_n = 0$ can be replaced by the one-sided condition

$$\underline{\lim}_{n\to\infty} a_n = 0.$$

A complete analogy to Littlewood's theorem! We see that (in this case) imposition of a condition in the complex domain leads to a weakening of the additional Tauberian condition (instead of $a_n = O(1/n)$ in Littlewood's theorem we require only $a_n \to 0$ in Fatou's theorem).

Korevaar [85] gave a quantitative form and generalization of Fatou's theorem. We give here a version of Korevaar's theorem that is somewhat less general than the original.

THEOREM. *Suppose a function represented by a power series $f(z) = \sum_0^\infty a_n z^n$ is analytic for $|z| < 1$ and analytic at the point $z = 1$. Suppose α is a real number, and let $\psi(n) = Kn^\alpha$, where $K > 0$ is a constant. Assume that $a_n \geq -\psi(n)$, $n = 1, 2, \ldots$. Then*

$$\sum_{k\leq n} a_k = f(1) + O(\psi(n)).$$

Korevaar proved this theorem by a method similar to the one used to prove Littlewood's theorem.

Fatou's theorem was extended to general Dirichlet series by M. Riesz [126].

THEOREM. *Let c be any fixed positive number. Suppose a function $f(s)$ is represented by a Dirichlet series $\sum_0^\infty a_n e^{-\lambda_n s}$ whose coefficients satisfy the condition*

$$\lim_{n\to\infty} \frac{a_0 e^{-\lambda_0 c} + \cdots + a_n e^{-\lambda_n c}}{e^{-\lambda_n c}} = 0,$$

as a result of which the function $f(s)$ is regular for $\operatorname{Re} s > 0$. Suppose also that $f(s)$ is regular at the point $s = s_0$, where $\operatorname{Re} s_0 = 0$. Then the series

$$\sum_{n=0}^{\infty} a_n e^{-\lambda_n s_0}$$

converges.

To make a connection between Fatou's theorem for a power series and Riesz's theorem, note (this is easy to prove) that the condition $a_n \to 0$ is equivalent to saying that for any $R > 1$

$$\lim_{n\to\infty} \frac{a_0 + a_1 R + \cdots + a_n R^n}{R^n} = 0.$$

Postnikov [115] proved a complex analogue of the Hardy-Littlewood Tauberian theorem.

THEOREM. *Suppose a power series* $f(z) = \sum_0^\infty a_n z^n$ *has radius of convergence equal to* 1. *Let* $z = re^{i\theta}$ $(r = |z|)$. *Assume that* $f(z) = 1/(1-z) + O(1)$ *for* $|\theta| \le c < \pi$. *Then*

$$\sum_{n \le N} a_n = N + O(\ln N).$$

Comparing Postnikov's theorem with Fatou's, we see that imposing conditions in the complex domain leads (in the given situation) to a significant improvement in the remainder term.

There is a complex Tauberian theorem for Dirichlet integrals, namely Ikehara's theorem, that is useful in analytic number theory.

Suppose $\psi(x)$ is a nondecreasing function on the interval $(0, \infty)$ such that $\psi(0) = 0$. Consider the integral

$$f(s) = \int_0^\infty e^{-sx}\, d\psi(x). \tag{6}$$

THEOREM. *Suppose the following conditions are satisfied:*

1) *The integral* (6) *converges for* $\operatorname{Re} s > 1$.

2) *Put* $s = 1 + \varepsilon + it$ $(\varepsilon > 0)$ *and* $h_\varepsilon(t) = f(s) - 1/(s-1)$. *The limit*

$$h(t) = \lim_{\varepsilon \to 0} h_\varepsilon(t)$$

exists uniformly in t *on any finite interval of the straight line* $\operatorname{Re} s = 1$.

Then

$$\lim_{x \to \infty} \frac{\psi(x)}{e^x} = 1.$$

The proof of this theorem can be found in many sources, so we will not give it here, referring the reader instead to Lang [93], Chapter VIII, §2.

This theorem has the following

COROLLARY. *Suppose* $f(s) = \sum_{n=1}^\infty a_n/n^s$ *is an ordinary Dirichlet series with nonnegative coefficients that converges for* $\operatorname{Re} s > 1$. *Suppose also that* $f(s)$ *is a regular function on the line* $\operatorname{Re} s = 1$ *except at the point* $s = 1$, *where it has a first-order pole with residue* 1. *Then*

$$\lim_{N \to \infty} \frac{1}{N} \sum_{n \le N} a_n = 1.$$

In Ikehara's theorem a condition is imposed on the behavior of a Dirichlet series in a neighborhood of the whole abscissa of convergence. Comparing this result with the Hardy-Littlewood theorem for an ordinary Dirichlet series, we see that imposing conditions in a neighborhood of the whole abscissa of convergence leads to a qualitatively stronger result.

Generalizations of Ikehara's theorem were given by Raĭkov [120] and Delange [27]. We state Delange's theorem.

THEOREM. *Suppose ρ is a real number, $\rho \neq 0, -1, -2, \dots$. Assume that for* $\operatorname{Re} s > a$

$$\int_0^\infty e^{-sx} d\varphi(x) = \frac{1}{(s-a)^\rho} g(s) + h(s),$$

where $g(s)$ and $h(s)$ are regular functions for $\operatorname{Re} s \geq a$ *and where $g(a) \neq 0$. Then as $x \to \infty$*

$$\varphi(x) \sim \frac{g(a)}{\Gamma(\rho)} e^{ax} x^{\rho-1}.$$

Tauberian theorems pertaining to rapidly increasing Laplace-Stieltjes transforms are important for additive number theory problems connected with partitions of natural numbers into an infinite number of summands. In this book we will prove the following theorem of Ingham (see [75]).

THEOREM. *Suppose that in the Laplace-Stieltjes transform*

$$f(s) = \int_0^\infty e^{-us}\, d\alpha(u) \tag{4}$$

the function $\alpha(u)$ satisfies the following conditions:

1) $\alpha(0) = 0$.

2) *$\alpha(u)$ is nondecreasing as $u \to \infty$.*

3) $$f(s) \sim C \left(\frac{M}{s}\right)^{m\beta - 1/2} \exp\left(\frac{1}{\beta}\left(\frac{M}{s}\right)^\beta\right)$$

(where C, M, m, $\beta > 0$ are real constants) uniformly as $s \to 0$ in any angle of the form $|\operatorname{Im} s| \leq \Delta \operatorname{Re} s$, $0 < \Delta < \infty$.

Then as $u \to \infty$

$$\alpha(u) \sim C\sqrt{\frac{1-\alpha}{2\pi}} (uM)^{m\alpha - 1/2} e^{(uM)^\alpha/\alpha}.$$

§1.2. The Hardy-Littlewood Tauberian theorem

We state Freud's theorem.

THEOREM. *Suppose $f(t)$ is a nonnegative function defined for* $0 \leq t < \infty$ *and*

$$F(s) = \int_0^\infty f(t) e^{-st}\, d\tau(t),$$

where $\tau(t)$ *is a monotone nondecreasing function defined for* $0 \le t < \infty$. *Let* $\alpha > 0$. *If for real* $s > 0$

$$F(s) = A\frac{\Gamma(1+\alpha)}{s^\alpha}(1+r(s)),$$

where $|r(s)| \le C_0 s^\varepsilon$ *and* $A > 0$, $C_0 > 0$, *and* $\varepsilon > 0$ *are constants, then*

$$\int_0^x f(t)\,d\tau(t) = Ax^\alpha(1+\rho(x)),$$

where $|\rho(x)| \le C_1/\ln x$, $C_1 > 0$ *being a constant.*

We will show how this theorem contains the Hardy-Littlewood theorem (with remainder term) for a power series.

Suppose that as $z \to 1$ a series with nonnegative coefficients a_n satisfies a relation

$$\sum_{n=0}^\infty a_n z^n = \frac{1}{(1-z)^\alpha}(1+O((1-z)^\varepsilon)),$$

where $\varepsilon > 0$. Then as $s \to 0$

$$\sum_{n=0}^\infty a_n e^{-ns} = \frac{1}{s^\alpha}(1+O(s^\varepsilon)),$$

where $\varepsilon > 0$. We choose any nonnegative function $f(t)$ such that $f(n) = a_n$, $n = 0, 1, \ldots$, and we define the function $\tau(t)$ by $\tau(0) = 0$ and $\tau(t) = [t] + 1$, $t > 0$. Then

$$\sum_{n=0}^\infty a_n e^{-ns} = \int_0^\infty f(t)e^{-ts}\,d\tau(t) = \frac{1}{s^\alpha}(1+O(s^\varepsilon)).$$

Freud's theorem (with $A = 1/\Gamma(1+\alpha)$) yields

$$\sum_{n\le x} a_n = \int_0^x f(t)\,d\tau(t) = \frac{1}{\Gamma(1+\alpha)}x^\alpha + O\left(\frac{x^\alpha}{\ln x}\right).$$

The proof of Freud's theorem is based on Markov's theory of one-sided approximation.

For a given $\alpha > 0$ we introduce the numbers

$$s_k = \frac{1}{\Gamma(\alpha)}\int_0^1 x^k\left(\ln\frac{1}{x}\right)^{\alpha-1}dx, \qquad k = 0, 1, 2, \ldots.$$

These numbers can be calculated: we make the substitution $\ln(1/x) = u$ and obtain

$$\begin{aligned} s_k &= \frac{1}{\Gamma(\alpha)}\int_0^\infty e^{-u(k+1)}u^{\alpha-1}\,du \\ &= \frac{1}{(k+1)^\alpha\Gamma(\alpha)}\int_0^\infty e^{-z}z^{\alpha-1}\,dz = \frac{1}{(k+1)^\alpha}. \end{aligned}$$

LEMMA 1. *For any integer* $m \geq 1$

$$c_m = \begin{vmatrix} s_0 & s_1 & \dots & s_{m-1} \\ s_1 & s_2 & \dots & s_m \\ \dots & \dots & \dots & \dots \\ s_{m-1} & s_m & \dots & s_{2m-2} \end{vmatrix} \neq 0.$$

If we assume $c_m = 0$, then there must be a linear relation among the rows, i.e., there must exist m numbers $\beta_0, \beta_1, \dots, \beta_{m-1}$, not all equal to zero, such that

$$\begin{aligned} \beta_0 s_0 + \beta_1 s_1 + \dots + \beta_{m-1} s_{m-1} &= 0, \\ \beta_0 s_1 + \beta_1 s_2 + \dots + \beta_{m-1} s_m &= 0, \\ \dots\dots\dots\dots\dots\dots\dots\dots\dots\dots \\ \beta_0 s_{m-1} + \beta_1 s_m + \dots + \beta_{m-1} s_{2m-2} &= 0. \end{aligned}$$

But the first equality can be written as follows:

$$\begin{aligned} \beta_0 \int_0^1 \frac{1}{\Gamma(\alpha)} \left(\ln \frac{1}{x}\right)^{\alpha-1} dx + \beta_1 \int_0^1 \frac{1}{\Gamma(\alpha)} x \left(\ln \frac{1}{x}\right)^{\alpha-1} dx \\ + \dots + \beta_{m-1} \int_0^1 \frac{1}{\Gamma(\alpha)} x^{m-1} \left(\ln \frac{1}{x}\right)^{\alpha-1} dx \\ = \frac{1}{\Gamma(\alpha)} \int_0^1 (\beta_0 + \beta_1 x + \dots + \beta_{m-1} x^{m-1}) \left(\ln \frac{1}{x}\right)^{\alpha-1} dx = 0. \end{aligned}$$

Thus,

$$\begin{aligned} \frac{1}{\Gamma(\alpha)} \int_0^1 (\beta_0 + \beta_1 x + \dots + \beta_{m-1} x^{m-1}) \left(\ln \frac{1}{x}\right)^{\alpha-1} dx &= 0, \\ \frac{1}{\Gamma(\alpha)} \int_0^1 (\beta_0 x + \beta_1 x^2 + \dots + \beta_{m-1} x^m) \left(\ln \frac{1}{x}\right)^{\alpha-1} dx &= 0, \\ \dots\dots\dots\dots\dots\dots\dots\dots\dots\dots\dots\dots \\ \frac{1}{\Gamma(\alpha)} \int_0^1 (\beta_0 x^{m-1} + \beta_1 x^m + \dots + \beta_{m-1} x^{2m-2}) \left(\ln \frac{1}{x}\right)^{\alpha-1} dx &= 0. \end{aligned}$$

Multiplying the first equality by β_0, the second by β_1, the third by $\beta_2, \dots$, and then adding, we obtain

$$\frac{1}{\Gamma(\alpha)} \int_0^1 (\beta_0 + \beta_1 x + \dots + \beta_{m-1} x^{m-1})^2 \left(\ln \frac{1}{x}\right)^{\alpha-1} dx = 0.$$

There exists an interval (a, b) in $(0, 1)$ on which $\beta_0 + \beta_1 x + \dots + \beta_{m-1} x^{m-1} \neq 0$. Thus on (a, b) we have

$$(\beta_0 + \beta_1 x + \dots + \beta_{m-1} x^{m-1})^2 \geq \rho > 0$$

and

$$\frac{1}{\Gamma(\alpha)}\int_0^1 (\beta_0+\beta_1 x+\cdots+\beta_{m-1}x^{m-1})^2\left(\ln\frac{1}{x}\right)^{\alpha-1} dx$$
$$\geq \frac{\rho}{\Gamma(\alpha)}\int_a^b \left(\ln\frac{1}{x}\right)^{\alpha-1} dx > 0,$$

since $(\ln(1/x))^{\alpha-1}>0$ on (a,b).

This contradiction proves our assertion.

We now introduce polynomials $t_m(x)$, $m=0,1,\ldots$, as follows: $t_0(x)=1$, and for $m\geq 1$

$$t_m(x)=L_0^{(m)}+L_1^{(m)}x+\cdots+L_{m-1}^{(m)}x^{m-1}+L_m^{(m)}x^m,$$

where the numbers $L_0^{(m)}, L_1^{(m)},\ldots,L_{m-1}^{(m)}$ are determined from the system of equations

$$\begin{gathered}
L_0^{(m)}s_0+L_1^{(m)}s_1+\cdots+L_{m-1}^{(m)}s_{m-1}+s_m=0,\\
L_0^{(m)}s_1+L_1^{(m)}s_2+\cdots+L_{m-1}^{(m)}s_m+s_{m+1}=0,\\
\cdots\cdots\cdots\cdots\cdots\cdots\cdots\cdots\cdots\cdots\cdots\cdots\\
L_0^{(m)}s_{m-1}+L_1^{(m)}s_m+\cdots+L_{m-1}^{(m)}s_{2m-2}+s_{2m-1}=0,
\end{gathered}$$

and we also put $L_m^{(m)}=1$. Since $c_m\neq 0$, the coefficients are uniquely determined:

$$t_m(x)=\frac{(-1)^m}{c_m}\begin{vmatrix} 1 & x & \cdots & x^m\\ s_0 & s_1 & \cdots & s_m\\ s_1 & s_2 & \cdots & s_{m+1}\\ \cdots & \cdots & \cdots & \cdots\\ s_{m-1} & s_m & \cdots & s_{2m-1}\end{vmatrix}.$$

LEMMA 2. *For each integer $m\geq 2$ there exist real numbers a_m and b_m such that*

$$t_m(x)=(x+b_m)t_{m-1}(x)-a_mt_{m-2}(x).$$

PROOF. We divide $t_m(x)$ by $t_{m-1}(x)$ with a remainder:

$$t_m(x)=(x+b_m)t_{m-1}(x)+R(x),$$

where $R(x)$ has degree at most $m-2$. We have

$$L_m^{(m)}-L_{m-1}^{(m-1)}=0,\qquad L_{m-1}^{(m)}-b_mL_{m-1}^{(m-1)}-L_{m-2}^{(m-1)}=0.$$

Let $R(x)=\sum_0^{m-2}b_kx^k$. We prove that the coefficients of $R(x)$ satisfy the system of equations

$$\sum_{k=0}^{m-2} l_k s_{k+j}=0,\qquad j=0,1,\ldots,m-1.$$

Indeed,

$$\begin{aligned}\sum_{k=0}^{m-2} l_k s_{k+j} &= s_j(L_0^{(m)} - b_m L_0^{(m-1)}) + \sum_{k=1}^{m-2}(L_k^{(m)} - b_m L_k^{(m-1)} - L_{k-1}^{(m-1)})s_{k+j} \\ &= s_j(L_0^{(m)} - b_m L_0^{(m-1)}) + \sum_{k=1}^{m-2}(L_k^{(m)} - b_m L_k^{(m-1)} - L_{k-1}^{(m-1)})s_{k+j} \\ &+ s_{j+m-1}(L_{m-1}^{(m)} - b_m L_{m-1}^{(m-1)} - L_{m-2}^{(m-1)}) + s_{j+m}(L_m^{(m)} - L_{m-1}^{(m-1)}) \\ &= \sum_{k=0}^{m} L_k^{(m)} s_{k+j} - b_m \sum_{k=0}^{m-1} L_k^{(m-1)} s_{k+j} - \sum_{k=1}^{m} L_{k-1}^{(m-1)} s_{k+j} = 0.\end{aligned}$$

By definition, the coefficients of the polynomial $t_{m-2}(x)$ satisfy the system

$$\sum_{k=0}^{m-2} L_k^{(m-2)} s_{k+j} = 0, \qquad j = 0, 1, \ldots, m-1.$$

Since

$$\begin{vmatrix} s_0 & \cdots & s_{m-1} \\ \cdots & \cdots & \cdots \\ s_{m-1} & \cdots & s_{2(m-1)} \end{vmatrix} \neq 0,$$

it follows that $R(x) = -a_m t_{m-2}(x)$. The lemma is proved.

LEMMA 3. *For $j = 0, 1, \ldots, m-1$*

$$\frac{1}{\Gamma(\alpha)} \int_0^1 x^j t_m(x) \left(\ln \frac{1}{x}\right)^{\alpha-1} dx = 0.$$

PROOF.

$$\begin{aligned}&\frac{1}{\Gamma(\alpha)} \int_0^1 x^j t_m(x) \left(\ln \frac{1}{x}\right)^{\alpha-1} dx \\ &= \frac{1}{\Gamma(\alpha)} \int_0^1 (L_0^m x^j + L_1^{(m)} x^{j+1} + \cdots + x^{m+j}) \left(\ln \frac{1}{x}\right)^{\alpha-1} dx \\ &= \frac{L_0^{(m)}}{\Gamma(\alpha)} \int_0^1 x^j \left(\ln \frac{1}{x}\right)^{\alpha-1} dx \\ &+ \cdots + L_{m-1}^{(m)} \frac{1}{\Gamma(\alpha)} \int_0^1 x^{m-1+j} \left(\ln \frac{1}{x}\right)^{\alpha-1} dx \\ &+ \frac{1}{\Gamma(\alpha)} \int_0^1 x^{m+j} \left(\ln \frac{1}{x}\right)^{\alpha-1} dx \\ &= L_0^{(m)} s_j + L_1^{(m)} s_{j+1} + \cdots + L_{m-1}^{(m)} s_{m-1+j} + s_{m+j} = 0\end{aligned}$$

for $j = 0, 1, \ldots, m-1$.

COROLLARY. *If $f(x)$ is any polynomial of degree at most $m-1$, then*

$$\frac{1}{\Gamma(\alpha)}\int_0^1 f(x)t_m(x)\left(\ln\frac{1}{x}\right)^{\alpha-1}dx=0.$$

In particular, if $i\neq j$, then

$$\frac{1}{\Gamma(\alpha)}\int_0^1 t_j(x)t_i(x)\left(\ln\frac{1}{x}\right)^{\alpha-1}dx=0.$$

Thus the polynomials 1, $t_1(x)$, $t_2(x),\ldots$ form an orthogonal system on $(0,1)$ with weight

$$\frac{1}{\Gamma(\alpha)}\left(\ln\frac{1}{x}\right)^{\alpha-1}.$$

LEMMA 4. *The roots of the polynomial $t_m(x)$ are real and distinct, and lie in the interval $(0,1)$.*

PROOF. Let $c_1,\ldots,c_k$ be the real roots of $t_m(x)$ of odd multiplicity. Assume $k<m$. Form the polynomial

$$\omega(x)=(x-c_1)\cdots(x-c_k)t_m(x).$$

It is clear from a factorization of $t_m(x)$ that $\omega(x)\geq 0$.

By assumption, $k<m$; hence

$$\frac{1}{\Gamma(\alpha)}\int_0^1 (x-c_1)\cdots(x-c_k)t_m(x)\left(\ln\frac{1}{x}\right)^{\alpha-1}dx=0,$$

$$\frac{1}{\Gamma(\alpha)}\int_0^1 \omega(x)\left(\ln\frac{1}{x}\right)^{\alpha-1}dx=0.$$

But this is impossible, since $(\ln(1/x))^{\alpha-1}>0$ on $(0,1)$ and $\omega(x)$ is a polynomial, not identically zero, that assumes only nonnegative values.

We will prove that for any $m\geq 1$ the roots of $t_m(x)$ all lie in $(0,1)$. Assume there is a root ξ of $t_m(x)$ that does not lie in $(0,1)$. Suppose $\xi\geq 1$ (the case $\xi\leq 0$ is handled analogously). Since $-t_m(x)/(x-\xi)$ is a polynomial of degree at most $m-1$, it follows that

$$\frac{1}{\Gamma(\alpha)}\int_0^1 t_m(x)\left(-\frac{t_m(x)}{x-\xi}\right)\left(\ln\frac{1}{x}\right)^{\alpha-1}dx=0.$$

But on (0,1) we have $-t_m^2(x)/(x-\xi)\geq 0$ and $(\ln(1/x))^{\alpha-1}>0$, which contradicts

$$\frac{1}{\Gamma(\alpha)}\int_0^1 \left(-\frac{t_m^2(x)}{x-\xi}\right)\left(\ln\frac{1}{x}\right)^{\alpha-1}dx=0.$$

Let

$$\psi_m(z)=\frac{1}{\Gamma(\alpha)}\int_0^1 \frac{t_m(x)-t_m(z)}{x-z}\left(\ln\frac{1}{x}\right)^{\alpha-1}dx.$$

This is a polynomial of degree at most $m-1$ (it is easy to see that the coefficient of its leading term is 1).

Using the recurrence relation for the polynomials $t_m(x)$, we can prove

LEMMA 5. *For* $m \geq 2$

$$\psi_m(z) = (z + b_m)\psi_{m-1}(z) - a_m\psi_{m-2}(z).$$

Consider the rational function $\psi_m(x)/t_m(x)$. Expand it into partial fractions. Since the roots of the polynomials $t_m(x)$ are real and simple, the expansion of $\psi_m(x)/t_m(x)$ has the form

$$\frac{\psi_m(x)}{t_m(x)} = \sum_{\xi} \frac{\rho_m(\xi)}{x - \xi}$$

(ξ stands for a root of the equation $t_m(x) = 0$; $\rho_m(\xi)$ is the notation for the coefficients).

LEMMA 6.

$$\rho_m(\xi_i) = \frac{\psi_m(\xi_i)}{t'_m(\xi_i)}.$$

The proof is obvious.

We will establish the following quadrature formula.

LEMMA 7. *For any polynomial* $\Omega(x)$ *of degree at most* $2m-1$

$$\frac{1}{\Gamma(\alpha)} \int_0^1 \Omega(x) \left(\ln \frac{1}{x}\right)^{\alpha-1} dx = \sum_{\xi} \rho_m(\xi)\Omega(\xi).$$

PROOF. Take any polynomial $\Omega(x)$ of degree at most $2m-1$ and divide it by $t_m(x)$ with a remainder:

$$\Omega(x) = t_m(x)\theta(x) + \Omega_0(x).$$

We will show that

$$\Omega_0(x) = \sum_{\xi_i} \frac{t_m(x)}{(x - \xi_i)t'_m(\xi_i)}\Omega(\xi_i).$$

The right- and left-hand sides are polynomials of degree at most $m-1$. To show that the right-hand side is equal to the left it suffices to show that the values of the right-hand side agree with those of the left at m points.

For $x = \xi_j$, $j = 1, \dots, m$,

$$\Omega_0(\xi_j) = \Omega(\xi_j) - t_m(\xi_j)\theta(\xi_j) = \Omega(\xi_j),$$

and

$$\sum_{\xi_j} \frac{t_m(x)}{(x - \xi_j)t'_m(\xi_j)}\Omega(\xi_j)\bigg|_{x=\xi_j} = \Omega(\xi_j).$$

This establishes the formula.

Consider

$$\begin{aligned}
&\frac{1}{\Gamma(\alpha)}\int_0^1 \Omega(x)\left(\ln\frac{1}{x}\right)^{\alpha-1}dx\\
&\quad=\frac{1}{\Gamma(\alpha)}\int_0^1 t_m(x)\theta(x)\left(\ln\frac{1}{x}\right)^{\alpha-1}dx+\frac{1}{\Gamma(\alpha)}\int_0^1 \Omega_0(x)\left(\ln\frac{1}{x}\right)^{\alpha-1}dx\\
&\quad=\frac{1}{\Gamma(\alpha)}\int_0^1 \Omega_0(x)\left(\ln\frac{1}{x}\right)^{\alpha-1}dx.
\end{aligned}$$

The first integral vanishes, since $\theta(x)$ is a polynomial of degree at most $m-1$. Also,

$$\begin{aligned}
\frac{1}{\Gamma(\alpha)}\int_0^1 \Omega(x)\left(\ln\frac{1}{x}\right)^{\alpha-1}dx&=\frac{1}{\Gamma(\alpha)}\int_0^1 \Omega_0(x)\left(\ln\frac{1}{x}\right)^{\alpha-1}dx\\
&=\frac{1}{\Gamma(\alpha)}\int_0^1\left(\sum_{\xi_i}\frac{t_m(x)}{(x-\xi_i)t'_m(\xi_i)}\right)\Omega(\xi_i)\left(\ln\frac{1}{x}\right)^{\alpha-1}dx\\
&=\sum_{\xi_i}\frac{\Omega(\xi_i)}{t'_m(\xi_i)}\frac{1}{\Gamma(\alpha)}\int_0^1\frac{t_m(x)-t_m(\xi_i)}{x-\xi_i}\left(\ln\frac{1}{x}\right)^{\alpha-1}dx\\
&=\sum_{\xi_i}\Omega(\xi_i)\frac{\psi_m(\xi_i)}{t'_m(\xi_i)}=\sum_{\xi_i}\rho_m(\xi_i)\Omega(\xi_i).
\end{aligned}$$

The lemma is proved.

We now obtain some corollaries.

COROLLARY 1. *$\rho_m(\xi_j)>0$ for $j=1,\dots,m$.*

Indeed, consider the polynomial

$$Q(x)=\prod_{i\neq j}(x-\xi_j)^2=\frac{t_m^2(x)}{(x-\xi_i)^2}.$$

Its degree is $2m-2$. On $(0,1)$ we have $Q(x)\geq 0$. Also,

$$\sum_{\xi}\rho_m(\xi)Q(\xi)=\frac{1}{\Gamma(\alpha)}\int_0^1 Q(x)\left(\ln\frac{1}{x}\right)^{\alpha-1}dx>0,$$

$$0<\sum_{\xi}\rho_m(\xi)Q(\xi)=\rho_m(\xi_i)Q(\xi_i).$$

But $Q(\xi_i)>0$, and hence $\rho_m(\xi_i)>0$, as required.

We will now denote the roots of the polynomial $t_m(x)$ by $\xi_i^{(m)}$, and the quantities $\rho_m(\xi_i^{(m)})$ will be written $\rho(\xi_i^{(m)})$.

COROLLARY 2. *For any continuous function $f(x)$ on $[0,1]$*

$$\lim_{m\to\infty}\sum_{i=1}^{m}\rho(\xi_i^{(m)})f(\xi_i^{(m)})=\frac{1}{\Gamma(\alpha)}\int_0^1 f(x)\left(\ln\frac{1}{x}\right)^{\alpha-1}dx.$$

PROOF. Let $f(x)$ be any continuous function on $[0,1]$. Given $\varepsilon>0$, by a classical theorem of Weierstrass, there exists a polynomial $\Omega(x)$ such that $|f(x)-\Omega(x)|\le\varepsilon$.

Let N be the degree of $\Omega(x)$. For $m\ge(N+2)/2$ we have, by the quadrature formula,

$$\sum_{i=1}^{m}\Omega(\xi_i^{(m)})\rho(\xi_i^{(m)})=\frac{1}{\Gamma(\alpha)}\int_0^1\Omega(x)\left(\ln\frac{1}{x}\right)^{\alpha-1}dx.$$

Also, for $m\ge(N+2)/2$

$$\begin{aligned}
&\left|\sum_{i=1}^{m}f(\xi_i^{(m)})\rho(\xi_i^{(m)})-\frac{1}{\Gamma(\alpha)}\int_0^1 f(x)\left(\ln\frac{1}{x}\right)^{\alpha-1}dx\right|\\
&\quad=\left|\sum_{i=1}^{m}\rho(\xi_i^{(m)})(f(\xi_i^{(m)})-\Omega(\xi_i^{(m)}))+\sum_{i=1}^{m}\Omega(\xi_i^{(m)})\rho(\xi_i^{(m)})\right.\\
&\qquad\qquad-\frac{1}{\Gamma(\alpha)}\int_0^1\Omega(x)\left(\ln\frac{1}{x}\right)^{\alpha-1}dx\\
&\qquad\qquad\left.+\frac{1}{\Gamma(\alpha)}\int_0^1(\Omega(x)-f(x))\left(\ln\frac{1}{x}\right)^{\alpha-1}dx\right|\\
&\quad\le\sum_{i=1}^{m}\rho(\xi_i^{(m)})|f(\xi_i^{(m)})-\Omega(\xi_i^{(m)})|\\
&\qquad\qquad+\frac{1}{\Gamma(\alpha)}\int_0^1|\Omega(x)-f(x)|\left(\ln\frac{1}{x}\right)^{\alpha-1}dx\\
&\quad\le\varepsilon\left(\sum_{i=1}^{m}\rho(\xi_i^{(m)})+\frac{1}{\Gamma(\alpha)}\int_0^1\left(\ln\frac{1}{x}\right)^{\alpha-1}dx\right)\\
&\quad=\frac{2\varepsilon}{\Gamma(\alpha)}\int_0^1\left(\ln\frac{1}{x}\right)^{\alpha-1}dx=2\varepsilon.
\end{aligned}$$

Thus for any $\varepsilon>0$

$$\overline{\lim_{m\to\infty}}\left|\sum_{i=1}^{m}f(\xi_i^{(m)})\rho(\xi_i^{(m)})-\frac{1}{\Gamma(\alpha)}\int_0^1 f(x)\left(\ln\frac{1}{x}\right)^{\alpha-1}dx\right|<2\varepsilon.$$

But ε can be arbitrarily small. Thus,

$$\lim_{m\to\infty}\sum_{i=1}^{m}f(\xi_i^{(m)})\rho(\xi_i^{(m)})=\frac{1}{\Gamma(\alpha)}\int_0^1 f(x)\left(\ln\frac{1}{x}\right)^{\alpha-1}dx.$$

COROLLARY 3. *Suppose η is any fixed number such that $0 < \eta < 1$. Suppose also that $\xi_\nu^{(m)} < \eta < \xi_{\nu+1}^{(m)}$ (ν depends on m). Then*

$$\lim_{m\to\infty} \xi_\nu^{(m)} = \eta, \qquad \lim_{m\to\infty} \xi_{\nu+1}^{(m)} = \eta.$$

We will prove that $\lim_{m\to\infty} \xi_\nu^{(m)} = \eta$ (the second equality is handled analogously).

Take any $\varepsilon > 0$. Consider a continuous function $f(x)$ such that $f(x) = 0$ on $[0, \eta-\varepsilon]$ and $[\eta, 1]$, $f(x) > 0$ on $(\eta-\varepsilon, \eta)$, and $f(x)$ is arbitrary elsewhere. Applying Corollary 2 to this function, we obtain

$$\lim_{m\to\infty} \sum_{\eta-\varepsilon<\xi_i^{(m)}<\eta} f(\xi_i^{(m)})\rho(\xi_i^{(m)}) = \frac{1}{\Gamma(\alpha)} \int_0^1 f(x) \left(\ln\frac{1}{x}\right)^{\alpha-1} dx > 0.$$

Thus, from some m on,

$$\sum_{\eta-\varepsilon<\xi_i^{(m)}<\eta} f(\xi_i^{(m)})\rho(\xi_i^{(m)}) > 0.$$

Consequently, the sum is not empty. Thus the interval $(\eta-\varepsilon, \eta)$ contains a root of $t_m(x)$. The corollary is proved.

LEMMA 8. *Suppose η is a fixed number such that $0 < \eta < 1$, and $\xi_\nu^{(m)}$ and $\xi_{\nu+1}^{(m)}$, $\xi_\nu^{(m)} \le \eta < \xi_{\nu+1}^{(m)}$, are two roots of $t_m(x)$ adjacent to η. Then*

$$\rho(\xi_\nu^{(m)}) \le C/m, \qquad \rho(\xi_{\nu+1}^{(m)}) \le C/m,$$

where the constant C depends on η.

PROOF. We will use the letter C to denote positive constants, not necessarily equal to each other. Let us prove the first inequality; the second can be obtained analogously. We will first prove an auxiliary assertion: For any polynomial $\pi(x)$ of degree at most $m-1$ such that $\pi(\xi_\nu^{(m)}) = 1$ we have

$$\frac{1}{\Gamma(\alpha)} \int_0^1 \pi^2(x) \left(\ln\frac{1}{x}\right)^{\alpha-1} dx \ge \rho(\xi_\nu^{(m)}).$$

Indeed, the degree of $\pi^2(x)$ is at most $2m-2$; therefore, applying the quadrature formula, we obtain

$$\frac{1}{\Gamma(\alpha)} \int_0^1 \pi^2(x) \left(\ln\frac{1}{x}\right)^{\alpha-1} dx = \sum_{i=1}^{m} \pi^2(\xi_i^{(m)})\rho(\xi_i^{(m)}) \ge \rho(\xi_\nu^{(m)}).$$

To obtain an estimate of $\rho(\xi_\nu^{(m)})$ we consider the polynomial

$$\frac{2}{m-1}\left(\frac{1}{2} + \sum_{k=1}^{m-2}\left(1 - \frac{k}{m-1}\right)\cos k\theta\right) = \frac{1}{(m-1)^2}\frac{\sin^2((m-1)\theta/2)}{\sin^2(\theta/2)},$$
$$-\pi/2 < \theta \le \pi/2.$$

The expansion

$$\cos nx = \frac{(\cos x + i \sin x)^n + (\cos x - i \sin x)^n}{2}$$
$$= \cos^n x - \frac{n(n-1)}{2} \cos^{n-2} x \sin^2 x + \cdots$$

contains only even powers of the sine, and any even power of the sine can be expressed as a polynomial in $\cos x$. Therefore,

$$\frac{2}{m-1}\left(\frac{1}{2} + \sum_{k=1}^{m-2}\left(1 - \frac{k}{m-1}\right)\cos k\theta\right)$$
$$= \frac{1}{(m-1)^2}\frac{\sin^2((m-1)\theta/2)}{\sin^2(\theta/2)} = P(\cos\theta).$$

$P(\cos\theta)$ is a polynomial of degree $m-2$. Obviously,

$$0 \le P(\cos\theta) \le \frac{2}{m-1}\left(\frac{1}{2} + \sum_{k=1}^{m-2}\left(1 - \frac{k}{m-1}\right)\right) = 1$$

and

$$P(\cos\theta) \le \frac{1}{(m-1)^2 \sin^2(\theta/2)} < \frac{C}{(m-1)^2\theta^2},$$

since $2/\pi \le \sin x/x \le 1$ when $0 \le x \le \pi/2$. Thus,

$$0 \le P(\cos\theta) \le \min(1, C/(m-1)^2\theta^2).$$

At $\theta = 0$

$$P(\cos\theta)|_{\theta=0} = \frac{2}{m-1}\left(\frac{1}{2} + \sum_{k=1}^{m-2}\left(1 - \frac{k}{m-1}\right)\right) = 1.$$

Consider the expression

$$\frac{P(\cos(\theta + \pi/2)) + P(\cos(\pi/2 - \theta))}{2}$$
$$= \frac{2}{m-1}\left(\frac{1}{2} + \sum_{k=1}^{m-2}\left(1 - \frac{k}{m-1}\right)\frac{\cos k(\pi/2 - \theta) + \cos k(\pi/2 + \theta)}{2}\right)$$
$$= \frac{2}{m-1}\left(\frac{1}{2} + \sum_{k=1}^{m-2}\left(1 - \frac{k}{m-1}\right)\cos\frac{k\pi}{2}\cos k\theta\right) = Q(\cos\theta).$$

The degree of $Q(\cos\theta)$ is at most $m-2$, and

$$Q(\cos\theta) \geq 0,$$

$$0 \leq Q(\cos\theta) \leq \min\left(1, \frac{C}{(m-1)^2(\pi/2+\theta)^2} + \frac{C}{(m-1)^2(\pi/2-\theta)^2}\right).$$

We now make a change of variable. For $0 \leq x \leq 1$ we put

$$\cos\theta = x - \xi_\nu^{(m)}, \qquad Q^*(x) = Q(x - \xi_\nu^{(m)}),$$

so that

$$0 \leq Q^{*2}(x) \leq \min\left(1, \left(\frac{C}{(m-1)^2(\pi/2 + \arccos(x - \xi_\nu^{(m)}))^2} + \frac{C}{(m-1)^2(\pi/2 - \arccos(x - \xi_\nu^{(m)}))^2}\right)^2\right).$$

Since $(a+b)^2 \leq 2a^2 + 2b^2$, it follows that

$$0 \leq Q^{*2}(x) \leq \min\left(1, \frac{C}{(m-1)^4\left(\pi/2 + \arccos(x - \xi_\nu^{(m)})\right)^4} + \frac{C}{(m-1)^4\left(\pi/2 - \arccos(x - \xi_\nu^{(m)})\right)^4}\right)$$

Since $-\xi_\nu^{(m)} \to -\eta$ as $m \to \infty$, for a sufficiently large m we have

$$x - \xi_\nu^{(m)} \geq -\xi_\nu^{(m)} \geq \eta' > -1$$

(the last inequality is strict). Therefore the quantity

$$\pi/2 + \arccos(x - \xi_\nu^{(m)})$$

is bounded away from zero by a finite quantity (not tending to zero), and

$$0 \leq Q^{*2}(x) \leq \min\left(1, \frac{C(\eta)}{(m-1)^4} + \frac{C}{(m-1)^4(\arcsin(x - \xi_\nu^{(m)}))^4}\right).$$

Since $|\arcsin x| \geq x$, it follows that

$$0 \leq Q^{*2}(x) \leq \min\left(1, \frac{C(\eta)}{(m-1)^4} + \frac{C}{(m-1)^4(x - \xi_\nu^{(m)})^4}\right).$$

We calculate $Q^*(\xi_\nu^{(m)})$, which we denote by d_m:

$$\begin{aligned}
d_m &= Q^*(\xi_\nu^{(m)}) = Q\left(\cos\frac{\pi}{2}\right) \\
&= \frac{2}{m-1}\left(\frac{1}{2} + \sum_{k=1}^{m-1}\left(1-\frac{k}{m-1}\right)\left(\cos\frac{k\pi}{2}\right)^2\right) \\
&= \frac{2}{m-1}\left(\frac{1}{2} + \sum_{\substack{k=1 \\ k\equiv 0 (\mathrm{mod}\, 2)}}^{m-2}\left(1-\frac{k}{m-1}\right)\right) \\
&= \frac{2}{m-1}\left(\frac{1}{2} + \left[\frac{m-2}{2}\right] - \frac{2}{m-1}\sum_{k'=1}^{[(m-2)/2]} k'\right) \\
&= \frac{2}{m-1}\left(\frac{1}{2} + \left[\frac{m-2}{2}\right] - \frac{1}{m-1}\left[\frac{m-2}{2}\right]\left(\left[\frac{m-2}{2}\right]+1\right)\right).
\end{aligned}$$

Obviously $d_m \to \frac{1}{2}$, since as $m \to \infty$

$$d_m \sim \frac{2}{m}\left(\frac{m}{2} - \frac{1}{m}\frac{m^2}{4}\right) \sim \frac{1}{2}.$$

We distinguish two cases.

I. $\alpha \geq 1$. Then

$$\frac{1}{\Gamma(\alpha)}\int_0^1 \left(\frac{Q^*(x)}{d_m}\frac{x}{\xi_\nu^{(m)}}\right)^2 \left(\ln\frac{1}{x}\right)^{\alpha-1} dx \geq \rho(\xi_\nu^{(m)}).$$

Since $d_m \to \frac{1}{2}$ and $\xi_\nu^{(m)} \to \eta$, it follows that

$$\rho(\xi_\nu^{(m)}) < C\int_0^1 Q^{*2}(x)x^2\left(\ln\frac{1}{x}\right)^{\alpha-1} dx.$$

Since $x^2(\ln(1/x))^{\alpha-1} < C$ on $[0,1]$,we have

$$\rho(\xi_\nu^{(m)}) < C\int_0^1 Q^{*2}(x)\, dx,$$

$$\begin{aligned}
\rho(\xi_\nu^{(m)}) < C\Bigg(&\int_{|x-\xi_\nu^{(m)}|\leq 1/m} dx + \frac{1}{(m-1)^4}\int_{|x-\xi_\nu^{(m)}|\geq 1/m,} dx \\
&+ \frac{1}{(m-1)^4}\int_{|x-\xi_\nu^{(m)}|\geq 1/m} \frac{dx}{(x-\xi_\nu^{(m)})^4}\Bigg)
\end{aligned}$$

$$= O\left(\frac{1}{m}\right) + O\left(\frac{1}{m^4}\int_0^1 dx\right) + O\left(\frac{1}{m^4}\int_{\substack{-\infty \\ |z|>1/m}}^{\infty} \frac{dz}{z^4}\right) = O\left(\frac{1}{m}\right).$$

II. $0 < \alpha < 1$. Then

$$\frac{1}{\Gamma(\alpha)} \int_0^1 \left(\frac{Q^*(x)}{d_m} \frac{1-x}{1-\xi_\nu^{(m)}} \right)^2 \left(\ln \frac{1}{x} \right)^{\alpha-1} dx \geq \rho(\xi_\nu^{(m)}),$$

$$\frac{1}{\Gamma(\alpha)} \int_0^1 \left(\frac{Q^*(x)}{d_m} \frac{1-x}{1-\xi_\nu^{(m)}} \right)^2 \left(\ln \frac{1}{x} \right)^{\alpha-1} dx \leq C \int_0^1 Q^{*2}(x)\, dx = O\left(\frac{1}{m} \right).$$

Lemma 8 is proved.

We now present Markov's construction for approximating a step function.

LEMMA 9. *Suppose η is a fixed number such that $0 < \eta < 1$, $\xi_1^{(m)} < \xi_2^{(m)} < \cdots < \xi_m^{(m)}$ are the roots of $t_m(x)$, and $\xi_\nu^{(m)}$ and $\xi_{\nu+1}^{(m)}$ are roots adjacent to η. Then we can construct a polynomial $P_m(x)$ of degree $2m-2$ such that*

$$P_m(\xi_1^{(m)}) = \cdots = P_m(\xi_{\nu+1}^{(m)}) = 1; \qquad P_m(\xi_{\nu+2}^{(m)}) = \cdots = P_m(\xi_m^{(m)}) = 0;$$
$$P'_m(\xi_1^{(m)}) = \cdots = P'_m(\xi_\nu^{(m)}) = 0; \qquad P'_m(\xi_{\nu+2}^{(m)}) = \cdots = P'_m(\xi_m^{(m)}) = 0.$$

This polynomial has the property that

$$P_m(x) \geq \begin{cases} 1, & \text{if } -\infty < x \leq \xi_{\nu+1}^{(m)}, \\ 0, & \text{if } \xi_{\nu+1}^{(m)} \leq x < \infty. \end{cases}$$

PROOF. Using the Lagrange interpolation formula, we can construct a polynomial $\tilde{P}_m(x)$ of degree at most $m-1$ such that

$$\tilde{P}_m(\xi_1^{(m)}) = \cdots = \tilde{P}_m(\xi_{\nu+1}^{(m)}) = 1; \qquad \tilde{P}_m(\xi_{\nu+2}^{(m)}) = \cdots = \tilde{P}_m(\xi_m^{(m)}) = 0.$$

We then construct a polynomial $A_m(x)$ of degree at most $m-2$ such that

$$A_m(\xi_i^{(m)}) = -\frac{\tilde{P}'_m(\xi_i^{(m)})}{t'_m(\xi_i^{(m)})}, \qquad i = 1, \ldots, m, \text{ but } i \neq \nu+1.$$

Finally we define $P_m(x)$ by

$$P_m(x) = \tilde{P}_m(x) + A_m(x) t_m(x).$$

The degree of this polynomial is at most $2m-2$. The conditions on the $P_m(\xi_i^{(m)})$ are satisfied by virtue of the construction of $\tilde{P}_m(x)$. Also,

$$P'_m(x) = \tilde{P}'_m(x) + t'_m(x) A_m(x) + t_m(x) A'_m(x).$$

The conditions on the $P'_m(\xi_i^{(m)})$ are satisfied by virtue of the construction of $A_m(x)$.

We will construct the graph of $P_m(x)$. Consider $P'_m(x)$. By Rolle's theorem, there is a point in each of the intervals

$$(\xi_i^{(m)}, \xi_{i+1}^{(m)}), \qquad i = 1, 2, \ldots, \nu, \nu+2, \ldots, m-1,$$

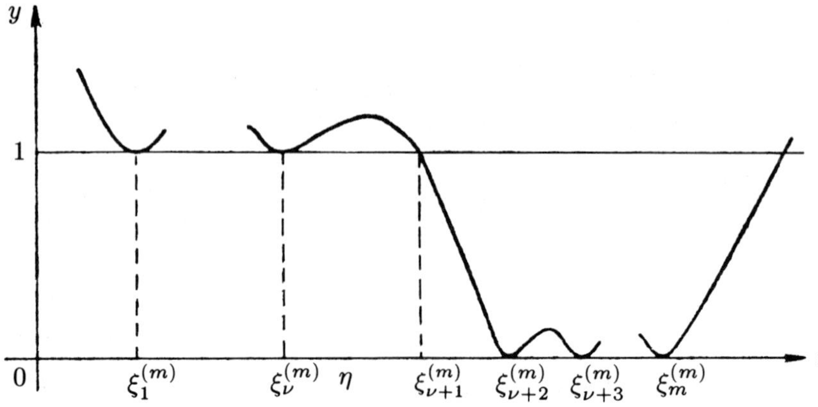

FIGURE 1

where $P'_m(x)$ vanishes. Moreover, there are $m-1$ other points where the derivative vanishes. Thus we have found $m-2+m-1=2m-3$ roots of $P'_m(x)$. But the degree of $P'_m(x)$ (being the derivative of $P_m(x)$) is at most $2m-3$. Therefore, these roots are all of the turning points of the graph of $P_m(x)$ (and the degree of $P'_m(x)$ is exactly $2m-3$).

In particular, $P'_m(x)$ has no roots in the half-open interval $[\xi_{\nu+1}^{(m)}, \xi_{\nu+2}^{(m)})$. The graph of $P_m(x)$ is shown in Figure 1. It is clear from the graph that

$$P_m(x) \geq \begin{cases} 1, & 0 \leq x \leq \xi_{\nu+1}^{(m)}, \\ 0, & \xi_{\nu+1}^{(m)} \leq x \leq 1. \end{cases}$$

Moreover, since $P'_m(x)$ has degree $2m-3$, it follows that $P_m(x)$ has degree exactly $2m-2$.

LEMMA $9'$. *Suppose $0<\eta<1$ and $\xi_1^{(m)} < \cdots < \xi_\nu^{(m)} < \eta < \xi_{\nu+1}^{(m)} < \cdots < \xi_m^{(m)}$. Then there exists a polynomial $p_m(x)$ of degree $2m-2$ such that*

$$p_m(\xi_1^{(m)}) = \cdots = p_m(\xi_{\nu-1}^{(m)}) = 1; \qquad p_m(\xi_\nu^{(m)}) = \cdots = p_m(\xi_m^{(m)}) = 0;$$
$$p'_m(\xi_1^{(m)}) = \cdots = p'_m(\xi_{\nu-1}^{(m)}) = 0; \qquad p'_m(\xi_{\nu+1}^{(m)}) = \cdots = p'_m(\xi_m^{(m)}) = 0$$

and

$$p_m(x) \leq \begin{cases} 1, & -\infty < x \leq \xi_\nu^{(m)}, \\ 0, & \xi_\nu^{(m)} \leq x < \infty. \end{cases}$$

The proof is analogous to that of Lemma 9. The graph of $p_m(x)$ is shown in Figure 2.

Consider the function $f(x)$ defined on $[0,1]$ as follows:

$$f(x) = \begin{cases} 1, & 0 \leq x < \eta, \\ 0, & \eta \leq x \leq 1. \end{cases}$$

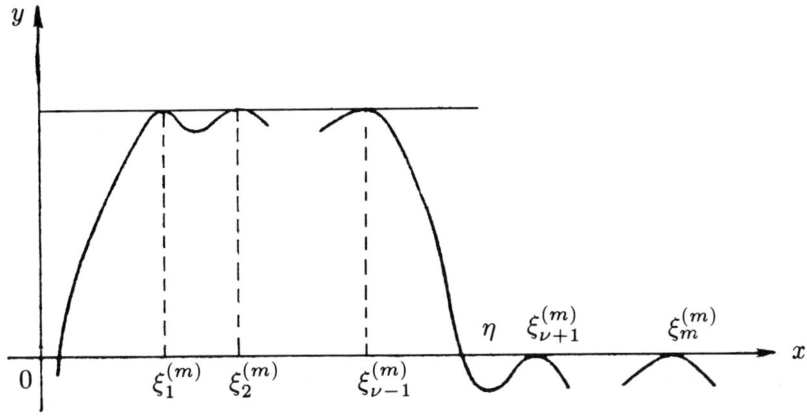

FIGURE 2

Obviously, $p_m(x) \leq f(x) \leq P_m(x)$. Also,

$$0 \leq \frac{1}{\Gamma(\alpha)} \int_0^1 (P_m(x) - p_m(x)) \left(\ln \frac{1}{x}\right)^{\alpha-1} dx$$
$$= \sum_{i=1}^{m} \rho(\xi_i^{(m)})[P_m(\xi_i^{(m)}) - p_m(\xi_i^{(m)})] = \rho(\xi_\nu^{(m)}) + \rho(\xi_{\nu+1}^{(m)}).$$

In view of Lemma 8,

$$0 \leq \frac{1}{\Gamma(\alpha)} \int_0^1 (P_m(x) - p_m(x)) \left(\ln \frac{1}{x}\right)^{\alpha-1} dx \leq \frac{C}{m}.$$

To prove Freud's theorem we must approximate the function $g(x)$ defined by

$$g(x) = \begin{cases} 0, & 0 \leq x < 1/e, \\ 1/x, & 1/e \leq x < 1 \end{cases}$$

(where e is the base of the natural logarithms). In what follows we take $\eta = 1/e$. We have $g(x) = h(x) - ef(x)$, where

$$h(x) = \begin{cases} e, & 0 \leq x < e^{-1}, \\ x^{-1}, & e^{-1} \leq x \leq 1. \end{cases}$$

The function $h(x)$ is continuous. It also satisfies the Lipschitz condition, i.e., there exists a constant $C > 0$ such that

$$|h(x_1) - h(x_2)| < C|x_1 - x_2|.$$

Let $\omega(\delta)$ be the modulus of continuity of $h(x)$.

By a theorem of Weierstrass, we can approximate $h(x)$ by a polynomial. Using Jackson's theorem (see Natanson [105], Part I, Chapter IV, §3), we see that for each natural number m there exists a polynomial $\chi_m(x)$ of degree m such that on $[0,1]$

$$|h(x) - \chi_m(x)| \le 12\omega(1/2m) \le C^*/m,$$

where C^* is a constant.

Consider the two polynomials

$$\begin{aligned}\varphi_m(x) &= \chi_m(x) - C^*/m - eP_m(x),\\ \Phi_m(x) &= \chi_m(x) + C^*/m - ep_m(x).\end{aligned}$$

It is clear that $\varphi_m(x) \le g(x) \le \Phi_m(x)$. We have

$$\begin{aligned}&\frac{1}{\Gamma(\alpha)}\int_0^1 \left(\ln\frac{1}{x}\right)^{\alpha-1} (\Phi_m(x) - \varphi_m(x))\,dx\\ &\quad= \frac{1}{\Gamma(\alpha)}\int_0^1 \frac{2C^*}{m}\left(\ln\frac{1}{x}\right)^{\alpha-1} dx\\ &\qquad+ e\frac{1}{\Gamma(\alpha)}\int_0^1 (P_m(x) - p_m(x))\left(\ln\frac{1}{x}\right)^{\alpha-1} dx\\ &\quad< \frac{C}{m} + \frac{C}{m} \le \frac{2C}{m}.\end{aligned}$$

We will require some properties of $\varphi_m(x)$ and $\Phi_m(x)$.

LEMMA 10. *There exists a constant $C > 0$ such that*

$$\max_{0\le x\le 1} |\Phi_m(x)| = O(m^C), \qquad \max_{0\le x\le 1} |\varphi_m(x)| = O(m^C).$$

PROOF. Let $\psi_m(x) = \Phi_m(x) - \varphi_m(x)$. Then $\psi_m(x)$ is a polynomial of degree at most $2m-2$, and $\psi_m(x) \ge 0$. Let $\max\psi_m(x) = \eta_m$ and suppose the maximum is attained at x_m: $\psi_m(x_m) = \eta_m$. By a well-known theorem of Markov (see Natanson [105], Part I, Chapter VI, §6),

$$|\psi_m'(x)| < 2(2m)^2\eta_m = 8m^2\eta_m.$$

If $|x - x_m| \le 1/16m^2$, then $\psi_m(x) \ge \eta_m/2$. Indeed, by Lagrange's finite difference theorem and Markov's theorem (see Natanson [105], loc. cit.),

$$\eta_m - \psi_m(x) \le |x - x_m| 8m^2\eta_m \le \eta_m/2.$$

Let X denote the set of values x for which $|x - x_m| \le 1/16m^2$.

If $\alpha \geq 1$, then

$$\begin{aligned}\frac{C}{m} &> \frac{1}{\Gamma(\alpha)} \int_0^1 (\Phi_m(x) - \varphi_m(x)) \left(\ln \frac{1}{x}\right)^{\alpha-1} dx \\ &\geq \frac{\eta_m}{2} \frac{1}{\Gamma(\alpha)} \int_X \left(\ln \frac{1}{x}\right)^{\alpha-1} dx \\ &\geq \frac{\eta_m}{2\Gamma(\alpha)} \int_{1-1/16m^2}^1 \left(\ln \frac{1}{x}\right)^{\alpha-1} dx \\ &> \frac{\eta_m}{2\Gamma(\alpha)} \int_{1-1/16m^2}^1 (1-x)^{\alpha-1}\, dx = \frac{\eta_m}{2\Gamma(\alpha)} \frac{1}{\alpha(16m^2)^\alpha}\end{aligned}$$

(since $\ln(1/x) > 1 - x$ for $0 < x \leq 1$).

If $\alpha < 1$, then

$$\begin{aligned}\frac{C}{m} &\geq \frac{1}{\Gamma(\alpha)} \int_0^1 \left(\ln \frac{1}{x}\right)^{\alpha-1} (\Phi_m(x) - \varphi_m(x))\, dx \\ &\geq \frac{\eta_m}{2} \frac{1}{\Gamma(\alpha)} \int_X \left(\ln \frac{1}{x}\right)^{\alpha-1} dx \geq \frac{\eta_m}{2} \frac{1}{\Gamma(\alpha)} \int_0^{1/16m^2} \left(\ln \frac{1}{x}\right)^{\alpha-1} dx \\ &= \frac{\eta_m}{2} \frac{1}{\Gamma(\alpha)} \int_{16m^2}^\infty \frac{1}{(\ln y)^{1-\alpha}} \frac{1}{y^2}\, dy \\ &> \frac{\eta_m}{2} \frac{1}{\Gamma(\alpha)} \int_{16m^2}^\infty \frac{dy}{y^{3-\alpha}} = \frac{\eta_m}{2(2-\alpha)} \cdot \frac{1}{\Gamma(\alpha)(16m^2)^{2-\alpha}}.\end{aligned}$$

In either case, $C_1 m^{C_2} > \eta_m > 0$ and

$$0 \leq \Phi_m(x) = \varphi_m(x) + \psi_m(x) \leq g(x) + \eta_m \leq e + O(m^C) = O(m^C).$$

On the other hand,

$$e \geq \varphi_m(x) = \Phi_m(x) - \psi_m(x) \geq g(x) - \eta_m \geq -\eta_m \geq -C_1 m^{C_2}.$$

This proves the lemma.

LEMMA 11. *Suppose $\pi_\nu(x) = \sum_0^\nu \alpha_k x^k$ is a polynomial of degree ν with real coefficients such that $|\pi_\nu(x)| \leq M$ on the interval $[0,1]$. Then its coefficients satisfy the estimate*

$$\max_{0 \leq k \leq \nu} |\alpha_k| \leq C \frac{(4e)^\nu}{\nu} M,$$

where C is an absolute constant.

PROOF. In this lemma the letter C will stand for absolute constants, not necessarily equal to each other. Consider the $\nu + 1$ points $x_i = i/\nu$,

$i = 0, 1, \ldots, \nu$, in $[0,1]$. Since a polynomial of degree ν is determined by its values at $\nu + 1$ points, it follows from the Lagrange interpolation formula that

$$\pi_\nu(x) = \sum_{i=0}^{\nu} \pi_\nu(x_i) \frac{(x - x_0)\cdots(x - x_{i-1})(x - x_{i+1})\cdots(x - x_\nu)}{(x_i - x_0)\cdots(x_i - x_{i-1})(x_i - x_{i+1})\cdots(x_i - x_\nu)}.$$

Let $\Phi_i^{(k)}$ denote the kth symmetric function of the quantities $x_0, \ldots, x_{i-1}$, $x_{i+1}, \ldots, x_\nu$. We have

$$\alpha_k = \sum_{i=0}^{\nu} (-1)^k \frac{\pi_\nu(x_i)\Phi_i^{(k)}}{(x_i - x_0)\cdots(x_i - x_{i-1})(x_i - x_{i+1})\cdots(x_i - x_\nu)}.$$

Since $0 \le x_i \le 1$, it follows that

$$|\Phi_i^{(k)}| \le C_\nu^k \le C_\nu^{[\nu/2]}.$$

By Stirling's formula, $C_\nu^{[\nu/2]} \le C 2^\nu / \sqrt{\nu}$, and hence

$$\begin{aligned} |\alpha_k| \le \sum_{i=0}^{\nu} C \frac{2^\nu}{\sqrt{\nu}} \cdot \frac{|\pi_\nu(x_i)|}{i!(\nu - i)!/\nu^\nu} &= C \frac{2^\nu}{\sqrt{\nu}} M \frac{\nu^\nu}{\nu!} \sum_{i=0}^{\nu} \frac{\nu!}{i!(\nu - i)!} \\ &= C \frac{4^\nu}{\sqrt{\nu}} \frac{\nu^\nu}{\nu!} M < C \frac{(4e)^\nu}{\nu} M. \end{aligned}$$

Lemma 11 is proved.

Suppose

$$\Phi_m(x) = \sum_{j=0}^{2m-2} B_j x^j \quad \text{and} \quad \varphi_m(x) = \sum_{j=0}^{2m-2} b_j x^j.$$

LEMMA 12. *There exists a constant $C > 0$ such that*

$$\sum_{j=0}^{2m-2} |B_j| = O(e^{Cm}) \quad \textit{and} \quad \sum_{j=0}^{2m-2} |b_j| = O(e^{Cm}).$$

PROOF. It follows from Lemma 11 that

$$\sum_{k=0}^{\nu} |\alpha_k| \le C(4e)^\nu M.$$

Applying this to $\Phi_m(x)$ with $\nu = 2m - 2$ and $M = O(m^C)$, we obtain

$$\sum_{j=0}^{2m-2} |B_j| = O(m^C (4e)^{2m-2}) = O(e^{Cm}), \qquad C > 0.$$

Analogously, for $\varphi_m(x)$ we obtain

$$\sum_{j=0}^{2m-2} |b_j| = O(e^{Cm}), \qquad C > 0.$$

Lemma 12 is proved.

We now complete the proof of Freud's theorem. By hypothesis,

$$\int_0^\infty f(t)e^{-st}\,d\tau(t) = \frac{A\Gamma(1+\alpha)}{s^\alpha}(1+r(s)).$$

We replace s by $s(k+1)$, $k=0,1,2,\ldots$. Then

$$\int_0^\infty f(t)e^{-s(k+1)t}\,d\tau(t) = \frac{A\Gamma(1+\alpha)}{s^\alpha(k+1)^\alpha}(1+r((k+1)s)).$$

Multiplying by B_k and summing over k from 0 to $2m-2$, we obtain

$$\int_0^\infty f(t)e^{-st}\left(\sum_{k=0}^{2m-2} B_k e^{-skt}\right) d\tau(t) = \int_0^\infty f(t)e^{-st}\Phi_m(e^{-st})\,d\tau(t)$$
$$= \frac{A\Gamma(1+\alpha)}{s^\alpha}\left(\sum_{k=0}^{2m-2}\frac{B_k}{(k+1)^\alpha} + \sum_{k=0}^{2m-2}\frac{B_k r((k+1)s)}{(k+1)^\alpha}\right).$$

Since

$$\int_0^\infty e^{-t}t^{\alpha-1}\Phi_m(e^{-t})\,dt = \sum_{k=0}^{2m-2} B_k\int_0^\infty t^{\alpha-1}e^{-(k+1)t}\,dt$$
$$= \sum_{k=0}^{2m-2} B_k\frac{\Gamma(\alpha)}{(k+1)^\alpha} = \frac{\Gamma(\alpha+1)}{\alpha}\sum_{k=0}^{2m-2}\frac{B_k}{(k+1)^\alpha},$$

it follows that

$$\int_0^\infty f(t)e^{-st}\Phi_m(e^{-st})\,d\tau(t)$$
$$= \frac{A\alpha}{s^\alpha}\int_0^\infty t^{\alpha-1}\Phi_m(e^{-t})e^{-t}\,dt + \frac{A\Gamma(\alpha+1)}{s^\alpha}\sum_{k=0}^{2m-2} B_k\frac{r((k+1)s)}{(k+1)^\alpha}$$
$$= \frac{A\alpha}{s^\alpha}\int_0^\infty t^{\alpha-1}e^{-t}\Phi_m(e^{-t})\,dt + O\left(\frac{s^\varepsilon}{s^\alpha}\sum_{k=0}^{2m-2}\frac{|B_k|}{(k+1)^{\alpha-\varepsilon}}\right)$$
$$= \frac{A\alpha}{s^\alpha}\int_0^\infty t^{\alpha-1}e^{-t}\Phi_m(e^{-t})\,dt + O\left(\frac{s^\varepsilon}{s^\alpha}m^C\sum_{k=0}^{2m-2}|B_k|\right)$$
$$= \frac{A\alpha}{s^\alpha}\int_0^\infty t^{\alpha-1}e^{-t}\Phi_m(e^{-t})\,dt + O\left(\frac{e^{Cm}}{s^{\alpha-\varepsilon}}\right).$$

Analogously,

$$\int_0^\infty f(t)e^{-st}\varphi_m(e^{-st})\,d\tau(t) = \frac{A\alpha}{s^\alpha}\int_0^\infty t^{\alpha-1}e^{-t}\varphi_m(e^{-t})\,dt + O\left(\frac{e^{Cm}}{s^{\alpha-\varepsilon}}\right).$$

In the integral

$$\int_0^\infty t^{\alpha-1}\Phi_m(e^{-t})e^{-t}\,dt$$

we make the change of variable $e^{-t} = x$:

$$\begin{aligned}
\int^{\infty} t^{\alpha-1}\Phi_m(e^{-t})e^{-t}\,dt &= \int_0^1 \left(\ln\frac{1}{x}\right)^{\alpha-1} \Phi_m(x)\,dx \\
&\le \int_0^1 \left(\ln\frac{1}{x}\right)^{\alpha-1} (g(x) + \Phi_m(x) - \varphi_m(x))\,dx \\
&= \int_0^1 g(x)\left(\ln\frac{1}{x}\right)^{\alpha-1} dx + \int_0^1 \left(\ln\frac{1}{x}\right)^{\alpha-1} (\Phi_m(x) - \varphi_m(x))\,dx \\
&\le \int_{1/e}^1 \left(\ln\frac{1}{x}\right)^{\alpha-1} \frac{dx}{x} + \frac{C\Gamma(\alpha)}{m} = \int_0^1 z^{\alpha-1}\,dz + \frac{\Gamma(\alpha)C}{m} \\
&= \frac{1}{\alpha} + \frac{C\Gamma(\alpha)}{m}.
\end{aligned}$$

Also,

$$\begin{aligned}
\int^{\infty} t^{\alpha-1}\varphi_m(e^{-t})e^{-t}\,dt &= \int_0^1 \left(\ln\frac{1}{x}\right)^{\alpha-1} \varphi_m(x)\,dx \\
&\ge \int_0^1 \left(\ln\frac{1}{x}\right)^{\alpha-1} (g(x) - (\Phi_m(x) - \varphi_m(x)))\,dx \\
&> \frac{1}{\alpha} - \frac{C\Gamma(\alpha)}{m}.
\end{aligned}$$

For $t \ge 0$

$$\varphi_m(e^{-st}) \le g(e^{-st}) \le \Phi_m(e^{-st}),$$

and hence

$$\begin{aligned}
\int_0^{\infty} f(t)e^{-st}g(e^{-st})\,d\tau(t) &\le \int_0^{\infty} f(t)e^{-st}\Phi_m(e^{-st})\,d\tau(t) \\
&= \frac{A\alpha}{s^{\alpha}} \int_0^{\infty} t^{\alpha-1}\Phi_m(e^{-t})e^{-t}\,dt + O\left(\frac{e^{Cm}}{s^{\alpha-\varepsilon}}\right) \\
&\le \frac{A\alpha}{s^{\alpha}}\left(\frac{1}{\alpha} + \frac{C}{m}\right) + O\left(\frac{e^{Cm}}{s^{\alpha-\varepsilon}}\right).
\end{aligned}$$

We have obtained the estimate

$$\int_0^{\infty} f(t)e^{-st}g(e^{-st})\,d\tau(t) \le \frac{A}{s^{\alpha}}\left(1 + \frac{C}{m}\right) + O\left(\frac{e^{C_0 m}}{s^{\alpha-\varepsilon}}\right).$$

Analogously,

$$\int_0^{\infty} f(t)e^{-st}g(e^{-st})\,d\tau(t) \ge \frac{A}{s^{\alpha}}\left(1 - \frac{C}{m}\right) + O\left(\frac{e^{C_0 m}}{s^{\alpha-\varepsilon}}\right).$$

Consider the integral

$$\int_0^{\infty} f(t)e^{-st}g(e^{-st})\,d\tau(t) = \int_0^{1/s} f(t)\,d\tau(t).$$

Since

$$g(e^{-st}) = \begin{cases} 0, & \text{if } t > 1/s, \\ e^{st}, & \text{if } t \le 1/s, \end{cases}$$

it follows that

$$\frac{A}{s^\alpha}\left(1 - \frac{C}{m}\right) + O\left(\frac{e^{C_0 m}}{s^{\alpha-\varepsilon}}\right) \le \int_0^{1/s} f(t)\,d\tau(t) \le \frac{A}{s^\alpha}\left(1 + \frac{C}{m}\right) + O\left(\frac{e^{C_0 m}}{s^{\alpha-\varepsilon}}\right).$$

Let $1/s = x$. As $s \to 0$ and $x \to \infty$ we have

$$\int_0^x f(t)\,d\tau(t) = Ax^\alpha + O\left(\frac{x^\alpha}{m}\right) + O(x^{\alpha-\varepsilon}e^{C_0 m}).$$

Choose $m = [(\varepsilon/2C_0)\ln x] + 1$. Then

$$\begin{aligned}\int_0^x f(t)\,d\tau(t) &= Ax^\alpha + O\left(\frac{x^\alpha}{\ln x}\right) + O\left(x^{\alpha-\varepsilon}\exp\left(C_0\left(\frac{\varepsilon}{2C_0}\ln x + 1\right)\right)\right) \\ &= Ax^\alpha + O\left(\frac{x^\alpha}{\ln x}\right) + O(x^{\alpha-\varepsilon+\varepsilon/2}) = Ax^\alpha + O\left(\frac{x^\alpha}{\ln x}\right),\end{aligned}$$

as required.

§1.3. Ingham's Tauberian theorem

Ingham's Tauberian theorem (Ingham [75]) pertains to rapidly increasing transforms.

We will use the following specific notation: $\nearrow$ signifies a monotone nondecreasing approach, and $\searrow$ a monotone nonincreasing approach. As usual, we denote a complex variable by $s = \sigma + it$.

THEOREM. *Suppose that the integral* $f(s) = \int_0^\infty e^{-us}\,dA(u)$ *converges for* $\sigma > 0$, *and that the following conditions hold:*

1) $A(u)$ *is nondecreasing,* $0 \le u < \infty$.

2) *We are given a region* D *lying entirely in the half-plane* $\operatorname{Re} s > 0$ *and containing the point* $s = 0$ *on its boundary and a function* $f_0(s) = \chi(s)e^{\varphi(s)}$, *where the functions* $\chi(s)$ *and* $\varphi(s)$ *(not Euler's function) satisfy the following conditions:*

a) $\chi(s)$ *and* $\varphi(s)$ *are analytic in* D.

b) $\chi(s)$ *and* $\varphi(s)$ *are real and positive on a half-open interval* $(0, h]$ *of the positive real axis lying in* D.

c) $-\sigma\varphi'(\sigma) \nearrow \infty$ *as* $\sigma \searrow 0$ *(and therefore* $(-\sigma\varphi'(\sigma))' \ge 0$; *hence* $\sigma\varphi''(\sigma) \ge -\varphi'(\sigma) > 0$ *for sufficiently small positive* σ*).*

d) *If* $\delta(\sigma)$ *denotes the distance from the point* σ *to the boundary of* D, *then as* $\sigma \searrow 0$

$$\frac{(\varphi''(\sigma))^{1/2}}{-\varphi'(\sigma)} = o\left(\frac{\delta(\sigma)}{\sigma}\right)$$

(since the origin does not lie in D, *it follows that* $\delta(\sigma) \le \sigma$*).*

e) $\varphi''(\sigma + z) = O(\varphi''(\sigma))$ *uniformly for* $|z| < \delta(\sigma)$ *as* $\sigma \searrow 0$.
f) $\chi(\sigma + z) = O(\chi(\sigma))$ *uniformly for* $|z| < \delta(\sigma)$ *as* $\sigma \searrow 0$.
g) *The asymptotic formula* $f(s) \sim f_0(s)$ *holds uniformly as* $s \to 0$ *in* D.
h) $f(s) = O(f_0(|s|))$ *when* s *lies in some fixed angle of the form* $|t| \leq \Delta\sigma$.
Then

$$\eta_1(\Delta) \leq \varliminf_{u\to\infty} \frac{A(u)}{A_0(u)} \leq \varlimsup_{u\to\infty} \frac{A(u)}{A_0(u)} \leq \eta_2(\Delta), \tag{1}$$

where

$$A_0(\omega) = \frac{\chi(\sigma)e^{\varphi(\sigma)+\omega\sigma}}{\sqrt{2\pi\sigma^2\varphi''(\sigma)}},$$

and $\sigma = \sigma_\omega$ *is the root of the equation* $-\varphi'(\sigma) = \omega$ *(this solution exists and is unique for large* ω*), the functions* $\eta_1(\Delta)$ *and* $\eta_2(\Delta)$ *depend only on* Δ*, are strictly positive and finite for any* Δ*,* $0 < \Delta < \infty$*, and* $\eta_1(\Delta) \to 1$*,* $\eta_2(\Delta) \to 1$ *as* $\Delta \to \infty$*. In particular, if* $f(s) = O(f_0(|s|))$ *for any* Δ*,* $0 < \Delta < \infty$*, then*

$$\lim_{u\to\infty} \frac{A(u)}{A_0(u)} = 1.$$

Before proving Ingham's Tauberian theorem, we give its concrete formulation in special cases.

Suppose $\varphi(s) = (1/\beta)(M/s)^\beta$ and $\chi(s) = C(M/s)^{m\beta-1/2}$, where β, M, and C are positive constants and m is a real constant:

$$f_0(s) = C(M/s)^{m\beta-1/2}(1/e^\beta)(M/s)^\beta.$$

As the region D we take a region the form $|t| \leq \Delta\sigma$, where $\Delta > 0$ is fixed. We verify the conditions: for small σ

$$\delta(\sigma) = \frac{\Delta}{\sqrt{1+\Delta^2}}\sigma,$$

$$\varphi'(\sigma) = -M^\beta\sigma^{-\beta-1}, \qquad \varphi''(\sigma) = (\beta+1)M^\beta\sigma^{-\beta-2}.$$

Since for $|z| \leq \delta(\sigma)$

$$2\sigma \geq |\sigma + z| \geq \sigma - \Delta\sigma/\sqrt{1+\Delta^2},$$

conditions e) and f) are satisfied. The equation $-\varphi'(\sigma) = \omega$ has the solution

$$\sigma_\omega = \frac{M^{\beta/(\beta+1)}}{\omega^{1/(\beta+1)}}.$$

A simple calculation yields

$$A_0(\omega) = \frac{C}{\sqrt{2\pi(\beta+1)}}(\omega M)^{m(\beta/\beta+1)-1/2}\exp\left(\frac{\beta+1}{\beta}(\omega M)^{\beta/(\beta+1)}\right).$$

Putting $\alpha = \beta/(\beta+1)$, we obtain

$$A_0(\omega) = C\sqrt{\frac{1-\alpha}{2\pi}}(\omega M)^{m\alpha-1/2}e^{(\omega,M)^\alpha/\alpha}.$$

Thus we have obtained the following special case.

COROLLARY 1. *Suppose $f(s) = \int_0^\infty e^{-us}\, dA(u)$ and the following conditions hold:*

1) *$A(0) = 0$, and $A(u)$ is nondecreasing as $u \to \infty$.*

2) *$f(s) \sim C(M/s)^{m\beta-1/2} e^{(1/\beta)(M/s)^\beta}$ (where C, M and β are positive constants and m is a real constant) uniformly as $s \to 0$ in some fixed angle of the form $|t| < \Delta_0\sigma$, $\Delta_0 > 0$.*

Then

$$\eta_1 \le \underline{\lim}_{u\to\infty} \frac{A(u)}{C\sqrt{(1-\alpha)/2\pi}(uM)^{m\alpha-1/2}e^{(uM)^\alpha/\alpha}} \le \overline{\lim}_{u\to\infty} \frac{A(u)}{C\sqrt{(1-\alpha)/2\pi}(uM)^{m\alpha-1/2}e^{(uM)^\alpha/\alpha}} \le \eta_2,$$

where η_1 and η_2 are constants, and $\alpha = \beta/(\beta+1)$.

COROLLARY 2. *Suppose $f(s) = \int_0^\infty e^{-us} dA(u)$, and the following conditions hold:*

1) *$A(0) = 0$, and $A(u)$ is nondecreasing as $u \to \infty$.*

2) *$f(s) \sim C(M/s)^{m\beta-1/2} e^{(1/\beta)(M/s)^\beta}$ (where C, M, and β are positive constants and m is a real constant) uniformly as $s \to 0$ in any angle of the form $|t| < \Delta\sigma$.*

Then as $u \to \infty$

$$A(u) \sim C\sqrt{\frac{1-\alpha}{2\pi}}(uM)^{m\alpha-1/2}e^{(uM)^\alpha/\alpha},$$

where $\alpha = \beta/(\beta+1)$.

PROOF OF INGHAM'S THEOREM. We will first prove the following lemmas.

LEMMA 1. *Uniformly for $|z| \le \tau(\sigma) = o(\delta(\sigma))$, as $\sigma \searrow 0$,*

$$\begin{aligned}\frac{f_0(\sigma+z)}{f_0(\sigma)} &= \exp\left[z\varphi'(\sigma) + \frac{z^2}{2}\varphi''(\sigma)(1+o(1)) + o(1)\right] \\ &= \exp\left[z\varphi'(\sigma) + \frac{z^2}{2}\varphi''(\sigma)\right] \\ &\quad \times (1 + o(|z|^2\varphi''(\sigma)+1)) \exp[\varepsilon|z|^2\varphi''(\sigma)]\end{aligned}$$

for any fixed $\varepsilon > 0$.

Indeed, by definition of $f_0(s)$,

$$\frac{f_0(\sigma+z)}{f_0(\sigma)} = \frac{\chi(\sigma+z)}{\chi(\sigma)}\exp[\varphi(\sigma+z)-\varphi(\sigma)]$$
$$= \left(1+\frac{\chi(\sigma+z)-\chi(\sigma)}{\chi(\sigma)}\right)\exp[\varphi(\sigma+z)-\varphi(\sigma)]. \tag{2}$$

But

$$\chi'(\sigma+z) = \frac{1}{2\pi i}\int_\Gamma \frac{\chi(\sigma+\varsigma)}{(\varsigma-z)^2}\,d\varsigma = O\left(\frac{\chi(\sigma)}{\delta(\sigma)}\right).$$

Here Γ is the circle $|\varsigma - z| = \delta(\sigma)/2$ (on this circle $\sigma+\varsigma$ does not fall outside D, since $|z| = o(\delta(\sigma))$). The estimate $\chi'(\sigma+z) = O(\chi(\sigma)/\delta(\sigma))$ is uniform for $|z| \le \tau(\sigma)$. Integration yields

$$\chi(\sigma+z)-\chi(\sigma) = O\left(|z|\frac{\chi(\sigma)}{\delta(\sigma)}\right) = O\left(\frac{\tau(\sigma)\chi(\sigma)}{\delta(\sigma)}\right) = o(\chi(\sigma)).$$

Similarly,

$$\varphi''(\sigma+z)-\varphi''(\sigma) = o(\varphi''(\sigma)).$$

Expanding the real and imaginary parts separately into Taylor series, we obtain

$$\varphi(\sigma+z)-\varphi(\sigma) = z\varphi'(\sigma) + \frac{z^2\varphi''(\sigma)}{2} + o(|z|^2\varphi''(\sigma)).$$

Substituting this into (2), we see that

$$\frac{f_0(\sigma+z)}{f_0(\sigma)} = (1+o(1))\exp\left[z\varphi'(\sigma)+\frac{z^2}{2}\varphi''(\sigma)+o(|z|^2\varphi''(\sigma))\right]$$
$$= \exp\left[z\varphi'(\sigma)+\frac{z^2}{2}\varphi''(\sigma)+o(|z|^2\varphi''(\sigma))+o(1)\right].$$

This yields the first equality of Lemma 1.

The second can be obtained from the first by using the relation $|e^w - 1| \le |w|e^{|w|}$:

$$\exp\left[z\varphi'(\sigma)+\frac{z^2}{2}\varphi''(\sigma)+o(|z|^2\varphi''(\sigma))+o(1)\right]$$
$$= \exp\left[z\varphi'(\sigma)+\frac{z^2}{2}\varphi''(\sigma)\right](1+\exp[o(|z|^2\varphi''(\sigma))+o(1)]-1)$$
$$= \exp\left[z\varphi'(\sigma)+\frac{z^2}{2}\varphi''(\sigma)\right]$$
$$\times\{1+o(1+|z|^2\varphi''(\sigma))\exp[o(|z|^2\varphi''(\sigma)+1)]\}$$

$$
\begin{aligned}
&= \exp\left[z\varphi'(\sigma) + \frac{z^2}{2}\varphi''(\sigma)\right] \\
&\qquad \times \{1 + o(1 + |z|^2\varphi''(\sigma))(1 + o(1)) \exp[o(|z|^2\varphi''(\sigma))]\} \\
&= \exp\left[z\varphi'(\sigma) + \frac{z^2}{2}\varphi''(\sigma)\right] \\
&\qquad \times \{1 + o(1 + |z|^2\varphi''(\sigma)) \exp[o(|z|^2\varphi''(\sigma))]\} \\
&= \exp\left[z\varphi'(\sigma) + \frac{z^2}{2}\varphi''(\sigma)\right] \\
&\qquad \times \{1 + o(1 + |z|^2\varphi''(\sigma)) \exp[\varepsilon|z|^2\varphi''(\sigma)] \\
&\qquad\qquad \times \exp[o(|z|^2\varphi''(\sigma)) - \varepsilon|z|^2\varphi''(\sigma)]\} \\
&= \exp\left[z\varphi'(\sigma) + \frac{z^2}{2}\varphi''(\sigma)\right] \\
&\qquad \times \{1 + o(1 + |z|^2\varphi''(\sigma))O(1) \exp[\varepsilon|z|^2\varphi''(\sigma)]\} \\
&= \exp\left[z\varphi'(\sigma) + \frac{z^2}{2}\varphi''(\sigma)\right] \\
&\qquad \times \{1 + o(1 + |z|^2\varphi''(\sigma)) \exp[\varepsilon|z|^2\varphi''(\sigma)]\},
\end{aligned}
$$

where ε is an arbitrarily small fixed number.

LEMMA 2. *$f_0(\sigma)$ increases as σ decreases (for sufficiently small positive values of σ).*

PROOF. We have

$$-\frac{f_0'(\sigma)}{f_0(\sigma)} = -\frac{\chi'(\sigma)}{\chi(\sigma)} - \varphi'(\sigma).$$

Using the relation $\chi'(\sigma+z) = O(\chi(\sigma)/\delta(\sigma))$ for $z = 0$, we see that as $\sigma \searrow 0$

$$-\frac{f_0'(\sigma)}{f_0(\sigma)} = -\varphi'(\sigma) + O\left(\frac{1}{\delta(\sigma)}\right) \sim -\varphi'(\sigma),$$

since, by the conditions of the theorem,

$$\frac{1}{\delta(\sigma)} = o\left(-\frac{\varphi'(\sigma)}{\sigma\sqrt{\varphi''(\sigma)}}\right) = o\left(\frac{-\varphi'(\sigma)}{\sqrt{\sigma\varphi'(\sigma)}}\right) = o(-\varphi'(\sigma)).$$

Consequently, $-f_0'(\sigma) > 0$ for small values of σ, and the lemma is proved.

Let us turn to the main part of the proof of the theorem. The principal variables are σ and ω, which are connected by the equation $-\varphi'(\sigma) = \omega$. We assume that σ is sufficiently small and that ω is sufficiently large. Limits are taken as $\sigma \searrow 0$ and $\omega \nearrow \infty$.

In the equality $f(s) = \int_0^\infty e^{-us}\, dA(u)$ we integrate by parts for $\sigma > 0$:

$$f(s) = s\int_0^\infty A(u)e^{-us}\, du$$

(for $\sigma > 0$, since the integral $\int_0^\infty e^{-us}\, dA(u)$ converges $A(u)e^{-us} \to 0$). Putting $s = \sigma + z$ in the last formula, we obtain

$$\int_0^\infty A(u)e^{-u(\sigma+z)}\, du = \frac{f(\sigma+z)}{\sigma+z}.$$

Take $T > 0$, a real ξ, and a function $k(\varsigma)$ that is regular on the interval $[0, i]$ of the imaginary axis of the ς-plane. Multiplying the last equality by $k(z/T)e^{\xi z}$, we obtain

$$k\left(\frac{z}{T}\right)e^{\xi z}\int_0^\infty A(u)e^{-u(\sigma+z)}\, du = k\left(\frac{z}{T}\right)e^{\xi z}\frac{f(\sigma+z)}{\sigma+z}.$$

We now integrate along the interval from 0 to iT of the imaginary axis of the z-plane and interchange the order of integration:

$$\int_0^\infty A(u)e^{-u\sigma}\left(\int_0^{iT} k\left(\frac{z}{T}\right)e^{-z(u-\xi)}\, dz\right) du = \int_0^{iT}\frac{f(\sigma+z)}{\sigma+z}k\left(\frac{z}{T}\right)e^{\xi z}\, dz.$$

Making the change of variable $z/T = \varsigma$ in the inner integral, we obtain

$$\int_0^\infty A(u)e^{-u\sigma}T\left(\int_0^{i} k(\varsigma)e^{-T\varsigma(u-\xi)}\, d\varsigma\right) du = \int_0^{iT}\frac{f(\sigma+z)}{\sigma+z}k\left(\frac{z}{T}\right)e^{\xi z}\, dz.$$

Let

$$R(v) = \int_0^i k(\varsigma)e^{-v\varsigma}\, d\varsigma.$$

Then

$$\int_0^\infty A(u)R(T(u-\xi))Te^{-u\sigma}\, du = \int_0^{iT}\frac{f(\sigma+z)}{\sigma+z}k\left(\frac{z}{T}\right)e^{\xi z}\, dz.$$

Put $T = \Delta\sigma$, where the parameter Δ defines the angle $|t| \le \Delta\sigma$ in condition h) of the theorem. We choose the function ξ to satisfy

$$\xi = \xi(\sigma) = \omega + O\left(\frac{1}{\sigma}\right) = -\varphi'(\sigma) + O\left(\frac{1}{\sigma}\right),$$

and we impose on $k(\varsigma)$ the additional condition $k(0) \ne 0$.

We introduce a parameter $\tau = \tau(\sigma)$ satisfying the following conditions:

1) $\tau(\sigma) = o(\delta(\sigma))$ (hence $\tau(\sigma) = o(\sigma)$).

2) $\sigma\sqrt{\varphi''(\sigma)}/(-\varphi'(\sigma)) = o(\tau(\sigma))$.

Such a choice is possible since

$$\frac{\sigma\sqrt{\varphi''(\sigma)}}{-\varphi'(\sigma)} = o(\delta(\sigma)).$$

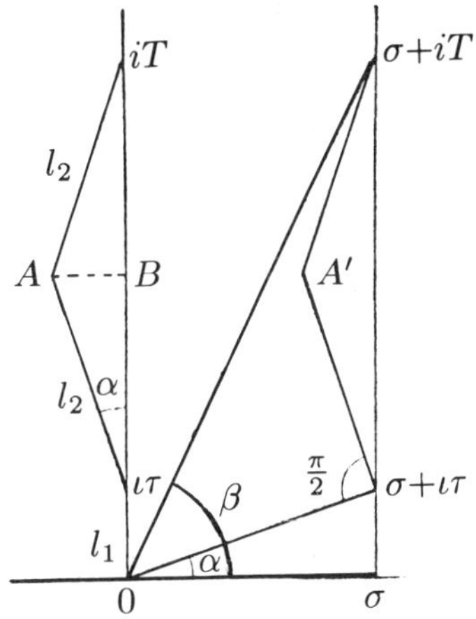

FIGURE 3

For example, we can take

$$\tau(\sigma) = \sqrt{\delta(\sigma)\frac{\sigma\sqrt{\varphi''(\sigma)}}{-\varphi'(\sigma)}}.$$

Let α be the acute angle defined by $\tan\alpha = \tau/\sigma$ (Figure 3). Using Cauchy's theorem, in the integral

$$\int_0^{iT} \frac{f(\sigma+z)}{\sigma+z} k\left(\frac{z}{T}\right) e^{\xi z}\, dz$$

we replace the path of integration $[0, iT]$ by the path $l = l(\sigma)$ consisting of the segment $[0, i\tau]$ of the imaginary axis (we denote this part of the new path by l_1) and the broken line l_2 made up of two segments: one from the point $i\tau$ to the point A with affix $-((T-\tau)\tan\alpha/2) + i(T+\tau)/2$, and the other from A to the point iT. Let B denote the point with affix $i(T+\tau)/2$, and let P denote the length of segment AB. We will prove that $\mathrm{Re}(\sigma+z) > 0$, i.e., $P < \sigma$, on l_2. We have

$$\frac{P}{\tan\alpha} = \frac{\Delta\sigma - \tau}{2}, \qquad \frac{P}{\tan\alpha} < \frac{\Delta\sigma}{2}, \qquad P < \frac{\Delta\sigma}{2}\frac{\tau}{\sigma} = o(\delta(\sigma));$$

hence $P = o(\sigma)$.

Let β denote the angle between the segment $[0, \sigma + iT]$ and the real axis. Since $T = \Delta\sigma$, it follows that $\tan\beta = \Delta$. Also $P/T = o(1)$, i.e., when σ is small, z/T lies in the region of analyticity of the function $k(\varsigma)$. Therefore the use of Cauchy's theorem is legitimate. Thus

$$T\int_0^\infty A(u)R(T(u-\xi))e^{-u\sigma}\, du = \int_l \frac{f(\sigma+z)}{\sigma+z} k\left(\frac{z}{T}\right) e^{\xi z}\, dz.$$

Let us examine the integral on the right-hand side of this equality. When $z = iy$ lies on l_1, the point $\sigma + z$ lies in D, since $\tau = o(\delta(\sigma))$, i.e., τ is less than $\delta(\sigma)$. Since in D we have $f(s) = f_0(s)e^{o(1)}$ uniformly in s, it follows that when z lies on l_1

$$f(\sigma + z) = f_0(\sigma + z)e^{o(1)}$$

uniformly in z. Also, when z lies on l_1,

$$k(z/T) = k(z/\Delta\sigma) = k(0)(1 + o(1)) = k(0)e^{o(1)}$$

uniformly in z, and

$$\left|\frac{\sigma}{\sigma + z}\right| \leq \frac{\sigma}{\sigma - \tau} \to 1,$$

i.e., $1/(\sigma + z) = e^{o(1)}/\sigma$ uniformly in z. On l_1 we have $dz = i dy$. Thus

$$\begin{aligned}\int_{l_1} \frac{f(\sigma + z)}{\sigma + z} k\left(\frac{z}{T}\right) e^{\xi z}\, dz &= \frac{ik(0)}{\sigma} \int_0^{\tau} f_0(\sigma + z)e^{\omega z + O(|z|/\sigma) + o(1)}\, dy \\ &= \frac{ik(0)}{\sigma} \int_0^{\tau} f_0(\sigma + z)e^{\omega z + o(1)}\, dy,\end{aligned}$$

since $|z|/\sigma \leq \tau/\sigma = o(1)$. By Lemma 1 with $\varepsilon = 1/4$ we have, uniformly in y,

$$\begin{aligned}f_0(\sigma + z) = f_0(\sigma) &\exp\left(z\varphi'(\sigma) + \frac{z^2\varphi''(\sigma)}{2}\right) \\ &\times \left(1 + o(1 + |z|^2\varphi''(\sigma)) \exp\left(\frac{|z|^2\varphi''(\sigma)}{4}\right)\right).\end{aligned}$$

But $\omega = -\varphi'(\sigma)$ and $z = iy$. Therefore

$$\begin{aligned}&\int_{l_1} \frac{f(\sigma + z)}{\sigma + z} k\left(\frac{z}{T}\right) e^{\xi z} dz \\ &\quad = \frac{if_0(\sigma)}{\sigma} k(0) \left(\int_0^{\tau} \exp\left(-\frac{y^2\varphi''(\sigma)}{2}\right)\, dy\right. \\ &\qquad\qquad \left. + o\left(\int_0^{\tau} (1 + y^2\varphi''(\sigma)) \exp\left(-\frac{y^2\varphi''(\sigma)}{4}\right)\, dy\right)\right).\end{aligned}$$

Making the change of variable $y\sqrt{\varphi''(\sigma)} = w$, we obtain

$$\begin{aligned}\int_{l_1} \frac{f(\sigma + z)}{\sigma + z} k\left(\frac{z}{T}\right) e^{\xi z}\, dz &= i\frac{f_0(\sigma)}{\sigma\sqrt{\varphi''(\sigma)}} k(0) \\ \times \left(\int_0^{\tau\sqrt{\varphi''(\sigma)}} e^{-w1/2}\, dw \right.&\left.+ o\left(\int_0^{\tau\sqrt{\varphi''(\sigma)}} (1 + w^2)e^{-w^2/4}\, dw\right)\right).\end{aligned}$$

Since $\sigma\sqrt{\varphi''(\sigma)}/(-\varphi'(\sigma)) = o(\tau)$, it follows that

$$\frac{\sigma\varphi''(\sigma)}{-\varphi'(\sigma)} = o(\tau\sqrt{\varphi''(\sigma)}).$$

But $\sigma\varphi''(\sigma)/(-\varphi'(\sigma)) \geq 1$; hence $1 = o(\tau\sqrt{\varphi''(\sigma)})$ and $1/\tau\sqrt{\varphi''(\sigma)} = o(1)$, i.e., $\tau\sqrt{\varphi''(\sigma)} \to \infty$ as $\sigma \searrow 0$, and

$$\begin{aligned}\int_{l_1} \frac{f(\sigma+z)}{\sigma+z} k\left(\frac{z}{T}\right) e^{\xi z}\, dz &= i\frac{f_0(\sigma)}{\sigma\sqrt{\varphi''(\sigma)}} k(0) \left(\int_0^\infty e^{-w^2/2}\, dw + o(1)\right) \\ &= \frac{i f_0(\sigma) k(0)}{\sigma\sqrt{\varphi''(\sigma)}} \sqrt{\frac{\pi}{2}}(1+o(1)).\end{aligned}$$

Since $\tan\alpha \to 0$ as $\sigma \searrow 0$ and Δ is a fixed angle, it follows that for a sufficiently small σ the broken line l_2 shifted σ units to the right lies entirely within the angle $|t| \leq \sigma\Delta$. Since the angle formed by the segment connecting 0 and $\sigma + i\tau$ and the segment connecting A' and $\sigma + i\tau$ is equal to $\pi/2$, it follows that $|\sigma + z| \geq \sigma$ when $z \in l_2$. Also, when $z \in l_2$, $k(z/T) = O(1)$. Finally, on l_2 we have, uniformly in z,

$$\begin{aligned}\frac{f(\sigma+z)}{\sigma+z} k\left(\frac{z}{T}\right) e^{\xi z} &= O\left(\frac{f_0(|\sigma+z|)}{\sigma} e^{\omega x + O(|z|/\sigma)}\right) \\ &= O\left(\frac{f_0(|\sigma+z|)}{\sigma} e^{\omega z}\right),\end{aligned}$$

since $O(|z|/\sigma) = O(T/\sigma) = O(1)$. But by Lemma 2 and the fact that $|\sigma+z| > \sigma$,

$$f_0(|\sigma+z|) < f_0(\sigma).$$

Thus on l_2, uniformly in z,

$$\frac{f(\sigma+z)}{\sigma+z} k\left(\frac{z}{T}\right) e^{\xi z} = O\left(\frac{f_0(\sigma)}{\sigma} e^{\omega x}\right).$$

Also on l_2, $x \leq 0$ and

$$\left|\frac{dz}{dx}\right| = \left|1 \pm \frac{\sigma}{\tau} i\right| = O\left(\frac{\sigma}{\tau}\right).$$

Consequently,

$$\begin{aligned}\int_{l_2} \frac{f(\sigma+z)}{\sigma+z} k\left(\frac{z}{T}\right) e^{\xi z}\, dz &= O\left(\frac{f_0(\sigma)\sigma}{\sigma\tau} \int_{-\infty}^0 e^{\omega x}\, dx\right) \\ &= O\left(\frac{f_0(\sigma)}{\tau\omega}\right) = O\left(\frac{f_0(\sigma)}{-\tau\varphi'(\sigma)}\right) = o\left(\frac{f_0(\sigma)}{\sigma}\frac{1}{\varphi''(\sigma)}\right),\end{aligned}$$

inasmuch as

$$-\frac{1}{\tau\varphi'(\sigma)} = o\left(\frac{1}{\sigma\sqrt{\varphi''(\sigma)}}\right).$$

Thus

$$\int_l \frac{f(\sigma+z)}{\sigma+z} k\left(\frac{z}{T}\right) e^{\xi z}\, dz = \frac{i f_0(\sigma) k(0)}{\sigma\sqrt{\varphi''(\sigma)}} \sqrt{\frac{\pi}{2}}(1+o(1)).$$

But

$$A_0(\omega) = \frac{1}{\sqrt{2\pi}} \frac{f_0(\sigma)}{\sigma\sqrt{\varphi''(\sigma)}} e^{\omega\sigma},$$

so

$$\int_l \frac{f(\sigma+z)}{\sigma+z} k\left(\frac{z}{T}\right) e^{\xi z}\, dz \sim \frac{ik(0)\pi A_0(\omega)}{e^{\omega\sigma}}.$$

Therefore as $\sigma \searrow 0$

$$\int_0^\infty A(u)R(T(u-\xi))Te^{-u\sigma}\, du \sim \frac{ik(0)\pi A_0(\omega)}{e^{\omega\sigma}}.$$

Making the change of variable $u = \xi + v/T$, we obtain

$$\int_{-\xi T}^\infty A\left(\xi + \frac{v}{T}\right) R(v) e^{-v\sigma/T}\, dv \sim ik(0)\pi A_0(\omega) e^{(\xi-\omega)\sigma}. \tag{3}$$

We have $T\xi = -\Delta\sigma\varphi'(\sigma) + O(1)$; therefore $T\xi \to \infty$ as $\sigma \searrow 0$.

To prove the estimates (1) we specify that function $k(\varsigma)$ and the quantity ξ in (3); this is done in different ways for the upper and lower bounds.

1. *Estimate of* $\overline{\lim}_{\omega\to\infty}(A(\omega)/A_0(\omega))$. Put $k(\varsigma) = 2(\varsigma - i)$, so that $k(0) = -2i$. Then

$$R(v) = 2\int_0^i (\varsigma - i)e^{-\varsigma v}\, d\varsigma = -\frac{2i}{v} + \frac{2}{v^2}(1 - e^{-iv}),$$

and hence

$$\operatorname{Re} R(v) = \frac{2}{v^2}(1 - \cos v) = \frac{4\sin^2(v/2)}{v^2} = \frac{\sin^2(v/2)}{(v/2)^2} \geq 0.$$

Taking real parts in the asymptotic formula (3), we obtain

$$\int_{-\xi T}^\infty A\left(\xi + \frac{v}{T}\right) e^{-v\sigma/T} \frac{\sin^2(v/2)}{(v/2)^2}\, dv \sim 2\pi A_0(\omega) e^{(\xi-\omega)\sigma}.$$

Let λ be a fixed positive number. Since $A(v)$ is nondecreasing and $e^{-v\sigma/T}$ is decreasing, it follows that

$$A\left(\xi - \frac{\lambda}{T}\right) e^{-\lambda\sigma/T} \int_{-\lambda}^{\lambda} \frac{\sin^2(v/2)}{(v/2)^2}\, dv \leq \int_{-\xi T}^\infty A\left(\xi + \frac{v}{T}\right) e^{-v\sigma/T} \frac{\sin^2(v/2)}{(v/2)^2}\, dv.$$

Therefore

$$A\left(\xi - \frac{\lambda}{T}\right) \leq e^{\lambda/\Delta} \left(\int_{-\lambda}^{\lambda} \frac{\sin^2(v/2)}{(v/2)^2}\, dv\right)^{-1} (1 + o(1)) 2\pi A_0(\omega) e^{(\xi-\omega)\sigma}.$$

Put $\xi = \omega + \lambda/T$; the condition $\xi = \omega + O(1/\sigma)$ is satisfied. We obtain

$$\frac{A(\omega)}{A_0(\omega)} \leq 2\pi e^{2\lambda/\Delta} \left(\int_{-\lambda}^{\lambda} \frac{\sin^2(v/2)}{(v/2)^2}\, dv\right)^{-1} (1 + o(1)),$$

$$\overline{\lim_{\omega\to\infty}} \frac{A(\omega)}{A_0(\omega)} \leq 2\pi e^{2\lambda/\Delta} \bigg/ \int_{-\lambda}^{\lambda} \frac{\sin^2(v/2)}{(v/2)^2}\, dv.$$

Taking $\lambda = \sqrt{\Delta}$, we see that

$$\varlimsup_{\omega\to\infty} \frac{A(\omega)}{A_0(\omega)} \le 2\pi e^{2/\sqrt{\Delta}} \left(\int_{-\sqrt{\Delta}}^{\sqrt{\Delta}} \frac{\sin^2(v/2)}{(v/2)^2}\, dv \right)^{-1}.$$

This relation and the formula

$$\int_{-\infty}^{\infty} \frac{\sin^2(v/2)}{(v/2)^2}\, dv = 2\pi$$

yield the upper bound in (1).

2. Now take

$$k(\varsigma) = 2(\varsigma - i) - \frac{e^{\mu\varsigma} - e^{-\mu\varsigma}}{\mu}; \qquad \mu > 0,\ \mu \equiv 0 \ (\mathrm{mod} 2\pi).$$

Then $k(0) = -2i$ and

$$\begin{aligned} R(v) &= 2\int_0^i (\varsigma - i)e^{-v\varsigma}\, d\varsigma - \frac{1}{\mu}\int_0^i (e^{\mu\varsigma} - e^{-\mu\varsigma})e^{-v\varsigma}\, d\varsigma \\ &= \frac{2}{v^2}(1 - e^{-iv} - iv) - \frac{1}{\mu}\left(\frac{e^{-vi} - 1}{\mu - v} + \frac{e^{-vi} - 1}{\mu + v} \right) \\ &= -\frac{2i}{v} + (1 - e^{-iv})\left(\frac{2}{v^2} + \frac{1}{\mu(\mu + v)} - \frac{1}{\mu(v - \mu)} \right). \end{aligned}$$

This implies

$$\begin{aligned} \operatorname{Re} R(v) &= (1 - \cos v)\left(\frac{2}{v^2} - \frac{2}{v^2 - \mu^2} \right) \\ &= \frac{\sin^2(v/2)}{(v/2)^2}\frac{\mu^2}{\mu^2 - v^2} \begin{cases} > 0 & \text{if } |v| < \mu, \\ < 0 & \text{if } |v| > \mu. \end{cases} \end{aligned}$$

Substituting into the main formula and taking real parts, we obtain

$$\int_{-\xi T}^{\infty} A\left(\xi + \frac{v}{T}\right) \frac{\sin^2(v/2)}{(v/2)^2}\frac{\mu^2}{\mu^2 - v^2} e^{-v\sigma/T}\, dv \sim 2\pi A_0(\omega)e^{(\xi - \omega)\sigma}(1 + o(1)),$$

$$\begin{aligned} \int_{-\mu}^{\mu} A\left(\xi + \frac{v}{T}\right) &\frac{\sin^2(v/2)}{(v/2)^2}\frac{\mu^2}{\mu^2 - v^2} e^{-v\sigma/T}\, dv \\ &\ge \int_{-\xi T}^{\infty} A\left(\xi + \frac{v}{T}\right) \frac{\sin^2(v/2)}{(v/2)^2}\frac{\mu^2}{\mu^2 - v^2} e^{-\sigma v/T}\, dv. \end{aligned}$$

Thus,

$$e^{\mu\sigma/T} A\left(\xi + \frac{\mu}{T}\right) \int_{-\mu}^{\mu} \frac{\sin^2(v/2)}{(v/2)^2}\frac{\mu^2}{\mu^2 - v^2}\, dv \ge 2\pi A_0(\omega)e^{(\xi - \omega)\sigma}(1 + o(1)).$$

Put $\xi = \omega - \mu/T$. Such a choice is possible since $\mu/T = O(1/\sigma)$. We obtain

$$e^{\mu/\Delta} A(\omega) \int_{-\mu}^{\mu} \frac{\sin^2(v/2)}{(v/2)^2} \frac{\mu^2}{\mu^2 - v^2} \, dv \geq 2\pi A_0(\omega) e^{-\mu/\Delta}(1 + o(1)),$$

and hence

$$\varliminf_{\omega \to \infty} \frac{A(\omega)}{A_0(\omega)} \geq 2\pi e^{-2\mu/\Delta} \left(\int_{-\mu}^{\mu} \frac{\sin^2(v/2)}{(v/2)^2} \frac{\mu^2}{\mu^2 - v^2} \, dv \right)^{-1}.$$

But

$$\begin{aligned}
&\int_{-\mu}^{\mu} 2\sin^2 \frac{v}{2} \left(\frac{2}{v^2} + \frac{1}{\mu(\mu - v)} + \frac{1}{\mu(\mu + v)} \right) dv \\
&\quad = \int_{-\mu}^{\mu} \frac{\sin^2(v/2)}{(v/2)^2} \, dv + 4 \int_{-\mu}^{\mu} \frac{\sin^2(v/2)}{\mu^2 - v^2} \, dv \\
&\quad = \int_{-\mu}^{\mu} \frac{\sin^2(v/2)}{(v/2)^2} \, dv + 8 \int_{0}^{\mu} \frac{\sin^2(v/2)}{\mu^2 - v^2} \, dv \\
&\quad = \int_{-\mu}^{\mu} \frac{\sin^2(v/2)}{(v/2)^2} \, dv + 8 \int_{0}^{\mu} \frac{\sin^2(w/2)}{w(2\mu - w)} \, dw \\
&\quad \leq \int_{-\mu}^{\mu} \frac{\sin^2(v/2)}{(v/2)^2} \, dv + \frac{8}{\mu} \int_{0}^{\mu} \frac{\sin^2(w/2)}{w} \, dw \\
&\qquad \leq \int_{-\mu}^{\mu} \frac{\sin^2(v/2)}{(v/2)^2} \, dv + \frac{8}{\mu} \int_{0}^{1} \frac{w^2}{4w} \, dw + \frac{8}{\mu} \int_{1}^{\mu} \frac{dw}{w}
\end{aligned}$$

(here we made the change of variable $\mu - v = w$ and used the inequality $1/(2\mu - w) < 1/\mu$). Thus,

$$\varliminf_{\omega \to \infty} \frac{A(w)}{A_0(w)} \geq 2\pi e^{-2\mu/\Delta} \left(\int_{-\mu}^{\mu} \frac{\sin^2(v/2)}{(v/2)^2} \, dv + \frac{1}{\mu} + \frac{8 \ln \mu}{\mu} \right)^{-1}.$$

Putting $\mu = 2\pi([\sqrt{\Delta}] + 1)$ establishes the lower bound in (1). The theorem is proved.

§1.4. The generalized Esseen inequality

Esseen's inequality (Esseen [43]; see also Gnedenko and Kolmogorov [62], §39) enables us to estimate the difference of two distribution functions in terms of the difference of their characteristic functions. Suppose $F(x)$ and $G(x)$ are distribution functions, and $f(t)$ and $g(t)$ their characteristic functions. If $G'(x)$ exists and $G'(x) \leq A$ for all x, then

$$\sup_x |F(x) - G(x)| < C \left(\frac{A}{T} + \int_0^T |f(t) - g(t)| \frac{dt}{t} \right).$$

For the needs of number theory the restriction "$G'(x)$ exists and is bounded" is sometimes inconvenient. Faĭnleĭb [45] gave an estimate of the difference of distribution functions in terms of the difference of their characteristic functions that is valid without any restriction.

THEOREM. *Suppose $F(x)$ and $G(x)$ are distribution functions, and $f(t)$ and $g(t)$ their characteristic functions. Then for $T > 0$*

$$\sup_x |F(x) - G(x)| < C_1 \left(S_G\left(\frac{1}{T}\right) + \int_0^T |f(t) - g(t)| \frac{dt}{t} \right), \tag{1}$$

where C_1 is an absolute constant and

$$S_G(h) = \sup_x \frac{1}{2h} \int_0^h (G(x+u) - G(x-u))\, du.$$

REMARK. It follows from the definition of $S_G(h)$ that

$$S_G(h) \le \tfrac{1}{2} Q_G(2h) \le Q_G(h),$$

where

$$Q_G(h) = \sup_x (G(x+h) - G(x))$$

is the concentration function for $G(x)$. If $G'(x)$ exists for each x and $G'(x) \le A$, then (1) becomes Esseen's inequality.

PROOF. Suppose $T > 0$. Consider a function $P_T(x)$ defined for all real x and having the following properties:

1) $0 \le P_T(x) \le C_1 T$ for all x;
2) $\int_{-\infty}^{\infty} P_T(x)\, dx = 1$;
3) $\int_0^{C_2/T} P_T(x)\, dx = A(T) \ge C_3 > \frac{1}{2}$;
4) $\varphi_T(t) = \int_{-\infty}^{\infty} e^{itx} P_T(x)\, dx = 0$ if $|t| > T$.

Here C_1, C_2, and C_3 are positive constants. It is easy to see that the function

$$P_T(x) = T \frac{1 - \cos(Tx - 3)}{\pi (Tx - 3)^2}$$

satisfies conditions 1)–4) with $C_1 = 1/2\pi$, $C_2 = 6$, and $C_3 = 1 - 4/3\pi > \frac{1}{2}$. Also,

$$\varphi_T(t) = \int_{-\infty}^{\infty} e^{itx} P_T(x)\, dx = \begin{cases} (1 - |t|/T) e^{3it/T}, & \text{if } |t| \le T, \\ 0, & \text{if } |t| > T. \end{cases}$$

Let $h = C_2/T$. Then, since $F(u)$ and $G(u)$ are distribution functions, it follows from conditions 1)–4) that

$$\begin{aligned}
F(x) &\le \frac{1}{A(T)}\int_x^{x+h} F(u)P_T(u-x)\,du \\
&= G(x) + \frac{1}{A(T)}\int_x^{x+h} (G(u)-G(x))P_T(u-x)\,du \\
&\quad + \frac{1}{A(T)}\int_x^{x+h} (F(u)-G(u))P_T(u-x)\,du \\
&\le G(x) + \frac{C_1 T}{A(T)}\int_0^{h} (G(x+u)-G(x))\,du \\
&\quad + \frac{1}{A(T)}\int_{-\infty}^{\infty} (F(u)-G(u))P_T(u-x)\,du \\
&\quad - \frac{1}{A(T)}\left(\int_{-\infty}^{x} + \int_{x+h}^{\infty}\right)(F(u)-G(u))P_T(u-x)\,du.
\end{aligned}$$

Let

$$F^*(x) = \int_{-\infty}^{\infty} F(u)P_T(u-x)\,du, \qquad G^*(x) = \int_{-\infty}^{\infty} G(u)P_T(u-x)\,du.$$

By the inversion formula, for $y < x$ we have

$$F^*(x) - F^*(y) = \frac{1}{2\pi i}\int_{-T}^{T} \frac{e^{-ity}-e^{-itx}}{t} f(t)\overline{\varphi}_T(t)\,dt$$

and

$$G^*(x) - G^*(y) = \frac{1}{2\pi i}\int_{-T}^{T} \frac{e^{-ity}-e^{-itx}}{t} g(t)\overline{\varphi}_T(t)\,dt$$

(since $\varphi_T(t) = 0$ for $|t| > T$); hence

$$\begin{aligned}
&F^*(x) - G^*(x) - (F^*(y) - G^*(y)) \\
&\quad = \frac{i}{2\pi}\int_{-T}^{T} \frac{f(t)-g(t)}{t}\overline{\varphi}_T(t)e^{-itx}\,dt \\
&\quad\quad - \frac{i}{2\pi}\int_{-T}^{T} \frac{f(t)-g(t)}{t}\overline{\varphi}_T(t)e^{-ity}\,dt. \qquad (2)
\end{aligned}$$

If $\int_0^1 |f(t)-g(t)|t^{-1}\,dt = \infty$, then the inequality being proved is trivial. If this is not the case, then, letting y tend to $-\infty$ in (2), we obtain (from the theorem on the Fourier coefficients tending to zero) that

$$\int_{-\infty}^{\infty} (F(u)-G(u))P_T(u-x)\,du = \frac{i}{2\pi}\int_{-T}^{T} \frac{f(t)-g(t)}{t}\overline{\varphi}_T(t)e^{-itx}\,dt,$$

and hence

$$\left|\int_{-\infty}^{\infty}(F(u)-G(u))P_T(u-x)\,du\right| \le \frac{1}{\pi}\int_0^T |f(t)-g(t)|\frac{dt}{t}. \tag{3}$$

Also,

$$\frac{C_1T}{A(T)}\int_0^h (G(x+u)-G(x))\,du$$
$$\le \frac{C_1}{C_3}T\int_0^h (G(x+u)-G(x-u))\,du \le \frac{2C_1C_2}{C_3}S_G(h). \tag{4}$$

It is clear from the definition of $S_G(h)$ that this is a nondecreasing function of h. Moreover, since

$$\begin{aligned}
&\frac{1}{4h}\int_0^{2h}(G(x+u)-G(x-u))\,du\\
&\quad= \frac{1}{4h}\int_0^h (G(x+u)-G(x-u))\,du\\
&\qquad+ \frac{1}{4h}\int_0^h (G(x+h+u)-G(x-h-u))\,du\\
&\quad= \frac{1}{4h}\int_0^h (G(x+u)-G(x-u))\,du\\
&\qquad+ \frac{1}{4h}\int_0^h (G(x+h+u)-G(x+h-u))\,du\\
&\qquad+ \frac{1}{4h}\int_0^h (G(x+h-u)-G(x-h+u))\,du\\
&\qquad+ \frac{1}{4h}\int_0^h (G(x-h+u)-G(x-h-u))\,du\\
&\quad= \frac{1}{2h}\int_0^h (G(x+u)-G(x-u))\,du\\
&\qquad+ \frac{1}{4h}\int_0^h (G(x+h+u)-G(x+h-u))\,du\\
&\qquad+ \frac{1}{4h}\int_0^h (G(x-h+u)-G(x-h-u))\,du,
\end{aligned}$$

it follows that $S_G(2h) \le 2S_G(h)$ and

$$S_G(h) = S_G(C_2/T) \le C_4S_G(1/T), \qquad C_4 = \text{const.} \tag{5}$$

Let $\alpha = \sup_x |F(x) - G(x)|$. We have

$$\begin{aligned}
&\left|\left(\int_{-\infty}^{x} + \int_{x+h}^{\infty}\right)(F(u) - G(u))P_T(u - x)\,du\right| \\
&\qquad \le \alpha\left(\int_{-\infty}^{x} + \int_{x+h}^{\infty}\right)P_T(u - x)\,du \\
&\qquad = \alpha\left(1 - \int_{x}^{x+h} P_T(u - x)\,du\right) = \alpha(1 - A(T)). \qquad (6)
\end{aligned}$$

It follows from (3)–(6) that

$$\begin{aligned}
F(x) \le G(x) &+ \frac{2C_1C_2C_4}{C_3} S_G\left(\frac{1}{T}\right) \\
&+ \frac{1}{\pi A(T)} \int_0^T |f(t) - g(t)| \frac{dt}{t} + \alpha\left(\frac{1}{A(T)} - 1\right),
\end{aligned}$$

i.e.

$$\begin{aligned}
F(x) - G(x) \le &\frac{2C_1C_2C_4}{C_3} S_G\left(\frac{1}{T}\right) \\
&+ \frac{1}{\pi C_3} \int_0^T |f(t) - g(t)| \frac{dt}{t} + \alpha\left(\frac{1}{C_3} - 1\right).
\end{aligned} \qquad (7)$$

In a completely analogous fashion, starting from the inequality

$$F(x) \ge \frac{1}{A(T)} \int_{x-h}^{x} F(u)P_T(x - u)\,du,$$

we arrive at a lower bound for $F(x) - G(x)$:

$$\begin{aligned}
F(x) - G(x) \ge -&\frac{2C_1C_2C_4}{C_3} S_G\left(\frac{1}{T}\right) \\
&+ \frac{1}{\pi C_3} \int_0^T |f(t) - g(t)| \frac{dt}{t} + \alpha\left(\frac{1}{C_3} - 1\right).
\end{aligned} \qquad (8)$$

Since inequalities (7) and (8) are valid for all x, it follows that

$$\alpha \le \frac{2C_1C_2C_4}{C_3} S_G\left(\frac{1}{T}\right) + \frac{1}{\pi C_3} \int_0^T |f(t) - g(t)| \frac{dt}{t} + \alpha\left(\frac{1}{C_3} - 1\right).$$

Therefore, since $C_3 > 1/2$, we obtain

$$\alpha \le \frac{2C_1C_2C_4}{C_3} S_G\left(\frac{1}{T}\right) + \frac{1}{\pi(2C_3 - 1)} \int_0^T |f(t) - g(t)| \frac{dt}{t}.$$

The theorem is proved.

COROLLARY 1. *Suppose $G(x)$ is a distribution function, $g(t)$ is its characteristic function, $h > 0$, and*

$$Q_G(h) = \sup_x (G(x+h) - G(x)),$$

$Q_G(h)$ being the concentration function. Then

$$Q_G(h) < C_5 \sup_{t \geq 1/h} \frac{1}{t} \int_0^t |g(u)|\, du, \tag{9}$$

where C_5 is an absolute constant.

For the proof we put $F(x) = G(x+h)$ in (1); then $f(t) = e^{-ith} g(t)$ and

$$Q_G(h) < C\left(S_G\left(\frac{1}{T}\right) + \int_0^T |g(t)(e^{-ith} - 1)| \frac{dt}{1}\right). \tag{10}$$

Here we take $T = 1/h$ and use the formula (see, for example, Cramér [22], Theorem 10)

$$\frac{1}{2h}\int_0^h (G(x+u) - G(x-u))\, du = \frac{1}{2\pi}\int_{-\infty}^{\infty} \left(\frac{\sin t}{t}\right)^2 e^{-2i(tx/h)} g\left(\frac{2t}{h}\right) dt,$$

from which it follows that

$$\begin{aligned} S_G(h) &\leq \frac{1}{\pi}\int_0^{\infty} \left(\frac{\sin t}{t}\right)^2 \left| g\left(\frac{2t}{h}\right)\right| dt \\ &\leq \frac{1}{\pi}\int_0^{1/2} \left| g\left(\frac{2t}{h}\right)\right| dt + \frac{1}{\pi}\int_{1/2}^{\infty} \left| g\left(\frac{2t}{h}\right)\right| \frac{dt}{t^2} \\ &= \frac{h}{2\pi}\int_0^{1/h} |g(t)|\, dt + \frac{2}{\pi h}\int_{1/h}^{\infty} |g(t)| \frac{dt}{t^2} \\ &\leq \frac{2}{\pi}\left(h\int_0^{1/h} |g(t)|\, dt + \frac{1}{h}\int_{1/h}^{\infty} \frac{1}{t^2} d\left(\int_0^t |g(u)|\, du\right)\right) \\ &= \frac{4}{\pi h}\int_{1/h}^{\infty} \left(\int_0^t |g(u)|\, du\right) \frac{dt}{t^3} \\ &\leq \frac{4}{\pi} \sup_{t\geq 1/h} \left(\frac{1}{t}\int_0^t |g(u)|\, du\right). \end{aligned} \tag{11}$$

This, together with the obvious inequality

$$\int_0^{1/h} |g(t)(e^{-ith} - 1)| \frac{dt}{t} \leq h\int_0^{1/h} |g(t)|\, dt \leq \sup_{t\geq 1/h} \frac{1}{t}\int_0^t |g(u)|\, du,$$

implies (9).

COROLLARY 2. *In the notation of the theorem,*

$$\sup_x |F(x) - G(x)| < C\left(\sup_{t\geq T}\frac{1}{t}\int_0^t |g(u)|\,du + \int_0^T |f(t)-g(t)|\frac{dt}{t}\right). \tag{12}$$

This follows immediately from (1) and (11).

CHAPTER 2

Additive Problems with an Increasing and an Infinite Number of Summands

§2.1. An additive sequence duplication problem

In the Introduction we mentioned a direct additive problem taken from Bachmann's book, *Die analytische Zahlentheorie* [6].

As another example we consider an additive problem (called to our attention by V. Tashbaev) in which the sequences to be added are equal (such problems are called *sequence duplication problems*).

Suppose we are given an infinite sequence of real numbers

$$a_1 \leq a_2 \leq a_3 \leq \cdots \tag{1}$$

We denote by $n(u)$ the number of terms of (1) not exceeding u, and by $q(u)$ the number of solutions of the inequality

$$a_i + a_j \leq u, \qquad i, j = 1, 2, \ldots. \tag{2}$$

Suppose we know that as $u \to \infty$

$$n(u) = Cu^\alpha + O(u^{\alpha_1}), \tag{3}$$

where $C > 0$ is a constant; $0 < \alpha \leq 1$, and $0 \leq \alpha_1 < \alpha$; α_1 and α are also constants.

THEOREM. *Under condition* (3),

$$q(N) = C^2 \frac{\Gamma^2(\alpha+1)}{\Gamma(2\alpha+1)} N^{2\alpha} + O(N^{\alpha+\alpha_1}).$$

PROOF. We first show that under condition (3)

$$a_i = C^{-1/\alpha} i^{1/\alpha} + O(i^{1/\alpha-(1-\alpha_1/\alpha)}).$$

Indeed, $n(\alpha_i) = i$. Therefore, by (3), $i = Ca_i^\alpha + O(a_i^{\alpha_1})$. It follows that $a_i^\alpha = O(i)$. We obtain

$$i = Ca_i^\alpha + O(i^{\alpha_1/\alpha}),$$

$$a_i = \left(\frac{i}{C} + O(i^{\alpha_1/\alpha})\right)^{1/\alpha} = \frac{i^{1/\alpha}}{C^{1/\alpha}}(1 + O(i^{-(1-\alpha_1/\alpha)}))^{1/\alpha}$$
$$= \frac{i^{1/\alpha}}{C^{1/\alpha}}(1 + O(i^{-(1-\alpha_1/\alpha)})) = \frac{i^{1/\alpha}}{C^{1/\alpha}} + O(i^{1/\alpha-(1-\alpha_1/\alpha)}).$$

We now prove the theorem. We have

$$\begin{aligned} q(N) &= \sum_{a_i \le N} n(N-a_i) \\ &= C\sum_{a_i\le N}(N-a_i)^\alpha + O\Big(\sum_{a_i\le N}(N-a_i)^{\alpha_1}\Big) \\ &= C \sum_{\substack{i\le n(N) \\ = CN^\alpha + O(N^{\alpha_1})}} \left(N - \left(\frac{i}{C}\right)^{1/\alpha} - O(i^{1/\alpha-(1-\alpha_1/\alpha)})\right)^\alpha + O(n(N)N^{\alpha_1}) \\ &= C \sum_{\substack{i\le n(N) \\ = CN^\alpha + O(N^{\alpha_1})}} \left(N - \left(\frac{i}{C}\right)^{1/\alpha} - O(i^{1/\alpha-(1-\alpha_1/\alpha)})\right)^\alpha + O(N^{\alpha+\alpha_1}). \end{aligned}$$

Since $0 < \alpha \le 1$, it follows from Lagrange's mean value theorem that

$$\begin{aligned} &(N - (i/C)^{1/\alpha} - O(i^{1/\alpha-(1-\alpha_1/\alpha)}))^\alpha - (N - (i/C)^{1/\alpha})^\alpha \\ &\qquad = O(i^{1/\alpha-(1-\alpha_1/\alpha)})O(1) = O(i^{1/\alpha-(1-\alpha_1/\alpha)}). \end{aligned}$$

Thus,

$$\begin{aligned} q(N) &= C \sum_{i\le CN^\alpha+O(N^{\alpha_1})} \left(N - \left(\frac{i}{C}\right)^{1/\alpha}\right)^\alpha \\ &\quad + O\Big(\sum_{i\le CN^\alpha+O(N^{\alpha_1})} i^{1/\alpha-(1-\alpha_1/\alpha)}\Big) + O(N^{\alpha_1+\alpha}) \\ &= C \sum_{i\le CN^\alpha+O(N^{\alpha_1})} \left(N - \left(\frac{i}{C}\right)^{1/\alpha}\right)^\alpha \\ &\quad + O\left(\int_1^{CN^\alpha+O(N^{\alpha_1})} x^{1/\alpha-(1-\alpha_1/\alpha)}\,dx\right) + O(N^{\alpha+\alpha_1}). \end{aligned}$$

It follows that

$$\begin{aligned}
q(N) &= C \sum_{i \le CN^\alpha + O(N^{\alpha_1})} \left(N - \left(\frac{i}{C}\right)^{1/\alpha}\right)^\alpha + O(N^{\alpha+\alpha_1}) \\
&= C \int_0^{CN^\alpha + O(N^{\alpha_1})} \left(N - \left(\frac{x}{C}\right)^{1/\alpha}\right)^\alpha dx + O(N^\alpha) + O(N^{\alpha+\alpha_1}) \\
&= C \int_0^{CN^\alpha} \left(N - \left(\frac{x}{C}\right)^{1/\alpha}\right)^\alpha dx + O(N^{\alpha_1} N^{(\alpha_1 - \alpha + 1)\alpha}) + O(N^{\alpha+\alpha_1}) \\
&= C \int_0^{CN^\alpha} \left(N - \left(\frac{x}{C}\right)^{1/\alpha}\right)^\alpha dx + O(N^{\alpha_1 + \alpha}),
\end{aligned}$$

since

$$\begin{aligned}
&\int_{CN^\alpha - O(N^{\alpha_1})}^{CN^\alpha} \left(N - \left(\frac{x}{C}\right)^{1/\alpha}\right)^\alpha dx \\
&\quad = O(N^{\alpha_1}((N^\alpha)^{1/\alpha} - (N^\alpha - O(N^{\alpha_1}))^{1/\alpha})^\alpha) \\
&\quad = O(N^{\alpha_1}(N^{\alpha_1} N^{\alpha(1/\alpha - 1)})^\alpha).
\end{aligned}$$

In the integral we make the change of variable

$$x = C(Nu)^\alpha, \qquad u = (1/N)(x/C)^{1/\alpha}.$$

We obtain

$$\begin{aligned}
q(N) &= C^2 N^{2\alpha} \alpha \int_0^1 u^{\alpha-1}(1-u)^\alpha \, du + O(N^{\alpha+\alpha_1}) \\
&= C^2 N^{2\alpha} \frac{\alpha \Gamma(\alpha)\Gamma(\alpha+1)}{\Gamma(2\alpha+1)} + O(N^{\alpha+\alpha_1}) \\
&= C^2 \frac{\Gamma^2(\alpha+1)}{\Gamma(2\alpha+1)} N^{2\alpha} + O(N^{\alpha+\alpha_1}),
\end{aligned}$$

as required.

§2.2. Schneider's lemma

As an example of a probabilistic interpretation of additive problems we state the lemma of Schneider [131] (Hilfsatz 1) mentioned in the Introduction.

LEMMA. *Suppose $r_1, \dots, r_n$ are natural numbers greater than 1; the n-dimensional parallelepiped $|x_l| \le r_l/2$, $l = 1, \dots, n$, is cut by the $(n-1)$-dimensional hyperplane $\sum_1^n x_l/r_l = t$ into two parts. Let $Q_n(t)$ denote the ratio of the number of lattice points (points with integral coordinates) in the smaller of the two parts to the total number of lattice points of the parallelepiped. Then $Q_n(t) \to 0$ for $n \to \infty$ and $t > \sqrt{(n \ln n)/2}$.*

To emphasize the "probabilistic" nature of Schneider's lemma we give the asymptotic law pertaining to this problem.

THEOREM. *Suppose $r_1, r_2, \dots$ are natural numbers greater than 1. Let $J_n(t)$ denote the number of solutions of the inequality $\sum_1^n x_l/r_l \le t$ in integers such that $|x_l| \le r_l/2$. Then, uniformly in x,*

$$\frac{J_n\left(x\sqrt{\frac{1}{3}\sum_{l=1}^n \frac{[r_l/2]([r_l/2]+1)}{r_l^2}}\right)}{\prod_{l=1}^n (2[r_l/2]+1)} = \frac{1}{\sqrt{2\pi}}\int_{-\infty}^{x} x^{-u^2/2}\,du + O(1/\sqrt{n}).$$

The idea on which the proof is based was due to Castelnuovo [18]; the theorem was proved by this author [116].

Consider a sequence of distribution laws $P_1(x), P_2(x), \dots$, where $P_l(x)$ is defined as follows:

$$P_l(x) = \begin{cases} \frac{1}{2[r_l/2]+1} & \text{for } x = n/r_l,\ n \text{ an integer such that } |n| \le r_l/2; \\ 0 & \text{for all other real } x. \end{cases}$$

Let us denote the corresponding distribution functions by $F_l(x)$, $l = 1, 2, \dots$. We calculate the expectation and variance of the law $P_l(x)$:

$$a_l = \frac{1}{2[r_l/2]+1}\sum_{|x_l|\le r_l/2}\frac{x_l}{r_l} = 0,$$

$$b_l^2 = \frac{1}{2[r_l/2]+1}\sum_{|x_l|\le r_l/2}\frac{x_l^2}{r_l^2} = \frac{[r_l/2]([r_l/2]+1)}{3r_l^2}.$$

Let $\Phi_n(x) = F_1(x) * \cdots * F_n(x)$ (the symbol $*$ denotes convolution). It is obvious that

$$\Phi_n(x) = \frac{J_n(x)}{\prod_{l=1}^n (2[r_l/2]+1)}.$$

By general theorems of probability theory, the expectation A_n and variance B_n^2 of the distribution function $\Phi_n(x)$ are

$$A_n = 0, \qquad B_n^2 = \frac{1}{3}\sum_{l=1}^n \frac{[r_l/2]([r_l/2]+1)}{r_l^2}.$$

Let us verify that the Lindeberg conditions are satisfied in this problem. For any fixed $\tau > 0$

$$\lim_{n\to\infty}\frac{1}{B_n^2}\sum_{l=1}^n \int_{|x|\ge \tau B_n} x^2\,dF_l(x) = \lim_{n\to\infty}\frac{1}{B_n^2}\sum_{l=1}^n \frac{1}{2[r_l/2]+1}\sum_{1/2\ge x_l/r_l\ge \tau B_n}\frac{x_l^2}{r_l^2}.$$

Obviously $b_l^2 \ge C$, where C is an absolute constant: if $r_l = 2$, then $b_l^2 = 2/(3\cdot 4) = \frac{1}{6}$; if $r_l \ge 3$, then

$$\left[\frac{r_l}{2}\right] \ge \frac{r_l}{3} \quad \text{and} \quad b_l^2 \ge \frac{1}{3r_l^2}\frac{r_l}{3}\frac{r_l}{2} = \frac{1}{18}.$$

Therefore $B_n^2 \geq Cn$, i.e. $B_n > \sqrt{Cn}$. For any $\tau > 0$ and a sufficiently large n, τB_n exceeds $1/2$ and the inner sum $\sum_{1/2 \geq x_l/r_l \geq \tau B_n} x_l^2/r_l^2$ is empty. Therefore

$$\lim_{n\to\infty} \frac{1}{B_n^2} \sum_{l=1}^{n} \int_{|x|\geq \tau B_n} x^2\, dF_l(x) = 0,$$

i.e., Lindeberg's conditions are satisfied. By the integral limit theorem we have, uniformly in x,

$$\lim_{n\to\infty} \frac{J_n(xB_n)}{\prod_{l=1}^{n}(2[r_l/2]+1)} = \frac{1}{\sqrt{2\pi}} \int_{-\infty}^{x} e^{-z^2/2}\, dz.$$

Substituting the value of B_n, we obtain the theorem.

Let us now calculate the remainder term in our theorem. We will use the method of characteristic functions and, in particular, Esseen's inequality.

We use the letter C to denote absolute constants. Let $\Phi(x)$ stand for Gauss's law:

$$\Phi(x) = \frac{1}{\sqrt{2\pi}} \int_{-\infty}^{x} e^{-u^2/2}\, du$$

We next introduce the following notation:

$$\beta_{2l} = \sigma_l^2 = \frac{1}{2[r_l/2]+1} \sum_{|x|\leq [r_l/2]} \frac{x^2}{r_l^2} = \frac{[r_l/2]([r_l/2]+1)}{3r_l^2},$$

$$\beta_{3l} = \frac{2}{2[r_l/2]+1} \sum_{x=0}^{[r_l/2]} \frac{x^3}{r_l^3} = \frac{([r_l/2]([r_l/2]+1))^2}{2(2[r_l/2]+1)r_l^3},$$

$$s_n^2 = \sum_{l=1}^{n} \sigma_l^2 = \sum_{l=1}^{n} \beta_{2l},$$

$$\rho_n = \frac{\sum_{l=1}^{n} \beta_{3l}}{s_n^3}\sqrt{n}, \qquad T_n = \frac{s_n^3}{4\left(\sum_{l=1}^{n} \beta_{3l}\right)}.$$

Let $f_l(t)$ and $\tilde{\varphi}_n(t)$ denote the characteristic functions of the distribution functions $F_l(x)$ and $\Phi_n(s_n x)$:

$$f_l(t) = \frac{1}{2[r_l/2]+1} \sum_{|x|\leq r_l/2} e^{itx/r_l},$$

$$\tilde{\varphi}_n(t) = \prod_{l=1}^{n} f_l\left(\frac{t}{s_n}\right) = \prod_{l=1}^{n} \frac{1}{2[r_l/2]+1} \left(\sum_{|x|\leq r_l/2} e^{itx/s_n r_l} \right).$$

REMARK 1. $\beta_{2l}^{1/2} \leq \beta_{3l}^{1/3}$.

Indeed, let $a_l = [r_l/2]$, $a_l \geq 1$. Then the stated inequality is equivalent to the obvious inequality $11(a_l^2 + a_l) \geq 4$.

REMARK 2. For any n, $\rho_n \geq 1$.

Indeed, by Hölder's inequality,

$$s_n^3 = \left(\sum_{l=1}^{n} \beta_{2l}\right)^{3/2} \leq \sqrt{n}\sum_{l=1}^{n} \beta_{2l}^{3/2}.$$

By Remark 2, $s_n^3 \leq \sqrt{n}\sum_{l=1}^{n} \beta_{3l}$, i.e., $\rho_n \geq 1$.

REMARK 3. $\rho_n \leq C$.

Since $[r_l/2] \geq 1$, it follows that $\sigma_l^2 \geq 1/12$ and $\beta_{3l} \leq C$. Therefore $\rho_n \leq C$. Of central importance is

LEMMA 1. *If* $|t| \leq \sqrt[3]{T_n}$, *then*

$$|\tilde{\varphi}_n(t) - e^{-t^2/2}| \leq C\frac{|t|^3}{T_n}e^{-t^2/2}$$

PROOF. If x is a real number, then

$$e^{ix} = 1 + ix - x^2/2 + \theta x^3/6, \qquad |\theta| \leq 1.$$

Consequently,

$$\begin{aligned} f_l\left(\frac{t}{s_n}\right) &= \frac{1}{2[r_l/2]+1}\sum_{|x|\leq r_l/2} e^{itx/s_n r_l} \\ &= \frac{1}{2[r_l/2]+1}\sum_{|x|\leq r_l/2}\left(1 + \frac{ixt}{r_l s_n} - \frac{x^2t^2}{2r_l^2 s_n^2} + \theta\frac{|x|^3|t|^3}{6r_l^3 s_n^3}\right) \\ &= 1 - \frac{1}{2}\beta_{2l}\frac{t^2}{s_n^2} + \frac{\theta}{6}\beta_{3l}\frac{|t|^3}{s_n^3}. \end{aligned}$$

From the inequality $\beta_{2l}^{1/2} \leq \beta_{3l}^{1/3}$ we obtain

$$\left|f_l\left(\frac{t}{s_n}\right) - 1\right| \leq \frac{1}{2}\frac{\beta_{3l}^{2/3}|t|^2}{s_n^2} + \frac{1}{6}\beta_{3l}\frac{|t|^3}{s_n^3} = \frac{1}{2}\left(\frac{\beta_{3l}|t|^3}{s_n^3}\right)^{2/3} + \frac{1}{6}\beta_{3l}\frac{|t|^3}{s_n^3}.$$

Since $\beta_{3l} \leq \sum_{i=1}^{n} \beta_{3i}$ and $|t| \leq \sqrt[3]{T_n}$, it follows that

$$\beta_{3l}\frac{|t|^3}{s_n^3} \leq \frac{\sum_{l=1}^{n}\beta_{3l}}{s_n^3}T_n = \frac{1}{4} < 1.$$

Therefore,

$$|f_l(t/s_n) - 1| \leq \tfrac{1}{2} + \tfrac{1}{6} = \tfrac{2}{3}.$$

Note that $f_l(t/s_n)$ is a real number. If $|\alpha| \leq 2/3$, it follows from Taylor's formula with two terms and a remainder that

$$|\ln(1+\alpha) - \alpha| \leq \tfrac{9}{2}\alpha^2.$$

Therefore,

$$\ln f_l(t/s_n) = f_l(t/s_n) - 1 + \tfrac{9}{2}\theta\,|f_l(t/s_n) - 1|^2.$$

But for $|t| \le \sqrt[3]{T_n}$

$$\left| f_l\left(\frac{t}{s_n}\right) - 1 \right| \le \left(\frac{\beta_{3l}|t|^3}{s_n^3}\right)^{1/2} \left(\frac{1}{2} + \frac{1}{6}\right) = \frac{2}{3}\left(\frac{\beta_{3l}|t|^3}{s_n}\right)^{1/2}.$$

Therefore,

$$\ln f_l\left(\frac{t}{s_n}\right) = f_l\left(\frac{t}{s_n}\right) - 1 + 2\theta\frac{\beta_{3l}|t|^3}{s_n^3} = -\frac{1}{2}\beta_{2l}\frac{t^2}{s_n^2} + \frac{13}{6}\theta\frac{\beta_{3l}|t|^3}{s_n^3}.$$

Summing and replacing 13/6 by 3, we obtain

$$\ln\tilde{\varphi}_n(t) = -\frac{t^2}{2} + 3\theta\left(\sum_{l=1}^n \beta_{3l}\right)\frac{|t|^3}{s_n^3},$$

or

$$\ln(\tilde{\varphi}_n(t)e^{t^2/2}) = 3\theta\left(\sum_{l=1}^n \beta_{3l}\right)\frac{|t|^3}{s_n^3}.$$

But for $\alpha > 0$ we have $e^\alpha = 1 + \alpha e^\alpha\theta$ for some θ, $0 < \theta < 1$, and hence

$$\tilde{\varphi}_n(t)e^{t^2/2} = 1 + 3\theta'\left(\sum_{l=1}^n \beta_{3l}\right)\frac{|t|^3}{s_n^3}\exp\left(3\theta\left(\sum_{l=1}^n \beta_{3l}\right)\frac{|t|^3}{s_n^3}\right).$$

Since for $|t| \le \sqrt[3]{T_n}$ we have

$$\left(\sum_{l=1}^n \beta_{3l}\right)\frac{|t|^3}{s_n^3} \le \left(\sum_{l=1}^n \beta_{3l}\right)\frac{T_n}{s_n^3} = \frac{1}{4},$$

it follows that

$$\tilde{\varphi}_n(t)e^{t^2/2} = 1 + 4C\theta\left(\sum_{l=1}^n \beta_{3l}\right)\frac{|t|^3}{s_n^3},$$

or

$$|\tilde{\varphi}_n(t)e^{t^2/2} - 1| \le C|t|^3/T_n,$$

as required.

LEMMA 2. $|\tilde{\varphi}_n(t)|^2 \le \exp(-t^2 + |t|^3/3T_n)$.

PROOF. We have

$$\begin{aligned}|f_l(t)|^2 &= \frac{1}{(2[r_l/2]+1)^2}\sum_{|x_1|\le r_l/2}\ \sum_{|x_2|\le r_l/2} e^{it(x_1-x_2)/r_l} \\ &= \frac{1}{(2[r_l/2]+1)^2}\sum_{|x_1|\le r_l/2}\ \sum_{|x_2|\le r_l/2} \cos t\frac{x_1-x_2}{r_l},\end{aligned}$$

since the left-hand side of the equality is a real number. Also,

$$\cos t\frac{x_1-x_2}{r_l} \le 1-\frac{t^2}{2}\frac{(x_1-x_2)^2}{r_l^2}+\frac{1}{6}|t|^3\left|\frac{x_1-x_2}{r_l}\right|^3$$
$$\le 1-\frac{1}{2}\frac{t^2}{r_l^2}(x_1^2-2x_1x_2+x_2^2)+\frac{2}{3}|t|^3\frac{|x_1|^3+|x_2|^3}{r_l^3}.$$

This yields

$$|f_l(t)|^2 \le 1-\frac{t^2}{(2[r_l/2]+1)^2\,2r_l^2}\left(\sum_{|x_1|\le r_l/2}\sum_{|x_2|\le r_l/2}x_1^2 - 2\sum_{|x_1|\le r_l/2}\sum_{|x_2|\le r_l/2}x_1x_2+\sum_{|x_1|\le r_l/2}\sum_{|x_2|\le r_l/2}x_2^2\right)$$
$$+\frac{2}{3}\frac{|t|^3}{(2[r_l/2]+1)^2r_l^3}\left(\sum_{|x_1|\le r_l/2}\sum_{|x_2|\le r_l/2}|x_1|^3+\sum_{|x_1|\le r_l/2}\sum_{|x_2|\le r_l/2}|x_2|^3\right)$$
$$=1-\frac{t^2}{2[r_l/2]+1}\sum_{|x|\le r_l/2}\frac{x^2}{r^2}+\frac{4|t|^3}{3(2[r_l/2]+1)}\sum_{|x|\le r_l/2}\frac{|x|^3}{r_l^3}$$
$$=1-t^2\beta_{2l}+\tfrac{4}{3}|t^3|\beta_{3l}.$$

We have the inequality $1+\alpha\le e^\alpha$ $(\alpha>0)$, which implies that

$$|f_l(t)|^2\le\exp\left[-t^2\beta_{2l}+\tfrac{4}{3}|t|^3\beta_{3l}\right].$$

Therefore,

$$|\tilde\varphi_n(t)|^2\prod_{l=1}^n\left|f_l\left(\frac{t}{s_n}\right)\right|^2=\exp\left[-\frac{t^2}{s_n^2}\sum_{l=1}^n\beta_{2l}+\frac{4}{3}\frac{|t|^3}{s_n^3}\sum_{l=1}^n\beta_{3l}\right]$$
$$=\exp[-t^2+|t|^3/3T_n].$$

This is what we had to prove.

COROLLARY. *Suppose* $|t|\le T_n$. *Then*

$$|\tilde\varphi_n(t)-e^{-t^2/2}|\le C\frac{|t|^3}{T_n}e^{t^2/3}.$$

PROOF. If $|t|\le\sqrt[3]{T_n}$, then our assertion follows from Lemma 1:

$$|\tilde\varphi_n(t)-e^{t^2/2}|\le\frac{C|t|^3}{T_n}e^{-t^2/2}<C\frac{|t|^3}{T_n}e^{-t^2/3}.$$

Suppose $\sqrt[3]{T_n}\le|t|\le T_n$. In this case we apply Lemma 2:

$$|\tilde\varphi_n(t)-e^{t^2/2}|\le|\tilde\varphi_n(t)|+e^{-t^2/2}\le Ce^{t^3/3}+e^{-t^2/2}\le Ce^{-t^2/3}\le C\frac{|t|^3}{T_n}e$$

(since $|t|\ge\sqrt[3]{T_n}$).

LEMMA 3.

$$\int_{-T_n}^{T_n} \left| \frac{\tilde{\varphi}_n(t) - e^{-t^2/2}}{t} \right| dt \leq \frac{C}{T_n}.$$

PROOF. We have

$$\begin{aligned} \int_{-T_n}^{T_n} \left| \frac{\tilde{\varphi}_n(t) - e^{-t^2/2}}{t} \right| dt &\leq \frac{C}{T_n} \int_{-T_n}^{T_n} t^2 e^{-t^2/3}\, dt \\ &\leq \frac{C}{T_n} \int_{-\infty}^{\infty} t^2 e^{-t^2/3}\, dt = O\left(\frac{1}{T_n}\right). \end{aligned}$$

The lemma is proved.

Let us complete the proof of the theorem. We apply Esseen's inequality:

$$\left| \frac{J_n(s_n x)}{\prod_{l=1}^{n}(2[r_l/2]+1)} - \Phi(x) \right| \leq C\left(\frac{A}{T_n} + \int_{-T_n}^{T_n} \left| \frac{\tilde{\varphi}_n(t) - e^{-t^2/2}}{t} \right| dt \right);$$

here $A = \max|\Phi'(x)| = 1/\sqrt{2\pi}$. By Lemma 3,

$$\left| \frac{J_n(s_n x)}{\prod_{l=1}^{n}(2[r_l/2]+1)} - \frac{1}{\sqrt{2\pi}} \int_{-\infty}^{x} e^{-u^2/2}\, du \right| \leq \frac{C}{T_n} \leq C\frac{\rho_n}{\sqrt{n}}.$$

But, by Remark 3, $\rho_n \leq C$. Therefore,

$$\left| \frac{J_n(x\sqrt{\sigma_1^2 + \cdots + \sigma_n^2})}{\prod_{l=1}^{n}(2[r_l/2]+1)} - \frac{1}{\sqrt{2\pi}} \int_{-\infty}^{x} e^{-u^2/2}\, du \right| \leq \frac{C}{\sqrt{n}},$$

as required.

Babaev (see [5], Chapter 1, §3) employed similar ideas in studying the distribution of lattice points on an equilateral hyperbola.

§2.3. The local limit theorem of probability theory and its applications to number theory

We say that a discrete random variable (i.e., a variable assuming a finite or countable set of values) ξ has a *lattice distribution* if there exist numbers a and $D > 0$ such that all values of this random variable belong to the progression $a + Dk$, $k = \dots, -1, 0, 1, \dots$. The number D is called the *spacing* of the distribution. We say that the spacing of the distribution is *maximal* if the values of ξ are not included in any progression $b + D_1 k$ with a larger difference $D_1 > D$.

If p_k denotes the probability of the equality $\xi = a + Dk$, then the characteristic function of ξ can be written in the form

$$f(t) = \sum_{k=-\infty}^{\infty} p_k e^{it(a+Dk)} = e^{ita} \sum_{k=-\infty}^{\infty} p_k e^{itDk}.$$

The characteristic function is a finite or infinite trigonometric series.

Suppose the random variable $\xi_1, \xi_2, \ldots,$ are mutually independent, are lattice distributed, and have the same distribution function $F(x)$. Consider their sum $\varsigma_n = \xi_1 + \cdots + \xi_n$; ς_n is a lattice random variable and its values can be written in the form $na + Dk$. We denote by $P_n(k)$ the probability of the equality $\varsigma_n = na + Dk$, and by A_n the expectation of ς_n. Let $B_n^2 = \operatorname{Var} \varsigma_n = n \operatorname{Var} \varsigma_1$, where Var denotes the variance, and put

$$z_{n,k} = (an + Dk - A_n)/B_n.$$

B. V. Gnedenko proved the following result, which is called the *local limit theorem* of probability theory.

THEOREM. *Suppose the independent random variables $\xi_1, \xi_2, \ldots$ have the same distribution function $F(x)$ and their expectations and variances are finite. Then in order that*

$$P_n(k) = \frac{D}{\sqrt{2\pi}\, B_n} e^{-z_{n,k}/2} + o\left(\frac{1}{B_n}\right)$$

uniformly in k $(-\infty < k < \infty)$ as $n \to \infty$ it is necessary and sufficient that the spacing D of the distribution be maximal.

The proof of Gnedenko's theorem ([61], §43) was effected by the method of characteristic functions.

Let us turn to the question of the remainder term in the local limit theorem for lattice random variables. The following theorem is given in [72] (Theorem 4.5.3).

THEOREM. *In order that*

$$\max_k \left| \frac{B_n P_n(k)}{D} - e^{-z_{n,k}^2/2} \right| = O(n^{-\delta/2}), \qquad 0 < \delta < 1,$$

it is necessary and sufficient that the following conditions be satisfied:

1) *The spacing D of the distribution is maximal.*

2) *As $z \to \infty$*

$$\int_{|x| \geq z} x^2 \, dF(x) = O(z^{-\delta}).$$

There have been many papers concerned with generalizations of the local limit theorem and, in particular, with extending it to variously distrubuted summands. We mention here the authors R. von Mises, G. M. Bavli (Bawly), B. V. Gnedenko, Yu. V. Prokhorov, G. A. Freĭman, Yu. A. Rozanov, V. V. Petrov, T. A. Azlarov, S. Kh. Sirazhdinov, A. A. Mitalauskas and V. A. Statulevičius, D. A. Moskvin, L. P. Postnikova, A. A. Yudin, and N. G. Gamkrelidze. Bibliographies can be found in the book of Ibragimov and Linnik

[72] and in the papers of Mitalauskas and Statulevičius [101] and Moskvin, Postnikova, and Yudin [102]. In Freĭman's paper [54], local limit theorems are studied by an elementary method.

Here are some examples of the use of local limit theorems of probability theory in arithmetic.

EXAMPLE 1. Let $r_{n,P}(N)$ denote the number of solutions of the equation $N = x_1 + \cdots + x_n$ in integers such that $0 \le x_i \le P$, $i = 1, \ldots, n$. Suppose P is fixed and $n \to \infty$. Consider the distribution law $P(x) = 1/(P+1)$ for $x = 0, 1, \ldots, P$. This is a lattice law with maximal spacing $D = 1$.

Consider also the distribution law $P_n(x) = P(x) * \cdots * P(x)$ (n factors). Obviously,

$$P_n(N) = \frac{r_{n,P}(N)}{(P+1)^n}.$$

The local limit theorem is applicable. Using this theorem we obtain, uniformly in integers N,

$$\sqrt{n\frac{P^2+2P}{12}}\,\frac{r_{n,P}(N)}{(P+1)^n} - \frac{1}{\sqrt{2\pi}}\exp\left(-\frac{(N-nP/2)^2}{2n(P^2+2P)/12}\right) \to 0.$$

Thus we see that for a fixed P and as $n \to \infty$ we have, uniformly in integers N,

$$r_{n,P}(N) = \frac{(P+1)^n}{\sqrt{\pi n(P^2+2P)/6}}\exp\left(-\frac{(N-nP/2)^2}{n(P^2+2P)/6}\right) + o\left(\frac{(P+1)^{n-1}}{\sqrt{n}}\right).$$

EXAMPLE 2. Suppose $r_{n,P}(N)$ has a different meaning than in Example 1; namely, here this symbol denotes the number of solutions of the Diophantine equation $N = x_1^2 + \cdots + x_n^2$ in integers such that $0 \le x_i \le P$, $i = 1, \ldots, n$. Suppose P is fixed and $n \to \infty$. Consider the distribution law $P(x) = 1/(P+1)$ for $x = 0^2, 1^2, 2^2, \ldots, P^2$. This is a lattice distribution law with maximal spacing $D = 1$. The expectation and variance of $P(x)$ are equal to

$$a = \frac{1}{P+1}(1^2 + 2^2 + \cdots + P^2) = \frac{P(2P+1)}{6},$$

$$\begin{aligned}\sigma^2 &= \frac{1}{P+1}(1^4 + 2^4 + \cdots + P^4) - \left(\frac{P(2P+1)}{6}\right)^2 \\ &= \frac{P(2P+1)(3P^2+3P-1)}{30} - \left(\frac{P(2P+1)}{6}\right)^2 \\ &= \frac{P(P+2)(2P+1)(8P-3)}{180}.\end{aligned}$$

We define the distrubiton law $P_n(x) = P(x) * \cdots * P(x)$ (n factors). It is clear that

$$P_n(N) = \frac{r_{n,P}(N)}{(P+1)^n}.$$

Applying the local limit theorem, we see that as $n \to \infty$ (P is fixed) we have, uniformly in integers N,

$$r_{n,P}(N) = \frac{(P+1)^n}{\sqrt{2\pi n P(P+2)(2P+1)(8P-3)/180}} \times \exp\left(-\frac{(N-nP(2P+1)/6)^2}{2nP(P+2)(2P+1)(8P-3)/180}\right) + o\left(\frac{(P+1)^{n-2}}{\sqrt{n}}\right).$$

EXAMPLE 3. Suppose $P \geq 1$ is a fixed natural number. Let $A_n(P)$ denote the number of solutions of the Diophantine equation $x_1 + \cdots + x_n = y_1 + \cdots + y_n$ in integers such that $0 \leq x_i \leq P$ and $0 \leq y_i \leq P$, $i = 1, \ldots, n$. Let

$$\sigma^2 = \frac{2}{(P+1)^2} \sum_{k=1}^{P} (P+1-k)k^2$$

(we can calculate this quantity exactly, but this will not be necessary).

THEOREM. *If P is fixed and $n \to \infty$, then*

$$\lim_{n\to\infty} \frac{A_n(P)}{(P+1)^{2n}/\sigma\sqrt{2\pi n}} = 1.$$

PROOF. Let $u(k)$ be the number of solutions of the Diophantine equation $x - y = k$ in integers such that $0 \leq x \leq P$ and $0 \leq y \leq P$. It is easy to find an explicit expression for $u(k)$:

$$\begin{aligned} u(k) &= P+1-|k| \quad &&\text{if } |k| \leq P, \\ u(k) &= 0 &&\text{if } |k| \geq P+1. \end{aligned}$$

Consider the distribution law defined as follows: $P(k) = u(k)/(P+1)^2$ for integers k. This is a lattice distribution law with maximal spacing $D = 1$. We calculate the mean and the variance σ^2 of $P(x)$:

$$a = \sum_{k=-P}^{P} \frac{u(k)}{(P+1)^2} k = 0,$$

$$\sigma^2 = 2\sum_{k=1}^{P} \frac{u(k)}{(P+1)^2} k^2 = \frac{2}{(P+1)^2} \sum_{k=1}^{P} (P+1-k)k^2.$$

We now form the convolution $P_n(x) = P(x) * \cdots * P(x)$ (n factors). Clearly,

$$P_n(0) = \frac{A_n(P)}{(P+1)^{2n}},$$

where $A_n(P)$ is the number of solutions of the Diophantine equation $x_1 - y_1 + \cdots + x_n - y_n = 0$ in integers such that $0 \leq x_i \leq P$ and $0 \leq y_i \leq P$, $i = 1, \ldots, n$.

Applying Gnedenko's local theorem, we see that

$$\sigma\sqrt{n}\frac{A_n(P)}{(P+1)^{2n}} - \frac{1}{\sqrt{2\pi}} \to 0$$

as $n \to \infty$, which was to be proved.

§2.4. Additive problems with an increasing number of summands

As we have seen, the local limit theorem of probability theory enables us to obtain an asymptotic formula for the number of solutions of the Diophantine equation $N = f(x_1) + \cdots + f(x_n)$, where $f(x)$ is an integral function of an integral argument and $0 \le x_1, \dots, x_n \le Q - 1$ as $n \to \infty$, Q being fixed. There arises the question of extending these asymptotic formulas to the case where Q can increase along with n.

It is this question on which we will now focus our attention.

In the probabilistic treatment of such problems we require that with the increase in the number of identically distributed lattice random variables $\xi_1, \xi_2, \dots$ the variable ξ itself could undergo specific changes, i.e., we need local limit theorems that are uniform relative to certain admissible changes.

Note that there are theorems of this kind in integral formulations of problems. Consider a sequence of identically distributed independent random variables $\xi_1, \xi_2, \dots$. Put $\mathsf{E}\xi_1 = a$ and $\operatorname{Var}\xi_1 = \sigma^2 > 0$. We denote the distribution function of the random variable

$$\eta_n = \frac{\xi_1 + \xi_2 + \cdots + \xi_n - na}{\sigma\sqrt{n}}$$

by $F_n(x)$. Assume the absolute third moment $\beta = \mathsf{E}|\xi_1 - a|^3$ exists. Then we have the estimate

$$\left|F_n(x) - \frac{1}{\sqrt{2\pi}}\int_{-\infty}^{x} e^{-t^2/2}\,dt\right| \le C\frac{\beta}{\sigma^3}\frac{1}{\sqrt{n}(1+|x|^3)}$$

with absolute constant C (see Nagaev [104]). This is a uniform estimate relative to the original distribution; it implies that the central limit theorem "remains valid" if the original distributions are altered so that $\beta/\sigma^3 = o(\sqrt{n})$.

In local limit theorems the question of uniform estimates is much more complicated, i.e., to establish a local limit theorem it is necessary to estimate a characteristic function not only in a neighborhood of zero (that is sufficiently well regulated by the first three moments), but also in the whole interval of variation of its argment. Thus if we wish to solve the problem of the number of representations of N in the form $N = f(x_1) + \cdots + f(x_n)$, $0 \le x_1, \dots, x_n \le Q - 1$, where n and Q increase, then we require estimates of trigonometric sums of the form

$$\frac{1}{Q}\sum_{x=0}^{Q-1} e^{2\pi i\alpha f(x)}$$

with $Q\to\infty$. Naturally, there are no universal theorems concerning estimates of sums of this kind. Therefore the most we can hope for is to distinguish classes of problems for which we can establish uniform local limit theorems.

Postnikov [117] considered the problem of representing N in the form $N = x_1+\cdots+x_n$, $0\le x_1,\ldots,x_n\le Q-1$, $n\to\infty$, the x_i integers, and solved it under the restriction $Q\le K^n$, where $K>1$. This restriction, as Sirazhdinov and Azlarov showed in [135], is not essential. An additive problem with a linear function can be solved by means of a special uniform local theorem; we state it here in the form in which it was proved by Zuparov [157].

THEOREM. *Suppose $\xi_1,\xi_2,\ldots$ is a sequence of identically distributed random variables. Let $\mathsf{P}(\xi = x) = p_x$, $x\in G$, where G is a set of Q distinct integers, and assume that the following conditions hold.*

1) *There exists a constant b, $1<b<\infty$, such that*

$$\max_{x\in G}\{x\} = bQ(1+o(1)).$$

2) $\min_{x\in G} p_x \ge C_1/Q$, *where $C_1>0$ is an absolute constant.*

3) $\max \mathsf{P}(\xi\equiv m \pmod h)\le \rho(h)$, *where $\rho(h)$ does not depend on Q and is strictly less than 1 for any $h\ge 2$.*

Let a denote the expectation and σ^2 the variance of the random variables $\xi_1,\xi_2,\ldots$. For a given integer N let

$$P_n(N) = \mathsf{P}(\xi_1+\xi_2+\cdots+\xi_n = N),$$

$$u_N = \frac{N-an}{\sigma\sqrt{n}},\qquad \Delta_n = \left|\sigma\sqrt{n}P_n(N) - \frac{1}{\sqrt{2\pi}}\exp\left(-\frac{u_N^2}{2}\right)\right|.$$

Then there exists a constant $C_2\ge 0$ such that for all $Q\ge 2$

$$\sup_N \Delta_n \le \frac{C_2}{\sqrt{n}}\left(1+\sigma\sum_{x\in G} p_x^2\right).$$

In the same paper [117] Postnikov considered the problem of representing N in the form $N = x_1^2+\cdots+x_n^2$ with nonnegative integers x_i not exceeding Q, $n\to\infty$, $Q\le K^n$, and $K>1$ a constant. A more general method, leading to a strengthening and generalization of Postnikov's result, was communicated by S. B. Stechkin at the Fourth All-Union Mathematical Congress (Leningrad, 1961). Sirazhdinov and Azlarov [135] considered the general equation $N = x_1^s+\cdots+x_n^s$, s an arbitrary natural number, where the restriction $Q\le K^n$, $K>1$, was dropped. We state the theorem in the form proved by Zuparov [157].

THEOREM. *Suppose $Q \geq 2$. Let $R_n(Q, N)$ denote the number of representations of N in the form $x_1^s + \cdots + x_n^s$ with integers x_i, $0 \leq x_1, \ldots, x_n \leq Q-1$. Assume $0 \leq N \leq nQ^s$. Let*

$$a = \frac{1}{Q}\sum_{j=0}^{Q-1} j^s, \qquad \sigma^2 = \frac{1}{Q}\sum_{j=0}^{Q-1} j^{2^s} - a^2, \qquad u_N = \frac{N - na}{\sigma\sqrt{n}}.$$

Then for any fixed k there exists an $n_0 = n_0(s, k)$ such that for $n \geq n_0$

$$R_N(Q, N) = \frac{Q^n}{\sigma\sqrt{n}}\left(\frac{1}{\sqrt{2\pi}}e^{-u_N^2/2} + \frac{\theta_{k,s}}{\sqrt{n(1 + |u_N|^k)}}\right), \qquad |\theta_{k,s}| \leq 1.$$

For the needs of quantum statistics, Khintchine ([82], Appendix IV) established an asymptotic formula for the number of solutions of the Diophantine equation $x_1 + \cdots + x_n = N$, where $n \to \infty$ and N/n remains constant. Khintchine's method can also be used to find an asymptotic formula for the number of solutions of the equation $x_1^s + \cdots + x_n^s = N$, where s is fixed, $n \to \infty$, and N/n remains constant. Combining Khintchine's idea with the method of trigonometric sums, Freĭman [56] removed the restriction that N/n remain constant and thereby solved Waring's problem with an increasing number of summands. Freĭman's theorem can be stated as follows.

THEOREM. *Let $r_n(N)$ denote the number of solutions of the equation $x_1^s + \cdots + x_n^s = N$. For $n \to \infty$ and $n < CN$, where $0 < C < 1$, we have the asymptotic formula*

$$r_n(N) = \frac{1}{\sqrt{2\pi n \operatorname{Var}\xi}} e^{-\sigma N}\Phi^n(\sigma)\left(1 + O\left(\frac{1}{n^{1/2-\varepsilon}}\right)\right)$$

where σ is determined from the equation

$$\frac{N}{n} = \frac{\sum_{x=1}^{\infty} x^s e^{-\sigma x^s}}{\sum_{x=1}^{\infty} e^{-\sigma x^s}}$$

(it is easy to prove this equation has a root when $n < CN$),

$$\Phi(\sigma) = \sum_{x=1}^{\infty} e^{-\sigma x^s},$$

ξ is a random variable that assumes the value x^s, $x = 1, 2, \ldots$, with probability $e^{-\sigma x^s}/\Phi(\sigma)$, $\operatorname{Var}\xi$ is its variance, and $\varepsilon > 0$ is an arbitarily small constant.

On the subject of representing numbers as the sum of a large number of squares see also Rankin [122].

Following Usol′tsev [142], we consider the additive problem with an increasing number of summands and exponential function $N = g^{x_1} + \cdots + g^{x_n}$, where $g \geq 4$ is an integer.

THEOREM. *Suppose $g \geq 4$, $Q \geq 3$, and n are natural numbers such that*

$$Q^2 \leq \frac{n}{9g^2 \ln g} - \sqrt{\frac{n}{9g^2 \ln g}} \ln n.$$

Let $r_{n,Q,g}(N)$ denote the number of solutions of the equations $N = g^{x_1} + \cdots + g^{x_n}$ in integers such that $0 \leq x_1, \ldots, x_n \leq Q-1$. Then for any natural number N such that $N \leq ng^{Q-1}$ and $N \equiv n \pmod{(g-1)}$ we have

$$r_{n,Q,g}(N) = \frac{(g-1)Q^n}{\sqrt{2\pi n}\,\sigma}\left(\exp\left[-\frac{(N-na)^2}{2n\sigma^2}\right] + 8\theta_{n,Q,g}\sqrt{Q/n}\right),$$

where

$$a = \frac{g^Q - 1}{(g-1)Q}, \quad \sigma^2 = \frac{g^{2Q}-1}{(g^2-1)Q} - \frac{(g^Q-1)^2}{(g-1)^2Q^2}, \quad |\theta_{n,Q,g}| \leq 1.$$

REMARK. Obviously $r_{n,Q,g}(N) = 0$ for $N > ng^{Q-1}$.

The following estimate of a trigonometric sum with exponential function plays an essential role in the proof of this theorem.

LEMMA. *If $g \geq 4$ and $A \geq 3$ are natural numbers and α is a real number such that*

$$1/g^Q \leq |\alpha| \leq 1/2(g-1),$$

then

$$\left|\sum_{x=0}^{Q-1} \exp(2\pi i \alpha g^x)\right| \leq Q - \frac{1}{9g^2}.$$

PROOF OF THE LEMMA. Since replacement of α by $-\alpha$ does not alter the value of the modulus of the sum, we may assume $\alpha > 0$, i.e.,

$$1/g^Q \leq \alpha \leq 1/2(g-1).$$

Let the g-adic representation of α be

$$\alpha = \frac{\varepsilon_1(\alpha)}{g} + \frac{\varepsilon_2(\alpha)}{g^2} + \cdots, \qquad 0 \leq \varepsilon_k(\alpha) \leq g-1 \quad (k = 1, 2, \ldots).$$

The g-adic expansion is not always unique, e.g.,

$$\frac{1}{g} = \frac{g-1}{g^2} + \frac{g-1}{g^3} + \cdots,$$

but if we agree to replace expansions in which $\varepsilon_k(\alpha) = g-1$ for $k \geq k_0$ by terminating ones, i.e., expansions in which $\varepsilon_k(\alpha) = 0$ for $k \geq k_1$, then the g-adic representation of α is unique. Since

$$\alpha \leq \frac{1}{2(g-1)} < \frac{1}{g}$$

(the latter inequality is strict inasmuch as $g \geq 4$), then $\varepsilon_1(\alpha) = 0$. Also,

$$0 \leq \alpha - \frac{0}{g} - \frac{\varepsilon_2(\alpha)}{g^2} - \cdots - \frac{\varepsilon_Q(\alpha)}{g^Q} < \frac{1}{g^Q}.$$

It is important to observe that the latter inequality is strict, by virtue of our agreement not to use an expansion in which $\varepsilon_k(\alpha) = g - 1$ for $k \geq Q$. Thus,

$$\alpha < \frac{\varepsilon_2(\alpha)}{g^2} + \cdots + \frac{\varepsilon_Q(\alpha)}{g^Q} + \frac{1}{g^Q}.$$

Since $\alpha \geq 1/g^Q$, it follows that

$$0 < \frac{\varepsilon_2(\alpha)}{g^2} + \cdots + \frac{\varepsilon_Q(\alpha)}{g^Q}.$$

Consequently, at least one of $\varepsilon_2(\alpha), \ldots, \varepsilon_Q(\alpha)$ is nonzero. Thus for numbers α such that $1/g^Q \leq \alpha \leq 1/2(g-1)$ we have $\varepsilon_1(\alpha) = 0$ and at least one of $\varepsilon_2(\alpha), \ldots, \varepsilon_Q(\alpha)$ is nonzero.

We will show that for each α in $[1/g^Q, 1/2(g-1)]$ there exist two numbers, $x_1 = x_1(\alpha)$ and $x_2 = x_2(\alpha)$, such that $x_1 \neq x_2$, $0 \leq x_1, x_2 \leq Q - 1$, and

$$\{g^{x_2}\alpha\} - \{g^{x_1}\alpha\} \geq 1/6g \quad \text{and} \quad 1 - \{g^{x_2}\alpha\} + \{g^{x_1}\alpha\} \geq 1/6g.$$

To prove this assertion we must consider several cases.

1) Among $\varepsilon_2(\alpha), \ldots, \varepsilon_Q(\alpha)$ there is a number l such that $2 \leq l \leq g - 2$. Suppose it appears in place x, i.e., $\varepsilon_x(\alpha) = l$. Then

$$\begin{aligned} \{g^{x-1}\alpha\} - \alpha &= \frac{l}{g} + \frac{\varepsilon_{x+1}(\alpha)}{g^2} + \cdots - \frac{\varepsilon_2(\alpha)}{g^2} - \cdots \\ &\geq \frac{l}{g} - \frac{g-1}{g^2} - \frac{g-1}{g^3} - \cdots = \frac{l-1}{g} \geq \frac{1}{g}. \end{aligned}$$

Representing 1 in the form

$$1 = \frac{g-1}{g} + \frac{g-1}{g^2} + \cdots,$$

we have

$$\begin{aligned} &1 + \alpha - \{g^{x-1}\alpha\} \\ &\quad = \frac{g-1}{g} + \frac{g-1+\varepsilon_2(\alpha)}{g^2} + \cdots - \frac{l}{g} - \frac{\varepsilon_{x+1}(\alpha)}{g^2} - \cdots \\ &\quad \geq \frac{g-1-l}{g} + \frac{g-1-\varepsilon_{x+1}(\alpha)}{g^2} + \cdots \\ &\quad \geq \frac{g-1-l}{g} \geq \frac{g-1-(g-2)}{g} = \frac{1}{g}. \end{aligned}$$

2) Among $\varepsilon_2(\alpha), \dots, \varepsilon_Q(\alpha)$ there is no number l such that $2 \le l \le g-2$, but the number $g-1$ does occur. Since for $g \ge 3$

$$\frac{g-1}{g^2} + \frac{g-1}{g^3} > \frac{1}{2(g-1)},$$

either $\varepsilon_2(\alpha)$ or $\varepsilon_3(\alpha)$ is different from $g-1$.

2a) Suppose $\varepsilon_2(\alpha) = g-1$. Then $\varepsilon(\alpha)$ is 0 or 1. We have

$$\begin{aligned}\{g\alpha\} - \alpha &= \frac{g-1}{g} + \frac{\varepsilon_3(\alpha)}{g^2} + \frac{\varepsilon_4(\alpha)}{g^3} + \cdots - \frac{g-1}{g^2} - \frac{\varepsilon_3(\alpha)}{g^3} - \cdots \\ &\ge \frac{g-1}{g} - \frac{g-1}{g^2} - \frac{g-1}{g^3} - \cdots = \frac{g-2}{g} \ge \frac{1}{g}\end{aligned}$$

(since $g \ge 3$).

Also,

$$\begin{aligned}1 + \alpha - \{\alpha g\} &= \frac{g-1}{g} + \frac{g-1}{g^2} + \cdots + \frac{g-1}{g^2} \\ &\quad + \frac{\varepsilon_3(\alpha)}{g^3} + \cdots - \frac{g-1}{g} - \frac{\varepsilon_3(\alpha)}{g^2} - \cdots \\ &\ge \frac{2(g-1) - \varepsilon_3(\alpha)}{g^2} + \frac{\varepsilon_3(\alpha)}{g^3} + \cdots \ge \frac{2(g-1)-1}{g^2} \ge \frac{1}{g},\end{aligned}$$

since $2g-3 \ge g$ when $g \ge 3$, hence also when $g \ge 4$.

2b) Suppose $\varepsilon_3(\alpha) = g-1$. Then $\varepsilon_2(\alpha)$ is either 0 or 1. We have

$$\begin{aligned}\{g^2\alpha\} - \{g\alpha\} &= \frac{g-1}{g} + \frac{\varepsilon_4(\alpha)}{g^2} + \cdots - \frac{\varepsilon_2(\alpha)}{g} - \frac{g-1}{g^2} - \frac{\varepsilon_4(\alpha)}{g^3} - \cdots \\ &\ge \frac{g-1-\varepsilon_2(\alpha)}{g} - \frac{g-1}{g^2} - \frac{g-1}{g^3} - \cdots \\ &= \frac{g-2-\varepsilon_2(\alpha)}{g} \ge \frac{g-3}{g} \ge \frac{1}{g},\end{aligned}$$

since $g \ge 4$. Also,

$$\begin{aligned}1 + \{g\alpha\} - \{g^2\alpha\} &= \frac{g-1}{g} + \frac{g-1}{g^2} + \cdots + \frac{\varepsilon_2(\alpha)}{g} + \frac{g-1}{g^2} \\ &\quad + \frac{\varepsilon_4(\alpha)}{g^3} + \cdots - \frac{g-1}{g} - \frac{\varepsilon_4(\alpha)}{g^2} - \cdots \\ &\ge \frac{2(g-1) - \varepsilon_4(\alpha)}{g^2} > \frac{g-1}{g^2}.\end{aligned}$$

2c) Suppose neither $\varepsilon_2(\alpha)$ nor $\varepsilon_3(\alpha)$ is $g-1$. Then there exists x, $4 \le x \le Q$, such that $\varepsilon_{x-1}(\alpha)$ is either 0 or 1 and $\varepsilon_x(\alpha) = g-1$. We have

$$\begin{aligned}\{g^{x-1}\alpha\} - \{g^{x-2}\alpha\} &= \frac{g-1}{g} + \frac{\varepsilon_{x+1}(\alpha)}{g^2} + \cdots - \frac{\varepsilon_{x-1}(\alpha)}{g} \\ &\quad - \frac{g-1}{g^2} - \frac{\varepsilon_{x+1}(\alpha)}{g^3} - \cdots \\ &\ge \frac{g-1-\varepsilon_{x-1}(\alpha)}{g} - \frac{g-1}{g^2} - \cdots \ge \frac{g-2-\varepsilon_{x-1}(\alpha)}{g} \\ &\ge \frac{g-2-1}{g} \ge \frac{1}{g},\end{aligned}$$

$$\begin{aligned}&1 + \{g^{x-2}\alpha\} - \{g^{x-1}\alpha\} \\ &\quad = \frac{g-1+\varepsilon_{x-1}(\alpha)}{g} + \frac{g-1+g-1}{g^2} + \frac{g-1+\varepsilon_{x+1}(\alpha)}{g^3} \\ &\quad + \cdots - \frac{g-1}{g} - \frac{\varepsilon_{x-1}(\alpha)}{g^2} - \cdots \\ &\quad \ge \frac{\varepsilon_{x-1}(\alpha)}{g} + \frac{2(g-1)-\varepsilon_{x+1}(\alpha)}{g^2} + \frac{\varepsilon_{x+1}(\alpha)}{g^3} + \frac{\varepsilon_{x+2}(\alpha)}{g^4} + \cdots \\ &\quad \ge \frac{g-1}{g^2} > \frac{1}{2g}.\end{aligned}$$

3) Suppose that each of $\varepsilon_2(\alpha), \dots, \varepsilon_Q(\alpha)$ is either 0 or 1. Let x be the smallest integer, $2 \le x \le Q$, for which $\varepsilon_x(\alpha) = 1$. We have

$$\begin{aligned}\{g^{x-1}\alpha\} - \alpha &= \frac{1}{g} + \frac{\varepsilon_{x+1}(\alpha)}{g^2} + \cdots - \frac{0}{g} - \cdots - \frac{0}{g^{x-1}} \\ &\quad - \frac{1}{g^x} - \cdots - \frac{1}{g^Q} - \frac{\varepsilon_{Q+1}(\alpha)}{g^{Q+1}} - \cdots \\ &> \frac{1}{g} - \sum_{r=x}^{Q} \frac{1}{g^r} - \sum_{r=Q+1}^{\infty} \frac{g-1}{g^r} \\ &= \frac{1}{g} - \frac{1}{g^{x-1}}\frac{1}{g-1} - \frac{1}{g^Q} \ge \frac{1}{g} - \frac{1}{g(g-1)} - \frac{1}{g^2}\end{aligned}$$

(since $x - 1 \ge 1$), i.e.,

$$\{g^{x-1}\alpha\} - \alpha \ge \frac{1}{g}\left(1 - \frac{1}{g-1} - \frac{1}{g}\right) \ge \frac{1}{g}\left(1 - \frac{1}{2} - \frac{1}{3}\right) = \frac{1}{6g}.$$

Also,

$$\begin{aligned}1 + \alpha - \{g^{x-1}\alpha\} &= \frac{g-1}{g} + \frac{g-1}{g^2} + \cdots + \frac{1}{g^x} + \cdots - \frac{1}{g} - \cdots \\ &\ge \frac{g-3}{g} \ge \frac{1}{g}.\end{aligned}$$

So if $g \geq 4$ and $Q \geq 3$, then for α in $[1/g^Q, 1/2(g-1)]$ there exist two numbers x_1 and x_2, $x_1 \neq x_2$, $0 \leq x_1, x_2 \leq Q-1$, such that

$$\{g^{x_2}\alpha\} - \{g^{x_1}\alpha\} \geq 1/6g \quad \text{and} \quad 1 - \{g^{x_2}\alpha\} + \{g^{x_1}\alpha\} \geq 1/6g.$$

Let φ denote the angle (not exceeding π) between the vectors $\exp(2\pi i g^{x_1}\alpha)$ and $\exp(2\pi i g^{x_2}\alpha)$. It has been shown that $\varphi \geq 2\pi/6g = \pi/3g$. We have

$$\begin{aligned}
&\left|\sum_{x=0}^{Q-1} \exp(2\pi i g^x \alpha)\right| \\
&\quad = \left|\exp(2\pi i g^{x_1}\alpha) + \exp(2\pi i g^{x_2}\alpha) + \sum_{\substack{x=0 \\ x\neq x_1,\ x\neq x_2}}^{Q-1} \exp(2\pi i g^x \alpha)\right| \\
&\quad \leq Q - 2 + |\exp(2\pi i g^{x_1}\alpha) + \exp(2\pi i g^{x_2}\alpha)| \\
&\quad \leq Q - 2 + 2|\cos\pi(\{g^{x_1}\alpha\} - \{g^{x_2}\alpha\})| \\
&\quad \leq Q - 2 + 2\cos(\pi/6g) = Q - 2(1 - \cos(\pi/6g)) \\
&\quad = Q - 4\sin^2(\pi/12g) \leq Q - 4(4/\pi^2)(\pi/12g)^2 = Q - 1/9g^2.
\end{aligned}$$

Thus,

$$\left|\frac{1}{Q}\sum_{x=0}^{Q-1} \exp(2\pi i g^x \alpha)\right| \leq 1 - \frac{1}{9g^2 Q}.$$

The lemma is proved.

We now turn to the proof of the theorem stated above. Note that

$$\frac{1}{Q}\sum_{x=0}^{Q-1} g^x = \frac{g^Q - 1}{(g-1)Q};$$

$$\frac{1}{Q}\sum_{x=0}^{Q-1}(g^x - a)^2 = \frac{g^{2Q} - 1}{(g^2-1)Q} - \frac{(g^Q-1)^2}{(g-1)^2 Q^2} = \sigma^2.$$

Observe also that

$$\left(1 - \frac{2}{Q}\right)\frac{g^{2Q}}{(g^2-1)Q} \leq \sigma^2.$$

Indeed, since for $g \geq 4$ and $Q \geq 3$ we have

$$\frac{Q}{g^{2Q}} + \frac{(g+1)(g^Q-1)^2}{(g-1)g^{2Q}} \leq \frac{3}{4^6} + \frac{5}{3} \leq 2,$$

it follows that

$$\sigma^2 = \frac{g^{2Q}}{(g^2-1)Q}\left(1 - \frac{1}{Q}\left[\frac{Q}{2g^Q} + \frac{(g-1)(g^Q-1)^2}{(g-1)g^{2Q}}\right]\right) \geq \frac{g^{2Q}}{(g^2-1)Q}\left(1 - \frac{2}{Q}\right).$$

Furthermore, it is obvious that

$$r_{n,Q,g}(N) = \int_{-1/2}^{1/2} \left(\sum_{x=0}^{Q-1} \exp\left(2\pi i \frac{\alpha}{g-1} g^x\right)\right)^n \exp\left(-2\pi i \frac{\alpha}{g-1} N\right) d\alpha,$$

and therefore

$$\begin{aligned}\frac{r_{n,Q,g}(N)}{(g-1)Q^n} &= \int_{-(1/2)(g-1)}^{(1/2)(g-1)} \left(\frac{1}{Q}\sum_{x=0}^{Q-1} \exp(2\pi i\alpha g^x)\right)^n \exp(-2\pi\alpha i N)\, d\alpha \\ &= I_1 + I_2 + I_3,\end{aligned}$$

where

$$\begin{aligned} I_1 &= \int_{-A/2\pi\sqrt{n}\,\sigma}^{A/2\pi\sqrt{n}\,\sigma} \left(\frac{1}{Q}\sum_{x=0}^{Q-1} \exp(2\pi i\alpha g^x)\right)^n \exp(-2\pi i\alpha N)\, d\alpha \\ &= \frac{1}{2\pi\sqrt{n}\,\sigma}\int_{-A}^{A} \left(\frac{1}{Q}\sum_{x=0}^{Q-1} \exp\left(\frac{iw}{\sqrt{n}\,\sigma}(g^x - a)\right)\right)^n \\ &\qquad\qquad \times \exp\left(\frac{-iw}{\sqrt{n}\,\sigma}(N - na)\right) dw, \\ I_2 &= \int_{A/2\pi\sqrt{n}\,\sigma \le |\alpha| \le 1/g^Q} \left(\frac{1}{Q}\sum_{x=0}^{Q-1} \exp(2\pi i\alpha g^x)\right)^n \exp(-2\pi i\alpha N)\, d\alpha \\ &= \frac{1}{2\pi\sqrt{n}\,\sigma}\int_{A\le |w| \le 2\pi\sqrt{n}\,\sigma/g^Q} \left(\frac{1}{Q}\sum_{x=0}^{Q-1} \exp\left(\frac{iw}{\sqrt{n}\,\sigma}(g^x - a)\right)\right)^n \\ &\qquad\qquad \times \exp\left(\frac{iw}{\sqrt{n}\,\sigma}(N - na)\right) dw, \\ I_3 &= \int_{1/g^Q \le |\alpha| \le 1/2(g-1)} \left(\frac{1}{Q}\sum_{x=0}^{Q-1} \exp(2\pi i\alpha g^x)\right)^n \exp(-2\pi i\alpha N)\, d\alpha, \end{aligned}$$

and the parameter A is such that $0 < A < 2\pi\sqrt{n}\,\sigma/g^Q$.

To estimate the integral I_3 we use the lemma proved above. According to the lemma,

$$\begin{aligned}|I_3| &\le \int_{1/g^Q \le |\alpha| \le 1/2(g-1)} \left|\frac{1}{Q}\sum_{x=0}^{Q-1} \exp(2\pi i\alpha g^x)\right|^n d\alpha \\ &\le \left(1 - \frac{1}{9g^2Q}\right)^n \int_{-1/2}^{1/2} d\alpha = \left(1 - \frac{1}{9g^2Q}\right)^n \le e^{-n/9g^2Q}\end{aligned}$$

Turning now to the integrals I_1 and I_2, we will use the symbols $\theta_1, \theta_2, \ldots$ to denote quantities depending on n, g, and Q with absolute value at most 1.

Since for any real t

$$|e^{it} - 1 - it + t^2/2| \le |t|^3/6,$$

it follows that

$$\begin{aligned}
&\frac{1}{Q}\sum_{x=0}^{Q-1} \exp\left(\frac{iw}{\sqrt{n}\,\sigma}(g^x - a)\right) \\
&\quad = 1 + \frac{iw}{\sqrt{n}\,\sigma}\frac{1}{Q}\sum_{x=0}^{Q-1}(g^x - a) - \frac{w^2}{2n\sigma^2}\frac{1}{Q}\sum_{x=0}^{Q-1}(g^x - a)^2 \\
&\qquad + \theta_1\frac{|w|^3}{6n^{3/2}\sigma^3}\frac{1}{Q}\sum_{x=0}^{Q-1}|g^x - a|^3 \\
&\quad = 1 - \frac{w^2}{2n} + \theta_2\frac{|w|^3 g^{Q-1}}{6n^{3/2}\sigma}.
\end{aligned}$$

If $|w| \le 2\pi\sqrt{n}\,\sigma/g^Q$, then

$$\frac{w^2}{n} \le \frac{4\pi^2}{g^{2Q}}\sigma^2 \le \frac{4\pi^2}{g^{2Q}}\frac{g^{2Q}}{(g^2-1)Q} \le \frac{4\pi^2}{(4^2-1)^3} \le 1$$

and

$$\frac{|w|^3 g^{Q-1}}{6n^{3/2}\sigma} \le \frac{w^2 2\pi\sqrt{n}\,\sigma g^{Q-1}}{6n^{3/2}\sigma g^Q} = \frac{\pi w^2}{3gn} \le \frac{\pi w^2}{12n},$$

i.e.,

$$\begin{aligned}
\left|\frac{1}{Q}\sum_{x=0}^{Q-1}\exp\left(\frac{iw}{\sqrt{n}\,\sigma}(g^x - a)\right)\right| &= \left|1 - \frac{w^2}{2n} + \theta_2\frac{|w|^3 g^{Q-1}}{6n^{3/2}\sigma}\right| \\
&\le 1 - \frac{w_2}{2n} + \frac{\pi w^2}{12n} \le 1 - \frac{w^2}{5n} \le e^{-w^2/5n},
\end{aligned}$$

from which it follows that

$$\begin{aligned}
|I_2| &\le \frac{1}{2\pi\sqrt{n}\,\sigma}\int_{A\le|w|\le 2\pi\sqrt{n}\,\sigma/g^Q}\left|\frac{1}{Q}\sum_{x=0}^{Q-1}\exp\left(\frac{iw}{\sqrt{n}\,\sigma}(g^x - a)\right)\right|^n dw \\
&\le \frac{2}{2\pi\sqrt{n}\,\sigma}\int_A^\infty e^{-w^2/5}\,dw \le \frac{5}{2\pi\sqrt{n}\,\sigma A}e^{-A^2/5}.
\end{aligned}$$

Since for $|w| \le 2\pi\sqrt{n}\,\sigma/g^Q$ we have

$$\left|\frac{w^2}{2n} - \theta_2\frac{|w|^3 g^{Q-1}}{6n^{3/2}\sigma}\right| \le \frac{w^2}{2n} + \frac{\pi w^2}{12n} \le \frac{16w^2}{21n} \le \frac{16}{21}$$

and for $|z| < 1$

$$|\ln(1+z) - z| \le \frac{|z|^2}{2(1-|z|)}$$

(where ln denotes the principal value of the logarithm), we see that for $|w| \le 2\pi\sqrt{n}\,\sigma/g^Q$

$$\begin{aligned}
&\ln\left(\frac{1}{Q}\sum_{x=0}^{Q-1}\exp\left(\frac{iw}{\sqrt{n}\,\sigma}(g^x-a)\right)\right) \\
&\quad= \ln\left(1-\frac{w^2}{2n}+\theta_2\frac{|w|^3 g^{Q-1}}{6n^{3/2}\sigma}\right) \\
&\quad= -\frac{w^2}{2n}+\theta_2\frac{|w|^3 g^{Q-1}}{6n^{3/2}}\frac{1}{\sigma}+\theta_3\frac{(16w^2/21n)^2}{2(1-16/21)} \\
&\quad= -\frac{w^2}{2n}+\theta_4\frac{|w|^3 g^{Q-1}}{6n^{3/2}}\frac{\sqrt{3(g^2-1)Q}}{g^Q}+\theta_3\frac{128|w|^3}{105n^{3/2}}\frac{|w|}{\sqrt{n}} \\
&\quad= -\frac{w^2}{2n}+\theta_5\frac{\sqrt{3}|w|^3\sqrt{Q}}{6n^{3/2}}+\theta_6\frac{128|w|^3}{105n^{3/2}} \\
&\quad= -\frac{w^2}{2n}+\theta_7\frac{|w|^3\sqrt{Q}}{n^{3/2}},
\end{aligned}$$

i.e.,

$$\left(\frac{1}{Q}\sum_{x=0}^{Q-1}\exp\left(\frac{iw}{\sqrt{n}\,\sigma}(g^x-a)\right)\right)^n = \exp\left(-\frac{w^2}{2}+\theta_7\frac{|w|^3\sqrt{Q}}{\sqrt{n}}\right).$$

Hence under the additional condition $A^6 \le n/Q$ we have

$$\begin{aligned}
I_1 &= \frac{1}{2\pi\sqrt{n}\,\sigma}\int_{-A}^{A}\left(-\frac{w^2}{2}+\theta_7\frac{|w|^3\sqrt{Q}}{\sqrt{n}}\right)\exp\left(-\frac{iw}{\sqrt{n}\,\sigma}(N-na)\right)dw \\
&= \frac{1}{2\pi\sqrt{n}\,\sigma}\int_{-A}^{A}\left(-\frac{w^2}{2}-iw\frac{N-na}{\sqrt{n}\,\sigma}\right)\left(1+\theta_8\frac{3|w|^3\sqrt{Q}}{\sqrt{n}}\right)dw \\
&= \frac{1}{2\pi\sqrt{n}\,\sigma}\int_{-\infty}^{\infty}\exp\left(-\frac{w^2}{2}-iw\frac{N-na}{\sqrt{n}\,\sigma}\right)dw \\
&\qquad+\theta_9\frac{1}{\pi\sqrt{n}\,\sigma}\int_A^{\infty}e^{-w^2/2}\,dw+\theta_{10}\frac{3\sqrt{Q}}{\pi n\sigma}\int_0^A w^3e^{-w^2/2}\,dw \\
&= \frac{1}{\sqrt{2\pi n}\,\sigma}\exp\left(-\frac{(N-na)^2}{2n\sigma^2}\right)+\theta_{11}\frac{1}{\pi\sqrt{n}\,\sigma A}e^{-A^2/2}+\theta_{12}\frac{6\sqrt{Q}}{\pi n\sigma}.
\end{aligned}$$

Combining the estimates for I_1, I_2, and I_3, and imposing the conditions $A < 2\pi\sqrt{n}\,\sigma/g^Q$ and $A^6 \le n/Q$, we obtain

$$\begin{aligned}
\frac{r_{n,Q,g}(N)}{(g-1)Q^n} &= I_1 + I_2 + I_3 \\
&= \frac{1}{\sqrt{2\pi n}\,\sigma}\exp\left(-\frac{(N-na)^2}{2n\sigma^2}\right) + \theta_{11}\frac{1}{\pi\sqrt{n}\,\sigma A}e^{-A^2/2} \\
&\quad + \theta_{12}\frac{6\sqrt{Q}}{\pi n\sigma} + \theta_{13}\frac{5}{2\pi\sqrt{n}\,\sigma A}e^{-A^2/5} + \theta_{14}e^{-n/9g^2Q} \\
&= \frac{1}{\sqrt{2\pi n}\,\sigma}\exp\left(-\frac{(N-an)^2}{2n\sigma^2}\right) + \theta_{15}\left(\frac{3}{A}e^{-A^2/5} + 7\sqrt{\frac{Q}{n}}\right),
\end{aligned}$$

since the condition

$$Q^2 \le \frac{n}{9g^2\ln g} - \sqrt{\frac{n}{9g^2\ln g}}\,\ln n$$

implies

$$\begin{aligned}
\sqrt{2\pi n}\,\sigma e^{-n/9g^2Q} &\le \frac{\sqrt{2\pi n}\,g^Q}{\sqrt{(g^2-1)}\,Q}e^{-n/9g^2Q} \\
&\le \frac{1}{\sqrt{n}}g^{-((n/9g^2\ln g)-Q^2-Q\ln n)(1/Q)} \le \frac{1}{\sqrt{n}}.
\end{aligned}$$

Taking $A = \sqrt{5\ln n}$, which does not violate the conditions $A < 2\pi\sqrt{n}\,\sigma/g^Q$ and $A^6 \le n/Q$, we obtain

$$r_{n,Q,g}(N) = \frac{(g-1)Q^n}{\sqrt{2\pi n}\,\sigma}\left(\exp\left(-\frac{(N-na)^2}{2n\sigma^2}\right) + 8\theta_{16}\sqrt{\frac{Q}{n}}\right),$$

which implies the assertion of the theorem.

§2.5. Bredikhin's theorem

Suppose G is a multiplicatively written free commutative semigroup with a countable system P of generators $\omega_1, \omega_2, \dots$, and suppose N is a homomorphism of G onto a multiplicative semigroup $\overline{G}$ consisting of positive real numbers such that for each real number x there exist in G only finitely many elements α such that $N(\alpha) \le x$. Let

$$\pi_G(x) = \sum_{\substack{N(\omega)\le x\\ \omega\in P}} 1 \quad\text{and}\quad \nu_G(x) = \sum_{\substack{N(\alpha)\le x\\ \alpha\in G}} 1.$$

THEOREM. *Suppose that as $x \to \infty$*

$$\pi_G(x) = \frac{ax}{\ln x} + O\left(\frac{x}{\ln^{1+\varepsilon} x}\right), \tag{1}$$

where $a > 0$ and $\varepsilon > 0$ are fixed. Then

$$\nu_G(x) = C_G x \ln^{a-1} x + O\left(\frac{x \ln^{a-1} x}{(\ln\ln x)^{\varepsilon_1}}\right),$$

where $\varepsilon_1 = \min(1, \varepsilon)$.

We first prove some lemmas.

LEMMA 1.

$$\nu_G(x) \le C_1(\varepsilon) x \ln^a x.$$

PROOF. Suppose ω_1 is an element such that $N(\omega_1) \le N(\omega)$. Then

$$\begin{aligned}\sum_{N(\alpha)\le x} \frac{1}{N(\alpha)} &\le \prod_{N(\omega)\le x} \left(1 - \frac{1}{N(\omega)}\right)^{-1} \\ &= \exp\left(\sum_{k\ge 1} \sum_{N(\omega)\le x} \frac{1}{kN^k(\omega)}\right) \\ &\le \exp\left(\sum_{N(\omega)\le x} \frac{1}{N(\omega)}\right. \\ &\qquad \left. + \left(\frac{1}{2} + \frac{1}{3N(\omega_1)} + \frac{1}{4N^2(\omega_1)} + \cdots\right) \sum_{N(\omega)\le x} \frac{1}{N^2(\omega)}\right).\end{aligned}$$

In view of condition (1), we have

$$\sum_{N(\omega)\le x} \frac{1}{N(\omega)} = a \ln\ln x + O(1), \qquad \sum_{N(\omega)\le x} \frac{1}{N^2(\omega)} = O(1).$$

Thus,

$$\sum_{N(\alpha)\le x} \frac{1}{N(\alpha)} \le C \ln^a x.$$

It follows that

$$\sum_{N(\alpha)\le x} 1 \le C_1 x \ln^a x.$$

LEMMA 2.

$$\nu_G(x) \ln x = \int_1^x \nu_G(u) \frac{du}{u} + \sum_{k\ge 1} \sum_{N(\omega)\le x} \nu_G\left(\frac{x}{N^k(\omega)}\right) \ln N(\omega).$$

PROOF. The following formula is a generalization of the well-known decomposition of $n!$ into primes:

$$\prod_{n(\alpha)\le x} N(\alpha) = \prod_{N(\omega)\le x} (N(\omega))^{\sum_{k\ge 1} \nu_G(x/N^k(\omega))}.$$

Taking the logarithm, we obtain

$$\sum_{N(\alpha)\le x} \ln N(\alpha) = \sum_{N(\omega)\le x} \left(\sum_{k\ge 1} \nu_G\left(\frac{x}{N^k(\omega)}\right) \ln N(\omega)\right).$$

But

$$\sum_{N(\alpha)\le x} \ln N(\alpha) = \sum_{N(\alpha)\le x} \int_1^{N(\alpha)} \frac{du}{u} = \int_1^x \frac{\sum_{u\le N(\alpha)\le x} 1}{u}\, du$$

$$= \nu_G(x)\ln x - \int_1^x \frac{\sum_{N(\alpha)\le u} 1}{u}\, du$$

$$= \nu_G(x)\ln x - \int_1^x \frac{\nu_G(u)}{u}\, du.$$

Lemma 3 strengthens the estimate of Lemma 1.

LEMMA 3.

$$\nu_G(x) \le C(\varepsilon) x \ln^{a-1} x.$$

PROOF. Substituting the estimate of Lemma 1 into the formula of Lemma 2, we obtain

$$\int_1^x \nu_G(u)\frac{du}{u} = O\left(\int_1^x \ln^a u\, du\right) = O(x\ln^a x).$$

Also,

$$\sum_{N(\omega)\le x} \nu_G\left(\frac{x}{N(\omega)}\right) \ln N(\omega) = \sum_{N(\alpha)\le x} \sum_{N(\omega)\le x/N(\alpha)} \ln N(\omega)$$

$$\le \sum_{N(\alpha)\le x} \ln \frac{x}{N(\alpha)} \sum_{N(\omega)\le x/N(\alpha)} 1$$

$$\le C(\varepsilon) \sum_{N(\alpha)\le x} \frac{x}{N(\alpha)} \le C(\varepsilon) x \ln^a x,$$

and, finally,

$$\sum_{k\ge 2} \sum_{N(\omega)\le x} \nu_G\left(\frac{x}{N^k(\omega)}\right) \ln N(\omega)$$

$$\le C(\varepsilon) \sum_{k\ge 2} \sum_{N(\omega)\le x} \frac{x\ln^a x}{N^k(\omega)} \ln N(\omega)$$

$$= O\left(x\ln^a x \sum_{k\ge 2} \sum_{N(\omega)\le x} \frac{\ln N(\omega)}{N^k(\omega)}\right) = O(x\ln^a x).$$

We now apply Lemma 2.

One should compare the statement of the following Lemma 4 with that of Lemma 2.

LEMMA 4.

$$\nu_G(x)\ln x = \sum_{N(\omega)\le x} \nu_G\left(\frac{x}{N(\omega)}\right)\ln N(\omega) + O(x\ln^{a-1} x).$$

The proof is analogous to that of Lemma 3, but uses Lemma 3 instead of Lemma 1.

LEMMA 5.

$$\nu_G(x) = a\frac{x}{\ln x}\sum_{N(\alpha)\le x}\frac{1}{N(\alpha)} + O\left(\frac{x\ln^{a-1} x}{(\ln\ln x)^{\varepsilon_1}}\right). \tag{2}$$

PROOF. We start from the formula of Lemma 4:

$$\nu_G(x)\ln x = \sum_{N(\omega)\le x} \nu_G\left(\frac{x}{N(\omega)}\right)\ln N(\omega) + O(x\ \ln^{a-1} x).$$

We have

$$\sum_{N(\omega)\le x} \nu_G\left(\frac{x}{N(\omega)}\right)\ln N(\omega) = \sum_{N(\alpha)\le x/N(\omega_1)}\ \sum_{N(\omega)\le x/N(\alpha)} \ln N(\omega).$$

But starting from these conditions of the theorem we obtain, by means of Abel's tranformation,

$$\sum_{N(\omega)\le x}\ln N(\omega) = ax + O\left(\frac{x}{\ln^{\varepsilon} x}\right).$$

It follows that

$$\sum_{N(\omega)\le x/N(\alpha)}\ln N(\omega) = a\frac{x}{N(\alpha)} + O\left(\frac{x}{N(\alpha)}\frac{x}{\ln^{\varepsilon}(x/N(\alpha))}\right),$$

and hence

$$\sum_{N(\omega)\le x} \nu_G\left(\frac{x}{N(\omega)}\right)\ln N(\omega)$$
$$= ax\sum_{N(\alpha)\le x/N(\omega_1)}\frac{1}{N(\alpha)} + O\left(x\sum_{N(\alpha)\le x/N(\omega_1)}\frac{1}{N(\alpha)\ln^{\varepsilon}(x/N(\alpha))}\right).$$

Note that

$$\sum_{x/N(\omega_1)<N(\alpha)\le x}\frac{1}{N(\alpha)} \le \frac{N(\omega_1)}{x}\sum_{N(\alpha)\le x} 1 = O(\ln^{a-1} x).$$

Therefore,

$$\sum_{N(\omega)\le x} \nu_G\left(\frac{x}{N(\omega)}\right)\ln N(\omega) = ax\sum_{N(\alpha)\le x}\frac{1}{N(\alpha)} + O(x\ln^{a-1} x)$$
$$+ O\left(x\sum_{N(\alpha)\le x/N(\omega_1)}\frac{1}{N(\alpha)\ln^{\varepsilon}(x/N(\alpha))}\right).$$

Now

$$
\begin{aligned}
&\sum_{N(\alpha)\le x/N(\omega_1)} \frac{1}{N(\alpha)\ln^{\varepsilon}(x/N(\alpha))} \\
&\quad= \sum_{N(\alpha)\le x/\sqrt{\ln x}} \frac{1}{N(\alpha)\ln^{\varepsilon}(x/N(\alpha))} \\
&\quad+ \sum_{x/\sqrt{\ln x}<N(\alpha)\le x/N(\omega_1)} \frac{1}{N(\alpha)\ln^{\varepsilon}(x/N(\alpha))} \\
&\quad= O\Bigg(\sum_{N(\alpha)\le x/\sqrt{\ln x}} \frac{1}{N(\alpha)(\ln\ln x)^{\varepsilon}}\Bigg) + O\Bigg(\frac{\sqrt{\ln x}}{x} \sum_{N(\alpha)\le x/N(\omega_1)} 1\Bigg) \\
&\quad= O\left(\frac{\ln^a x}{(\ln\ln x)^{\varepsilon}}\right) + O(\ln^{a-1/2} x) = O\left(\frac{\ln^a x}{(\ln\ln x)^{\varepsilon}}\right).
\end{aligned}
$$

Thus,

$$
\sum_{N(\omega)\le x} \nu_G\left(\frac{x}{N(\omega)}\right)\ln N(\omega) = ax \sum_{N(\alpha)\le x} \frac{1}{N(\alpha)} + O\left(\frac{x\ln^a x}{(\ln\ln x)^{\varepsilon}}\right).
$$

Lemma 5 is proved.

We introduce the zeta-function of the semigroup:

$$
\zeta_G(s) = \sum_{\alpha\in G} \frac{1}{N^s(\alpha)}.
$$

We will consider $\zeta_G(s)$ for real values of s.

LEMMA 6. *If*

$$
\pi_G(x) = ax/\ln x + O(x/\ln^{1+\varepsilon} x),
$$

then as $s\to 1+0$

$$
\zeta_G(s) = \frac{C}{(s-1)^a}(1+r(s)),
$$

where $|r(s)| \le C_1(s-1)^{\varepsilon}$.

PROOF. If $s>1$,

$$
\zeta_G(s) = \prod_{\omega\in P}\left(1-\frac{1}{N(\omega)}\right)^{-1}.
$$

Therefore

$$
\zeta_G(s) = \exp\left(\sum_{\omega\in P}\frac{1}{N^s(\omega)} + \varphi_1(s)\right),
$$

where $\varphi_1(s) = C_2 + O(s-1)$. Let $R(x) = \pi_G(x) - ax/\ln x$. Then

$$\sum_{\omega \in P} \frac{1}{N^s(\omega)} = s \sum_{\omega \in P} \int_{N(\omega)}^{\infty} \frac{dx}{x^{s+1}} = s \int_{N(\omega_1)}^{\infty} \frac{\sum_{N(\omega) \le x} 1}{x^{s+1}}\, dx$$
$$= sa \int_{N(\omega_1)}^{\infty} \frac{dx}{x^s \ln x} + s \int_{N(\omega_1)}^{\infty} \frac{R(x)}{x^{s+1}}\, dx$$

We have

$$sa \int_{N(\omega_1)}^{\infty} \frac{dx}{x^s \ln x} = sa \int_{\ln N(\omega_1)}^{\infty} \frac{e^{-(s-1)y}}{y}\, dy = sa \int_{(s-1)\ln N(\omega_1)}^{\infty} \frac{e^{-z}}{z}\, dz$$
$$= sa \int_{(s-1)\ln N(\omega_1)}^{1} \frac{e^{-z}}{z}\, dz + sa \int_{1}^{\infty} \frac{e^{-z}}{z}\, dz$$
$$= sa \int_{(s-1)\ln N(\omega_1)}^{1} \frac{e^{-z}}{z}\, dz + C_3 + O(s-1)$$
$$= sa \int_{(s-1)\ln N(\omega_1)}^{1} \frac{dz}{z} + O(s-1) + C_3 + O(s-1)$$
$$= a \ln \frac{1}{s-1} + C_4 + O(s-1).$$

Thus,

$$\sum_{\omega \in P} \frac{1}{N^s(\omega)} = a \ln \frac{1}{s-1} + s \int_{N(\omega_1)}^{\infty} \frac{R(x)}{x^{s+1}}\, dx + \varphi_2(s),$$

where $\varphi_2(s) = C_4 + O(s-1)$. Now

$$\int_{N(\omega_1)}^{\infty} \frac{R(x)}{x^{s+1}}\, dx = \int_{N(\omega_1)}^{\infty} \frac{R(x)}{x^2}\, dx + \int_{N(\omega_1)}^{\infty} \frac{R(x)}{x^2} \left(\frac{1}{x^{s-1}} - 1 \right) dx.$$

By the hypothesis of the theorem, all integrals in this formula are absolutely convergent. Also,

$$\int_{N(\omega_1)}^{\infty} \frac{R(x)}{x^2} \left(\frac{1}{x^{s-1}} - 1 \right) dx$$
$$= \int_{N(\omega_1)}^{e^{1/(s-1)}} \frac{R(x)}{x^2} \left(\frac{1}{x^{s-1}} - 1 \right) dx + \int_{e^{1/(s-1)}}^{\infty} \frac{R(x)}{x^2} \left(\frac{1}{x^{s-1}} - 1 \right) dx$$
$$= \int_{N(\omega_1)}^{e^{1/(s-1)}} \frac{R(x)(1 - x^{s-1})}{x^{s+1}}\, dx + O\left(\int_{e^{1/(s-1)}}^{\infty} \frac{dx}{x \ln^{1+\varepsilon} x} \right)$$
$$= \int_{N(\omega_1)}^{e^{1/(s-1)}} \frac{R(x)(1 - x^{s-1})}{x^{s+1}}\, dx + O((s-1)^{\varepsilon})$$
$$= O\left(\int_{N(\omega_1)}^{e^{1/(s-1)}} \frac{x^{s-1} - 1}{x^s \ln^{1+\varepsilon} x}\, dx \right) + O((s-1)^{\varepsilon}).$$

Since $e^z - 1 \le ze^z$ for $z > 0$, by taking $z = (s-1)\ln x$ we obtain

$$\int_{N(\omega_1)}^{e^{1/(s-1)}} \frac{x^{s-1}-1}{x^s \ln^{1+\varepsilon} x}\,dx \le (s-1)\int_{N(\omega_1)}^{e^{1/(s-1)}} \frac{dx}{x\ln^{\varepsilon} x} = (s-1)\int_{\ln N(\omega_1)}^{1/(s-1)} \frac{dy}{y^{\varepsilon}}.$$

In condition (1) we may assume $\varepsilon < 1$, because if $\varepsilon > 1$ we have

$$\frac{x}{\ln^{1+\varepsilon} x} < \frac{x}{\ln^{1+\varepsilon_0} x},$$

where ε_0 is less than 1. Then

$$(s-1)\int_{\ln N(\omega_1)}^{1/(s-1)} \frac{dy}{y^{\varepsilon}} = O((s-1)^{\varepsilon}) + O(s-1) = O((s-1)^{\varepsilon}).$$

Thus,

$$\begin{aligned} \varsigma_G(s) &= \exp\left(a\ln\frac{1}{s-1} + C_4 + O((s-1)^{\varepsilon})\right) \\ &= \frac{C_5}{(s-1)^a}(1 + O((s-1)^{\varepsilon})). \end{aligned}$$

The lemma is proved.

We now prove the theorem. We represent $\varsigma_G(s+1)$ as a Stieltjes integral:

$$\varsigma_G(s+1) = \int_0^{\infty} e^{-st}\,d\beta(t),$$

where

$$\beta(t) = \sum_{N(\alpha)\le e^t} \frac{1}{N(\alpha)};$$

$\beta(t)$ is a nondecreasing function. By Lemma 6,

$$\varsigma_G(s+1) = \frac{C}{s^a}(1 + r(s)),$$

where $|r(s)| \le C_1 s^{\varepsilon}$.

We now apply the Hardy-Littlewood Tauberian theorem with Freud's remainder term. We obtain

$$\beta(t) = \frac{C}{\Gamma(a+1)} t^a \left(1 + O\left(\frac{1}{\ln t}\right)\right).$$

Putting $t = \ln x$, we see that

$$\sum_{N(\alpha)\le x} \frac{1}{N(\alpha)} = \frac{C}{\Gamma(a+1)} \ln^a x \left(1 + O\left(\frac{1}{\ln\ln x}\right)\right).$$

This formula in conjunction with (2) proves the theorem.

COROLLARY. *Suppose that as* $x \to \infty$

$$\pi_G(x) = \frac{ax^\theta}{\theta \ln x} + O\left(\frac{x^\theta}{\ln^{1+\varepsilon} x}\right)$$

with some $\theta > 0$. *Then*

$$\nu_G(x) = C_G x^\theta \ln^{a-1} x + O\left(\frac{x^\theta \ln^{a-1} x}{(\ln\ln x)^{\varepsilon_1}}\right),$$

where $\varepsilon_1 = \min(1, \varepsilon)$.

PROOF. We define a homomorphism of the semigroup $\overline{G}$ onto the semigroup $\overline{\overline{G}}$ defined by the correspondence $x^\theta \to x$, and we put $\overline{N}(\alpha) = N^\theta(\alpha)$. We have

$$\pi_G(x^{1/\theta}) = \sum_{N(\omega) \le x^{1/\theta}} 1 = \sum_{\overline{N}(\omega) \le x} 1 = \frac{ax}{\ln x} + O\left(\frac{x}{\ln^{1+\varepsilon} x}\right).$$

Applying the theorem, we obtain

$$\nu_G(y^{1/\theta}) = \sum_{\overline{N}(\alpha) \le y} 1 = Cy \ln^{a-1} y + O\left(\frac{y \ln^{a-1} y}{(\ln\ln y)^{\varepsilon_1}}\right).$$

We now obtain the desired result by making the change of variable $x = y^{1/\theta}$.

§2.6. The asymptotic law for the distribution of basis elements for free semigroups

In this section we will present the content of Bredikhin's paper [9], in which he extended the elementary method of proving the asymptotic law of the distribution of primes to the study of the distribution of basis elements of free semigroups.

We will adopt the same notation introduced at the beginning of §2.5. In the Introduction we observed that if for each real number x there are only finitely many elements $\alpha \in G$ such that $N(\alpha) \le x$, then each element of G except the identity satisfies $N(\alpha) > 1$. If the semigroup has no identity, we formally adjoin one. Obviously, $N(1) = 1$.

We will omit the G from the notation $\nu_G(x) = \sum_{N(\alpha) \le x} 1$ and $\pi_G(x) = \sum_{N(\omega) \le x} 1$.

THEOREM. *Suppose*

$$\nu(x) = Gx^\theta + O(x^{\theta_1}), \tag{1}$$

where $\theta_1 < \theta$ *and* $C > 0$. *Then*

$$\lim_{x\to\infty} \frac{\pi(x)}{x^\theta / \ln x} = \frac{1}{\theta}. \tag{2}$$

Before proving this theorem, we introduce analogues of some classical functions of number theory on the semigroup G.

1. The Möbius function on G:

$$\mu(\alpha) = \begin{cases} 1, & \text{if } \alpha = 1, \\ (-1)^k, & \text{if } \alpha = \omega_1 \cdots \omega_k,\ \omega_i \neq \omega_j, \\ 0, & \text{if } \alpha \text{ is divisible by the square} \\ & \quad \text{of a nonidentity element.} \end{cases}$$

2. The von Mangoldt function on G:

$$\Lambda(\alpha) = \begin{cases} \ln N(\omega), & \text{if } \alpha = \omega^x \text{ with } x \geq 1, \\ 0, & \text{if } \alpha \neq \omega^x. \end{cases}$$

3. The Tchebycheff functions:

$$\psi(x) = \sum_{N(\alpha) \leq x} \Lambda(\alpha), \qquad \vartheta(x) = \sum_{N(\omega) \leq x} \ln N(\omega).$$

We will establish some elementary properties of these functions.

REMARK 1.

$$\sum_{\delta | \alpha} \mu(\delta) = \begin{cases} 0, & \alpha \neq 1, \\ 1, & \alpha = 1. \end{cases} \tag{3}$$

If $\alpha = 1$, the remark follows immediately from the definition of the Möbius function. Suppose $\alpha \neq 1$. Since the Möbius function is multiplicative, for $\alpha = \omega_1^{x_1} \cdots \omega_s^{x_s}$ we have, assuming not all of the x_i are zero,

$$\sum_{\delta | \alpha} \mu(\delta) = \prod_{j=1}^{s} (1 + \mu(\omega_j) + \cdots + \mu(\omega_j^{x_j})) = \prod_{j=1}^{s} (1 + \mu(\omega)) = 0.$$

REMARK 2. *The Tchebycheff-Möbius inversion formula.* If $f(\alpha) = \sum_{\delta|\alpha} \Phi(\delta)$, then

$$\Phi(\alpha) = \sum_{\delta | \alpha} \mu(\delta) f\left(\frac{\alpha}{\delta}\right).$$

REMARK 3.

$$\sum_{\delta | \alpha} \Lambda(\delta) = \ln N(\alpha). \tag{4}$$

Indeed, if $\alpha = \omega_1^{x_1} \cdots \omega_s^{x_s}$, then

$$\sum_{\delta | \alpha} \Lambda(\delta) = \sum_{j=1}^{s} x_j \ln \omega_j = \ln N(\alpha).$$

REMARK 4.

$$\Lambda(\alpha) = -\sum_{\delta | \alpha} \mu(\delta) \ln N(\delta). \tag{5}$$

For $\alpha = 1$ this identity can be verified directly. Suppose $\alpha \neq 1$. By the Tchebycheff-Möbius inversion formula and (4),

$$\Lambda(\alpha) = \sum_{\delta|\alpha} \mu(\delta) \ln N\left(\frac{\alpha}{\delta}\right) = \ln N(\alpha) \sum_{\delta|\alpha} \mu(\delta) - \sum_{\delta|\alpha} \mu(\delta) \ln N(\delta)$$
$$= -\sum_{\delta|\alpha} \mu(\delta) \ln N(\delta).$$

The proof of the theorem is based on several lemmas.

LEMMA 1. *If*

$$f(x) = \ln x \sum_{N(\alpha)\le x} h\left(\frac{s}{N(\alpha)}\right), \tag{6}$$

then

$$\sum_{N(\alpha)\le x} \mu(\alpha) f\left(\frac{x}{N(\alpha)}\right) = h(x)\ln x + \sum_{N(\alpha)\le x} \Lambda(\alpha) h\left(\frac{x}{N(\alpha)}\right). \tag{7}$$

PROOF. We have

$$\sum_{N(\alpha)\le x} \mu(\alpha) f\left(\frac{x}{N(\alpha)}\right)$$
$$= \sum_{N(\alpha)\le x} \mu(\alpha) \ln \frac{x}{N(\alpha)} \sum_{N(\beta)\le x/N(\alpha)} h\left(\frac{x}{N(\alpha)N(\beta)}\right)$$
$$= \ln x \sum_{N(\alpha)\le x} \mu(\alpha) \sum_{N(\beta)\le x/N(\alpha)} h\left(\frac{x}{N(\alpha\beta)}\right)$$
$$- \sum_{N(\alpha)\le x} \mu(\alpha) \ln N(\alpha) \sum_{N(\beta)\le x/N(\alpha)} h\left(\frac{x}{N(\alpha\beta)}\right)$$
$$= \ln x \sum_{N(\gamma)\le x} h\left(\frac{x}{N(\gamma)}\right) \sum_{\alpha|\gamma} \mu(\alpha) - \sum_{N(\gamma)\le x} h\left(\frac{x}{N(\gamma)}\right) \sum_{\alpha|\gamma} \mu(\alpha) \ln N(\alpha)$$
$$= \ln x h(x) + \sum_{N(\gamma)\le x} h\left(\frac{x}{N(\gamma)}\right) \Lambda(\gamma),$$

according to (3) and (5). In what follows we assume condition (1) holds.

LEMMA 2. *The following asymptotic formula holds*:

$$\sum_{N(\alpha)\le x} \frac{1}{N^\theta(\alpha)} = C\theta \ln x + C_0 + O(x^{\theta_1 - \theta}).$$

PROOF. Using Abel's transformation, we obtain

$$\sum_{N(\alpha)\le x}\frac{1}{N^\theta(\alpha)}=\theta\int_1^x\frac{\sum_{N(\alpha)\le u}1}{u^{\theta+1}}\,du+\frac{\sum_{N(\alpha)\le x}1}{x^\theta}$$
$$=\theta\int_1^x\frac{Cu^\theta+R(u)}{u^{\theta+1}}\,du+C+O(x^{\theta_1-\theta})$$
$$=C\theta\ln x+\int_1^\infty\frac{R(u)}{u^{\theta+1}}\,du-\int_x^\infty\frac{R(u)}{u^{\theta+1}}\,du+C+O(x^{\theta_1-\theta}).$$

Since $R(u)=O(u^{\theta_1})$, where $\theta_1<\theta$, the last two integrals converge. Thus,

$$\sum_{N(\alpha)\le x}\frac{1}{N^\theta(\alpha)}=C\theta\ln x+C'+O\left(\int_x^\infty\frac{du}{u^{1+\theta-\theta_1}}\right)+O(x^{\theta_1-\theta})$$
$$=C\theta\ln x+C'+O(x^{\theta_1-\theta}).$$

LEMMA 3.

$$\sum_{N(\alpha)\le x}\ln N(\alpha)=Cx^\theta\ln x-\frac{C}{\theta}x^\theta+O(x^{\theta_1}\ln x).$$

PROOF. Using Abel's transformation, we obtain

$$\sum_{N(\alpha)\le x}\ln N(\alpha)=-\int_1^x\frac{\sum_{N(\alpha)\le u}1}{u}\,du+\Big(\sum_{N(\alpha)\le x}1\Big)\ln x$$
$$=\nu(x)\ln x-\int_1^x\frac{\nu(u)}{u}\,du$$
$$=Cx^\theta\ln x-C\int_1^x u^{\theta-1}\,du+O\left(\int_1^x u^{\theta_1-1}\,du\right)+O(x^{\theta_1}\ln x)$$
$$=Cx^\theta\ln x-\frac{C}{\theta}x^\theta+O(x^{\theta_1}\ln x).$$

LEMMA 4.

$$\sum_{N(\alpha)\le x}\frac{\Lambda(\alpha)}{N^\theta(\alpha)}=\ln x+O\bigg(\frac{1}{x^{\theta-\theta_1}}\sum_{N(\alpha)\le x}\frac{\Lambda(\alpha)}{N^{\theta_1}(\alpha)}\bigg)+O(1).$$

PROOF. According to the formula generalizing the decomposition of $n!$ into prime factors,

$$\prod_{N(\alpha)\le x}N(\alpha)=\prod_{N(\omega)\le x}(N(\omega))^{\nu(x/N(\omega))+\nu(x/N^2(\omega))+\cdots}.$$

Therefore,

$$\begin{aligned}\sum_{N(\alpha)\le x} \ln N(\alpha) &= \sum_{N(\omega^r)\le x} \nu\left(\frac{x}{N(\omega^r)}\right) \ln N(\omega)\\ &= \sum_{N(\alpha)\le x} \nu\left(\frac{x}{N(\alpha)}\right) \Lambda(\alpha)\\ &= Cx^\theta \sum_{N(\alpha)\le x} \frac{\Lambda(\alpha)}{N^\theta(\alpha)} + O\left(x^{\theta_1} \sum_{N(\alpha)\le x} \frac{\Lambda(\alpha)}{N^{\theta_1}(\alpha)}\right).\end{aligned}$$

But, by Lemma 3,

$$Cx^\theta \ln x - \frac{C}{\theta}x^\theta + O(x^{\theta_1}\ln x) = Cx^\theta \sum_{N(\alpha)\le x} \frac{\Lambda(\alpha)}{N^\theta(\alpha)} + O\left(x^{\theta_1} \sum_{N(\alpha)\le x} \frac{\Lambda(\alpha)}{N^{\theta_1}(\alpha)}\right).$$

This implies the desired result:

$$\sum_{N(\alpha)\le x} \frac{\Lambda(\alpha)}{N^\theta(\alpha)} = \ln x + O\left(\frac{x^{\theta_1}}{x^\theta} \sum_{N(\alpha)\le x} \frac{\Lambda(\alpha)}{N^{\theta_1}(\alpha)}\right) + O(1).$$

LEMMA 5.

$$\sum_{N(\alpha)\le x} \psi\left(\frac{x}{N(\alpha)}\right) = Cx^\theta \ln x - \frac{C}{\theta}x^\theta + O(x^{\theta_1}\ln x).$$

PROOF. Indeed,

$$\begin{aligned}\sum_{N(\alpha)\le x} \psi\left(\frac{x}{N(\alpha)}\right) &= \sum_{N(\alpha)\le x}\ \sum_{N(\beta)\le x/N(\alpha)} \Lambda(\beta)\\ &= \sum_{N(\gamma)\le x} \sum_{\beta|\gamma} \Lambda(\beta)\\ &= \sum_{N(\gamma)\le x} \ln N(\gamma),\end{aligned}$$

according to (4). It now suffices to apply Lemma 3.

LEMMA 6.

$$\psi(x)\ln x + \sum_{N(\alpha)\le x} \Lambda(\alpha)\psi\left(\frac{x}{N(\alpha)}\right) = \frac{2}{\theta}x^\theta \ln x + O(x^\theta). \tag{8}$$

PROOF. In Lemma 1 put $h(x) = \theta\psi(x) - x^\theta + \lambda$, where λ is to be chosen later. Then

$$\begin{aligned} f(x) &= \ln x \sum_{N(\alpha)\le x} \left(\theta\psi\left(\frac{x}{N(\alpha)}\right) - \left(\frac{x}{N(\alpha)}\right)^\theta + \lambda\right) \\ &= \theta \ln x \sum_{N(\alpha)\le x} \psi\left(\frac{x}{N(\alpha)}\right) - x^\theta \ln x \sum_{N(\alpha)\le x} \frac{1}{N^\theta(\alpha)} + \lambda \ln x \nu(x) \\ &= C\theta x^\theta \ln^2 x - Cx^\theta \ln x + O(x^{\theta_1}\ln^2 x) - C\theta x^\theta \ln^2 x \\ &\quad - C_0 x^\theta \ln x + O(x^{\theta_1}\ln x) + C\lambda x^\theta \ln x + O(x^{\theta_1}\ln x) \\ &= (-C_0 - C + C\lambda)x^\theta \ln x + O(x^{\theta_1}\ln^2 x). \end{aligned}$$

If we take $\lambda = -(C + C_0)/C$, then

$$f(x) = O(x^{\theta_1}\ln^2 x). \tag{9}$$

We now transform the right-hand side of (7):

$$\begin{aligned} &h(x)\ln x + \sum_{N(\alpha)\le x} \Lambda(\alpha) h\left(\frac{x}{N(\alpha)}\right) \\ &\quad = \theta\psi(x)\ln x - x^\theta \ln x + O(\ln x) \\ &\qquad + \theta \sum_{N(\alpha)\le x} \Lambda(\alpha)\psi\left(\frac{x}{N(\alpha)}\right) \\ &\qquad - x^\theta \sum_{N(\alpha)\le x} \frac{\Lambda(\alpha)}{N^\theta(\alpha)} + O\Bigg(\sum_{N(\alpha)\le x} \Lambda(\alpha)\Bigg) \\ &\quad = \theta\Bigg(\psi(x)\ln x + \sum_{N(\alpha)\le x} \Lambda(\alpha)\psi\left(\frac{x}{N(\alpha)}\right)\Bigg) \\ &\qquad - x^\theta \ln x - x^\theta \ln x + O\Bigg(x^{\theta_1} \sum_{N(\alpha)\le x} \frac{\Lambda(\alpha)}{N^{\theta_1}(\alpha)}\Bigg) + O(\psi(x)) \\ &\quad = \theta\Bigg(\psi(x)\ln x + \sum_{N(\alpha)\le x} \Lambda(\alpha)\psi\left(\frac{x}{N(\alpha)}\right)\Bigg) - 2x^\theta \ln x \\ &\qquad + O\Bigg(x^{\theta_1} \sum_{N(\alpha)\le x} \frac{\Lambda(\alpha)}{N^{\theta_1}(\alpha)}\Bigg). \end{aligned}$$

This leads to the inequality

$$\psi(x)\ln x + \sum_{N(\alpha)\le x} \Lambda(\alpha)\psi\left(\frac{x}{N(\alpha)}\right) - \frac{2}{\theta}x^\theta \ln x + O\Bigg(x^{\theta_1} \sum_{N(\alpha)\le x} \frac{\Lambda(\alpha)}{N^{\theta_1}(\alpha)}\Bigg)$$
$$= \frac{1}{\theta} \sum_{N(\alpha)\le x} \mu(\alpha) f\left(\frac{x}{N(\alpha)}\right)$$
$$= O\Bigg(\sum_{N(\alpha)\le x} \left(\frac{x}{N(\alpha)}\right)^{\theta_1} \ln^2 \frac{x}{N(\alpha)}\Bigg) = O(x^{\theta_1}\ln^3 x),$$

and hence

$$\psi(x)\ln x + \sum_{N(\alpha)\le x} \Lambda(\alpha)\psi\left(\frac{x}{N(\alpha)}\right)$$
$$= \frac{2}{\theta}x^\theta \ln x + O\Bigg(x^{\theta_1} \sum_{N(\alpha)\le x} \frac{\Lambda(\alpha)}{N^{\theta_1}(\alpha)}\Bigg) + O(x^\theta). \tag{10}$$

It follows that

$$\psi(x) = O(x^\theta) + O\Bigg(\frac{x^{\theta_1}}{\ln x} \sum_{N(\alpha)\le x} \frac{\Lambda(\alpha)}{N^{\theta_1}(\alpha)}\Bigg)$$
$$= O(x^\theta) + O\Bigg(x^{\theta_1} \sum_{N(\alpha)\le x} \frac{1}{N^{\theta_1}(\alpha)}\Bigg).$$

Using Abel's transformation and the argument of Lemma 2, we obtain

$$\sum_{N(\alpha)\le x} \frac{1}{N^{\theta_1}(\alpha)} = O\left(\int_1^x \frac{u^\theta}{u^{\theta+1}}\,du\right) + O\left(\frac{x^\theta}{x^{\theta_1}}\right) = O(x^{\theta-\theta_1}).$$

Thus, $\psi(x) = O(x^\theta)$. In view of this,

$$x^{\theta_1} \sum_{N(\alpha)\le x} \frac{\Lambda(\alpha)}{N^{\theta_1}(\alpha)} = O\left(x^{\theta_1} \int_1^x \frac{\psi(u)\,du}{u^{\theta_1+1}}\right) + O\left(x^{\theta_1} \frac{\psi(x)}{x^{\theta_1}}\right)$$
$$= O\left(x^{\theta_1} \int_1^x \frac{u^\theta}{u^{\theta_1+1}}\,du\right) + O(x^\theta) = O(x^\theta),$$

and so it follows from (10) that

$$\psi(x)\ln x + \sum_{N(\alpha)\le x} \Lambda(\alpha)\psi\left(\frac{x}{N(\alpha)}\right) = \frac{2}{\theta}x^\theta \ln x + O(x^\theta).$$

Lemma 6 is proved.

Since we have established that

$$x^{\theta_1} \sum_{N(\alpha)\le x} \frac{\Lambda(\alpha)}{N^{\theta_1}(\alpha)} = O(x^\theta),$$

Lemma 4 can be restated as follows:

LEMMA 4′.

$$\sum_{N(\alpha)\le x} \frac{\Lambda(\alpha)}{N^\theta(\alpha)} = \ln x + O(1).$$

Also,

$$\sum_{N(\alpha)\le x} \frac{\Lambda(\alpha)}{N^\theta(\alpha)} = \sum_{N(\omega)\le x} \frac{\ln N(\omega)}{N^\theta(\omega)} + O\Bigg(\sum_{N(\omega)\le\sqrt{x}} \frac{\ln N(\omega)}{N^{2\theta}(\omega)}\left(1+\frac{1}{N^\theta(\omega_1)}+\frac{1}{N^{2\theta}(\omega_1)}+\cdots\right)\Bigg).$$

We have

$$1+\frac{1}{N^\theta(\omega_1)}+\frac{1}{N^{2\theta}(\omega_1)}+\cdots = O(1)$$

and

$$\sum_{N(\alpha)\le x} \frac{\Lambda(\alpha)}{N^\theta(\alpha)} = \sum_{N(\omega)\le x} \frac{\ln N(\omega)}{N^\theta(\omega)} + O\Bigg(\sum_{N(\alpha)\le\sqrt{x}} \frac{\Lambda(\alpha)}{N^{2\theta}(\alpha)}\Bigg).$$

But using Abel's transformation we obtain

$$\sum_{N(\alpha)\le\sqrt{x}} \frac{\Lambda(\alpha)}{N^{2\theta}(\alpha)} = O\left(\int_1^{\sqrt{x}} \frac{\psi(u)}{u^{2\theta+1}}\,du\right) + O\left(\frac{\psi(x)}{x^{2\theta}}\right) = O(1).$$

Thus we have yet another form of Lemma 4:

LEMMA 4″.

$$\sum_{N(\omega)\le x} \frac{\ln N(\omega)}{N^\theta(\omega)} = \ln x + O(1).$$

Also,

$$\begin{aligned}\psi(x) &= \sum_{N(\alpha)\le x} \Lambda(\alpha)\\ &= \sum_{N(\omega)\le x} \ln N(\omega) + O\Bigg(\frac{\ln x}{\ln N(\omega_1)} \sum_{N(\omega)\le\sqrt{x}} \ln N(\omega)\Bigg)\\ &= \vartheta(x) + O\left(\ln x\,\psi(\sqrt{x})\right) = \vartheta(x) + O(x^{\theta/2}\ln x).\end{aligned}$$

In what follows, the letters x, y, and z will stand for positive real numbers, and R, or more explicitly R_x, will denote the interval $(\ln x, x/\ln x)$.

It follows directly from Lemma 2 that

$$\sum_{y<N(\alpha)\le z} \frac{1}{N^\theta(\alpha)} = \theta C \ln\frac{z}{y} + O(y^{\theta_1-\theta}). \qquad (11)$$

By Lemma 4″,

$$\sum_{N(\omega)\in R} \frac{\ln N(\omega)}{N^\theta(\omega)} = \ln x + O(1). \tag{12}$$

The following modification of Lemma 6 is the basis of the proof of the asymptotic law.

LEMMA 6′.

$$\vartheta(x)\ln x + \sum_{N(\omega)\in R} \vartheta\left(\frac{x}{N(\omega)}\right)\ln N(\omega) = \frac{2}{\theta}x^\theta \ln x + O(x^\theta \ln\ln x).$$

PROOF. Since

$$\psi(x) = \vartheta(x) + O(x^{\theta/2}\ln x) = \vartheta(x) + O(x^{\theta/2+\varepsilon}),$$

where $0 < \varepsilon < \theta/2$, we have

$$\begin{aligned}\vartheta(x)\ln x + O(x^{\theta/2+\varepsilon}) &+ \sum_{N(\omega)\le x} \vartheta\left(\frac{x}{N(\omega)}\right)\ln N(\omega)\\ &+ O\Bigg(x^{\theta/2+\varepsilon}\sum_{N(\omega)\le x}\frac{\ln N(\omega)}{N^{\theta/2+\varepsilon}(\omega)}\Bigg) = \frac{2}{\theta}x^\theta \ln x + O(x^\theta),\end{aligned}$$

because, as we have seen in the proof of Lemma 6,

$$x^{\theta/2+\varepsilon}\sum_{N(\omega)\le x}\frac{\ln N(\omega)}{N^{\theta/2+\varepsilon}(\omega)} = O(x^\theta).$$

Thus,

$$\vartheta(x)\ln x + \sum_{N(\omega)\le x} \vartheta\left(\frac{x}{N(\omega)}\right)\ln N(\omega) = \frac{2}{\theta}x^\theta \ln x + O(x^\theta). \tag{13}$$

Now

$$\begin{aligned}\sum_{N(\omega)\le \ln x} \vartheta\left(\frac{x}{N(\omega)}\right)\ln N(\omega) &= O\Bigg(x^\theta \sum_{N(\omega)\le \ln x}\frac{\ln N(\omega)}{N^\theta(\omega)}\Bigg)\\ &= O(x^\theta \ln\ln x)\end{aligned}$$

and

$$\begin{aligned}\sum_{x/\ln x\le N(\omega)\le x} \vartheta\left(\frac{x}{N(\omega)}\right)\ln N(\omega) &= O\Bigg(x^\theta \sum_{x/\ln x\le N(\omega)\le x}\frac{\ln N(\omega)}{N^\theta(\omega)}\Bigg)\\ &= O\left(x^\theta\left(\ln x - \ln\frac{x}{\ln x}\right)\right) = O(x^\theta \ln\ln x).\end{aligned}$$

This proves Lemma 6′.

LEMMA 7. *If $y < z \le 2y$, then*

$$\vartheta(z) - \vartheta(y) \le \frac{2}{\theta}(z^\theta - y^\theta) + O\left(\frac{z^\theta}{\ln z}\right). \tag{14}$$

PROOF. From (13) with $x = z$ we subtract (13) with $x = y$:

$$\vartheta(z)\ln z - \vartheta(y)\ln y \le \frac{2}{\theta}(z^\theta \ln z - y^\theta \ln y) + O(z^\theta).$$

But

$$\vartheta(y)\ln y = \vartheta(y)\ln z - \vartheta(y)\ln\frac{z}{y} = \vartheta(y)\ln z + O(y^\theta),$$

$$(\vartheta(z) - \vartheta(y))\ln z \le \frac{2}{\theta}\left(z^\theta \ln z - y^\theta \ln z - y^\theta \ln\frac{y}{z}\right) + O(z^\theta),$$

and hence, in view of the fact that $\ln(y/z) = O(1)$,

$$\vartheta(z) - \vartheta(y) \le \frac{2}{\theta}(z^\theta - y^\theta) + O\left(\frac{z^\theta}{\ln z}\right).$$

LEMMA 8. *If $1 \le y < z \le x$, then*

$$\left|\sum_{y < N(\alpha) \le z} \vartheta\left(\frac{x}{N(\alpha)}\right) - Cx^\theta \ln\frac{z}{y}\right| \le Bx^\theta, \tag{15}$$

where the constant B does not depend on x, y, or z.

PROOF. We have

$$\begin{aligned}
\sum_{N(\alpha)\le z} \vartheta\left(\frac{x}{N(\alpha)}\right) &= \sum_{N(\alpha)\le z}\ \sum_{N(\omega)\le x/N(\alpha)} \ln N(\omega) \\
&= \sum_{N(\omega)\le x/z} \ln N(\omega) \sum_{N(\alpha)\le z} 1 + \sum_{x/z\le N(\omega)\le x} \ln N(\omega) \sum_{N(\alpha)\le x/N(\omega)} 1 \\
&= \nu(z)\vartheta\left(\frac{x}{z}\right) + \sum_{x/z\le N(\omega)\le x} \ln N(\omega)\nu\left(\frac{x}{N(\omega)}\right) \\
&= Cx^\theta \sum_{x/z\le N(\omega)\le x} \frac{\ln N(\omega)}{N^\theta(\omega)} + O(x^\theta) = Cx^\theta \ln z + O(x^\theta).
\end{aligned}$$

Analogously,

$$\sum_{N(\alpha)\le y} \vartheta\left(\frac{x}{N(\alpha)}\right) = Cx^\theta \ln y + O(x^\theta).$$

The desired result is obtained by subtracting these equalities.

Observe that, as a consequence of (15),

$$\sum_{N(\alpha)\in R} \vartheta\left(\frac{x}{N(\alpha)}\right) = Cx^\theta \ln x + O(x^\theta \ln\ln x) \tag{16}$$

(put $y = \ln x$ and $z = x/\ln x$).

Let

$$A = \overline{\lim_{x\to\infty}}\, \frac{\vartheta(x)}{x^\theta}, \qquad a = \underline{\lim_{x\to\infty}}\, \frac{\vartheta(x)}{x^\theta}.$$

LEMMA 9. $a + A = 2/\theta$.

PROOF. For $x \geq x_0(\varepsilon)$ we have $\vartheta(x)/x^\theta < A + \varepsilon$, and there exists an infinite sequence of numbers x tending to infinity such that $\vartheta(x)/x^\theta < a + \varepsilon$. From (13) and Lemma 4″ we obtain

$$\begin{aligned}\frac{2}{\theta} + O\left(\frac{1}{\ln x}\right) &\leq A + \varepsilon + (a+\varepsilon)\frac{1}{\ln x}\sum_{N(\omega)\leq x}\frac{\ln N(\omega)}{N^\theta(\omega)}\\ &= A + a + 2\varepsilon + O\left(\frac{1}{\ln x}\right),\end{aligned}$$

and hence $2/\theta \leq A + a + 2\varepsilon$. But ε is arbitrarily small. Therefore $A + a \geq 2/\theta$. The reverse inequality $A + a \leq 2/\theta$ can be proved analogously.

Let $a = 1/\theta - \gamma$ and $A = 1/\theta + \gamma$, where $\gamma \geq 0$. Our goal is to prove that $A = a = 1/\theta$, i.e., to prove that $\gamma = 0$. We will prove this by the method of contradiction: assume $\gamma > 0$.

Let h be a fixed positive number, with

$$0 < h < \min\left(\frac{3\gamma^3 C\theta}{400B}, h_1, h_2, \ldots,\right),$$

where B is the constant in (15) and $h_1, h_2, \ldots$ are sufficiently small positive constants, to be chosen in the course of the proof, depending only on C, θ, and γ. There will also appear in the proof a sufficiently large number x_0 depending only on h, C, θ, and γ.

Suppose $x > x_0$. We divide the interval R into half-open intervals I_t of the form

$$(e^{(t-1)h/\theta}\ln x, e^{th/\theta}\ln x],$$

where $t = 1, \ldots, m$ and $m = [(\ln x - 2\ln\ln x)\theta/h]$; the last interval I_{m+1} can have smaller length.

It follows from (11) that for $x \geq x_0$

$$\sum_{N(\alpha)\in I_t} \frac{1}{N^\theta(\alpha)} < Ch + h^2. \tag{a}$$

It follows from (12) that for $x \geq x_0$

$$(1-h^2)\ln x < \sum_{N(\omega)\in R} \frac{\ln N(\omega)}{N^\theta(\omega)} < (1+h^2)\ln x. \tag{b}$$

By the definition of γ, for $y \geq \ln x$ $(x \geq x_0)$ we have

$$\left(\frac{1}{\theta} - \gamma - h^2\right) y^\theta < \vartheta(y) < \left(\frac{1}{\theta} + \gamma + h^2\right) y^\theta. \tag{c}$$

By Lemma 6′, for $x \geq x_0$

$$\left|\vartheta(x)\ln x + \sum_{N(\omega)\in R} \vartheta\left(\frac{x}{N(\omega)}\right) \ln N(\omega) - \frac{2}{\theta} x^\theta \ln x\right| < h^2 x^\theta \ln x. \tag{d}$$

By Lemma 7, for $\ln x \leq y < z < 2y$ (and $x \geq x_0$)

$$\vartheta(z) - \vartheta(y) < \frac{2}{\theta}(z^\theta - y^\theta) + \frac{1}{\theta} h^2 z^\theta. \tag{e}$$

In view of (16), for $x \geq x_0$

$$\sum_{N(\alpha)\in R} \vartheta\left(\frac{x}{N(\alpha)}\right) > (1-h) C x^\theta \ln x. \tag{f}$$

There exists a sequence of numbers x tending to infinity such that, in accordance with the definition of γ,

$$\vartheta(x) > (1/\theta + \gamma - h^2) x^\theta. \tag{g}$$

We will henceforth assume tht $x \geq x_0$ and x is such that (g) holds.

LEMMA 10. *Suppose ω' is a basis element for which*

$$\vartheta\left(\frac{x}{N(\omega')}\right) < \left(\frac{1}{\theta} - \gamma + h\right) \frac{x^\theta}{N^\theta(\omega')}.$$

Then

$$\sum_{N(\omega')\in R} \frac{\ln N(\omega')}{N^\theta(\omega')} > \left(1 - \left(3 + \frac{1}{\theta}\right) h\right) \ln x. \tag{17}$$

Indeed, it follows from (d) that

$$\vartheta(x)\ln x + \sum_{N(\omega)\in R} \vartheta\left(\frac{x}{N(\omega)}\right) \ln N(\omega) - \frac{2}{\theta} x^\theta \ln x < h^2 x^\theta \ln x.$$

For ω not equal to ω' we have, by the hypothesis of the lemma,

$$\vartheta\left(\frac{x}{N(\omega)}\right) \geq \left(\frac{1}{\theta} - \gamma + h\right) \frac{x^\theta}{N^\theta(\omega)}.$$

For ω', according to (c),

$$\begin{aligned}\vartheta\left(\frac{x}{N(\omega')}\right) &\geq \left(\frac{1}{\theta}-\gamma-h^2\right)\frac{x^\theta}{N^\theta(\omega')}\\ &= \left(\frac{1}{\theta}-\gamma+h\right)\frac{x^\theta}{N^\theta(\omega')}-(h+h^2)\frac{x^\theta}{N^\theta(\omega')}.\end{aligned}$$

Finally, using (g), we have

$$\begin{aligned}&\vartheta(x)\ln x+\sum_{N(\omega)\in R}\vartheta\left(\frac{x}{N(\omega)}\right)\ln N(\omega)-\frac{2}{\theta}x^\theta\ln x\\ &\geq \left(\frac{1}{\theta}+\gamma-h^2\right)x^\theta\ln x+\left(\frac{1}{\theta}-\gamma+h\right)x^\theta\sum_{N(\omega)\in R}\frac{\ln N(\omega)}{N^\theta(\omega)}\\ &\quad -x^\theta(h+h^2)\sum_{N(\omega')\in R}\frac{\ln N(\omega')}{N^\theta(\omega')}-\frac{2}{\theta}x^\theta\ln x.\end{aligned}$$

Dividing by x^θ, we obtain

$$\begin{aligned}&(h+h^2)\sum_{N(\omega')\in R}\frac{\ln N(\omega')}{N^\theta(\omega')}\\ &> \left(\frac{1}{\theta}+\gamma-h^2\right)\ln x+\left(\frac{1}{\theta}-\gamma+h\right)\sum_{N(\omega)\in R}\frac{\ln N(\omega)}{N^\theta(\omega)}-\left(\frac{2}{\theta}+h^2\right)\ln x.\end{aligned}$$

It follows from (b) that

$$\begin{aligned}&\sum_{N(\omega')\in R}\frac{\ln N(\omega')}{N^\theta(\omega')}\\ &\geq \frac{\ln x}{h+h^2}\left[\frac{1}{\theta}+\gamma-h^2+\left(\frac{1}{\theta}-\gamma+h\right)(1-h^2)-\frac{2}{\theta}-h^2\right]\\ &= \frac{1-h-h/\theta+h\gamma-h^2}{1+h}\ln x\\ &\geq \frac{1-h-h/\theta-h^2}{1+h}\ln x\\ &\geq \left(\frac{1-h-h^2}{1+h}-\frac{h}{\theta}\right)\ln x \geq \left(1-3h-\frac{h}{\theta}\right)\ln x.\end{aligned}$$

The lemma is proved.

LEMMA 11.

$$\sum_{N(\omega)\in I_t}\frac{\ln N(\omega)}{N^\theta(\omega)}<\frac{2}{\theta}h+\frac{4}{\theta}h^2. \tag{18}$$

PROOF. In the interval $I_t = (y, ye^{h/\theta})$, where $y = e^{(t-1)h/\theta}\ln x$, we have $ye^{h/\theta} < 2y$ when $h < h_1$. Applying (e), we get

$$\sum_{N(\omega)\in I_t} \frac{\ln N(\omega)}{N^\theta(\omega)} \le \frac{1}{y^\theta} \sum_{y\le N(\omega)\le ye^{h/\theta}} \ln N(\omega) < \frac{2}{\theta}(e^h - 1) + \frac{h^2}{\theta}e^h,$$

which easily implies (18).

LEMMA 12. *The number of intervals I_t containing the numbers $N(\omega')$ must be greater than M, where $M = \frac{1}{2}m(1-(5+1/\theta)h)$.*

PROOF. Assume the number of intervals I_t containing the numbers $N(\omega')$ does not exceed M. Then

$$\begin{aligned}\sum_{N(\omega')\in R} \frac{\ln N(\omega')}{N^\theta(\omega')} &\le M \max_t \sum_{N(\omega')\in I_t} \frac{\ln N(\omega')}{N^\theta(\omega')} \\ &\le M \max_t \sum_{N(\omega)\in I_t} \frac{\ln N(\omega)}{N^\theta(\omega)} \\ &\le M\left(\frac{2}{\theta}h + \frac{4}{\theta}h^2\right) \\ &\le (1+2h)\left(1-\left(5+\frac{1}{\theta}\right)h\right)\ln x < \left(1-\left(3+\frac{1}{\theta}\right)h\right)\ln x.\end{aligned}$$

This inequality contradicts (17).

LEMMA 13. *If $N(\alpha_1) \in R$, $N(\alpha_2) \in R$, and $e^{-h/\theta} \le N(\alpha_1)/N(\alpha_2) \le e^{h/\theta}$, then*

$$\left|\vartheta\left(\frac{x}{N(\alpha_1)}\right)\left(\frac{N(\alpha_1)}{x}\right)^\theta - \vartheta\left(\frac{x}{N(\alpha_2)}\right)\left(\frac{N(\alpha_2)}{x}\right)^\theta\right| < \frac{6h}{\theta}.$$

PROOF. We may assume without loss of generality that

$$\vartheta\left(\frac{x}{N(\alpha_1)}\right)\left(\frac{N(\alpha_1)}{x}\right)^\theta \ge \vartheta\left(\frac{x}{N(\alpha_2)}\right)\left(\frac{N(\alpha_2)}{x}\right)^\theta.$$

In view of inequality (e), for $h < h_2$ we have

$$\begin{aligned}\left|\vartheta\left(\frac{x}{N(\alpha_1)}\right) - \vartheta\left(\frac{x}{N(\alpha_2)}\right)\right| &\le \frac{2}{\theta}\left|\frac{x^\theta}{N^\theta(\alpha_1)} - \frac{x^\theta}{N^\theta(\alpha_2)}\right| + \frac{e^h h^2}{\theta}\frac{x^\theta}{N^\theta(\alpha_1)} \\ &\le \frac{1}{\theta}\frac{2x^\theta}{N^\theta(\alpha_1)}\left((e^h-1) + \frac{h^2}{2}e^h\right) \\ &< \frac{1}{\theta}\left(\frac{x}{N(\alpha_1)}\right)^\theta 3h.\end{aligned}$$

Hence, according to (c),

$$\begin{aligned}\vartheta\left(\frac{x}{N(\alpha_1)}\right)\left(\frac{N(\alpha_1)}{x}\right)^{\theta} &\leq \frac{3h}{\theta}+\vartheta\left(\frac{x}{N(\alpha_2)}\right)\left(\frac{N(\alpha_1)}{x}\right)^{\theta}\\ &\leq \frac{3h}{\theta}+\vartheta\left(\frac{x}{N(\alpha_2)}\right)\left(\frac{N(\alpha_2)}{x}\right)^{\theta}+\frac{\vartheta(x/N(\alpha_2))}{x^{\theta}}|N^{\theta}(\alpha_1)-N^{\theta}(\alpha_2)|\\ &\leq \frac{3h}{\theta}+\vartheta\left(\frac{x}{N(\alpha_2)}\right)\left(\frac{N(\alpha_2)}{x}\right)^{\theta}\\ &\quad+\frac{1}{N^{\theta}(\alpha_2)}\left(\frac{1}{\theta}+\gamma+h^2\right)|N^{\theta}(\alpha_1)-N^{\theta}(\alpha_2)|.\end{aligned}$$

But for a sufficiently small h

$$\left(\frac{1}{\theta}+\gamma+h^2\right)(e^h-1)\leq\frac{3h}{\theta}\quad\left(\text{since } a\geq 0 \text{ and } \gamma\leq\frac{1}{\theta}\right).$$

LEMMA 13′. *If $N(\alpha)$ belongs to the same interval I_t as one of the numbers $N(\omega')$, then*

$$\vartheta\left(\frac{x}{N(\alpha)}\right)<\left(\frac{1}{\theta}-\gamma+\left(1+\frac{6}{\theta}\right)h\right)\left(\frac{x}{N(\alpha)}\right)^{\theta}. \tag{19}$$

PROOF. If $N(\alpha)$ and $N(\omega')$ belong to I_t, then the hypothesis of Lemma 13 is satisfied. Therefore,

$$\vartheta\left(\frac{x}{N(\alpha)}\right)\left(\frac{N(\alpha)}{x}\right)^{\theta}<\vartheta\left(\frac{x}{N(\omega')}\right)\left(\frac{N(\omega')}{x}\right)^{\theta}+\frac{6h}{\theta}.$$

Then inequality (19) follows from the definition of ω' given in the statement of Lemma 10.

LEMMA 14. *There are at least $m\gamma^2C\theta/100B$ intervals I_t such that for each number $N(\alpha)$ belonging to these intervals*

$$\left(\frac{1}{\theta}-\gamma+\left(1+\frac{6}{\theta}\right)h\right)\frac{x^{\theta}}{N^{\theta}(\alpha)}<\vartheta\left(\frac{x}{N(\alpha)}\right)<\left(\frac{1}{\theta}-\frac{\gamma}{2}\right)\frac{x^{\theta}}{N^{\theta}(\alpha)}.$$

PROOF. Suppose there are N successive intervals I_t in which

$$\vartheta\left(\frac{x}{N(\alpha)}\right)\left(\frac{N(\alpha)}{x}\right)^{\theta}<\frac{1}{\theta}-\frac{\gamma}{2}. \tag{20}$$

By Lemma 8,

$$\sum_{y<N(\alpha)\leq ye^{Nh/\theta}}\vartheta\left(\frac{x}{N(\alpha)}\right)>\frac{Cx^{\theta}Nh}{\theta}-Bx^{\theta}. \tag{21}$$

On the other hand, according to inequality (a),

$$\sum_{y<N(\alpha)\le ye^{Nh/\theta}} \vartheta\left(\frac{x}{N(\alpha)}\right) \le x^\theta\left(\frac{1}{\theta}-\frac{\gamma}{2}\right) \sum_{y<N(\alpha)\le ye^{Nh/\theta}} \frac{1}{N^\theta(\alpha)}$$
$$< \left(\frac{1}{\theta}-\frac{\gamma}{2}\right) x^\theta N(Ch+h^2). \tag{22}$$

It follows from (21) and (22) that there can be (for $h<h_4$) at most $[3B/C\gamma h]$ successive intervals I_t for which (20) holds.

By Lemma 13′, if $N(\alpha)$ belongs to an interval I_t containing an $N(\omega')$, then

$$\vartheta\left(\frac{x}{N(\alpha)}\right)\left(\frac{N(\alpha)}{x}\right)^\theta < \frac{1}{\theta}-\gamma+\left(1+\frac{6}{\theta}\right)h < \frac{1}{\theta}-\frac{\gamma}{2}.$$

So for each run of at most $[3B/C\gamma h]$ successive intervals I_t in which there is an $N(\omega')$ (except perhaps the last) there must be intervals in which $\vartheta(x/N(\alpha))(N(\alpha)/x)^\theta$ increases from a quantity not exceeding $1/\theta-\gamma+(1+6/\theta)h$ to at least $1/\theta-\gamma/2$, i.e., by an amount greater than

$$\gamma/2-(1+6/\theta)h.$$

By Lemma 13, such an increase in $\vartheta(x/N(\alpha))(N(\alpha)/x)^\theta$ requires (for $h<h_5$) at least

$$\frac{\gamma/2-(1+6/\theta)h}{6h/\theta} > \frac{\gamma\theta}{13h}$$

intervals I_t. Of these intervals, at least $\gamma\theta/13h-2>\gamma\theta/14h$ (for $h<h_6$) must be such that for each $N(\alpha)$ we have

$$\left(\frac{1}{\theta}-\gamma+\left(1+\frac{6}{\theta}\right)h\right)\left(\frac{x}{N(\alpha)}\right)^\theta < \vartheta\left(\frac{x}{N(\alpha)}\right) < \left(\frac{1}{\theta}-\frac{\gamma}{2}\right)\left(\frac{x}{N(\alpha)}\right)^\theta$$

on these intervals.

By Lemma 12, there exist at least

$$\frac{M}{[3B/C\gamma h]}-1 > M\frac{C\gamma h}{3B}-1$$

such sets of increasing relations. Consequently, the number of intervals I_t on which (19) holds must be at least

$$\left(\frac{MC\gamma h}{3B}-1\right)\frac{\gamma\theta}{14h} = \frac{m\gamma^2C\theta}{B}\frac{1-(5+1/\theta)h}{84}\left(1-\frac{3B}{MC\gamma h}\right) > \frac{m\gamma^2C\theta}{100B}$$

for sufficiently large x (for which inequalities (a)–(g) hold) and $h<h_7$.

LEMMA 15.

$$\lim_{x\to\infty}\frac{\vartheta(x)}{x^\theta}=\frac{1}{\theta}.$$

PROOF. We will show that $\gamma=0$. Assume $\gamma>0$. Then for sufficiently small h and sufficiently large x satisfying (a)–(g) Lemmas 10–14 hold true. Let N_1 denote the number of intervals I_t such that for each α with $N(\alpha)$ belonging to any one of them we have

$$\vartheta\left(\frac{x}{N(\alpha)}\right)<\left(\frac{1}{\theta}-\gamma+\left(1+\frac{h}{\theta}\right)h\right)\frac{x^\theta}{N^\theta(\alpha)};$$

let N_2 denote the number of intervals I_t such that for each α with $N(\alpha)$ belonging to any one of them

$$\left(\frac{1}{\theta}-\gamma+\left(1+\frac{6}{\theta}\right)h\right)\frac{x^\theta}{N^\theta(\alpha)}<\vartheta\left(\frac{x}{N(\alpha)}\right)<\left(\frac{1}{\theta}-\frac{\gamma}{2}\right)\frac{x^\theta}{N^\theta(\alpha)};$$

there are $m+1-N_1-N_2$ other intervals I_t in R.

By virtue of (a) and (c),

$$\begin{aligned}\sum_{N(\alpha)\in R}&\vartheta\left(\frac{x}{N(\alpha)}\right)\\
&<x^\theta(Ch+h^2)\left(N_1\left(\frac{1}{\theta}-\gamma+\left(1+\frac{6}{\theta}\right)h\right)+N_2\left(\frac{1}{\theta}-\frac{\gamma}{2}\right)\right.\\
&\qquad\left.+(m+1-N_1-N_2)\left(\frac{1}{\theta}+\gamma+h^2\right)\right)\\
&=x^\theta(Ch+h^2)\left((m+1)\left(\frac{1}{\theta}+\gamma+h^2\right)\right.\\
&\qquad\left.-N_1\left(2\gamma-\left(1+\frac{6}{\theta}\right)h\right)-N_2\left(\frac{3}{2}\gamma+h^2\right)\right).\end{aligned}$$

By Lemmas 12 and 14, inasmuch as $m\le(\theta/h)\ln x$, we obtain

$$\begin{aligned}\sum_{N(\alpha)\in R}\vartheta\left(\frac{x}{N(\alpha)}\right)&<C\left(1-\frac{3\gamma^3C\theta}{200B}+\frac{1}{2}(\theta+6)h+(5\theta+1)\gamma h+h^2\theta\right.\\
&\qquad+\frac{1}{m}(1+\theta\gamma+\theta h^2)+\frac{h}{C}-\frac{h}{C}\frac{3\gamma^3C\theta}{200B}\\
&\qquad+\frac{1}{2}(\theta+6)\frac{h^2}{C}+(5\theta+1)\frac{h^2}{C}+\frac{h^3\theta}{C}\\
&\qquad\left.+\frac{h}{Cm}(1+\theta\gamma+\theta h^2)\right)x^\theta\ln x.\end{aligned}$$

Take $h<h_8$ and then a sufficiently large x (hence m will also be sufficiently large) so that all summands within the parentheses except the first two amount

to less than $\frac{1}{2} \cdot 3\gamma^3 C\theta/200B$. Then we have

$$\sum_{N(\alpha)\in R} \vartheta\left(\frac{x}{N(\alpha)}\right) < \left(1 - \frac{3\gamma^3 C\theta}{400B}\right) Cx^\theta \ln x.$$

Since $h < 3\gamma^3 C\theta/400B$, it follows that

$$\sum_{N(\alpha)\in R} \vartheta\left(\frac{x}{N(\alpha)}\right) < (1-h)Cx^\theta \ln x.$$

But this is in violation of inequality (f). Thus the assumption $\gamma > 0$ has led to a contradiction. Lemma 15 is proved.

LEMMA 16.

$$\lim_{x\to\infty} \frac{\psi(x)}{x^\theta} = \frac{1}{\theta}.$$

This follows immediately from the relation $\psi(x) = \vartheta(x) + O(\vartheta(x^{1/2})\ln x)$ and Lemma 15.

We will now complete the proof of the asymptotic law of distribution of basis elements of a free semigroup. It follows from the definition of $\vartheta(x)$ that

$$\vartheta(x) \le \pi(x)\ln x.$$

On the other hand, for any fixed a, $0 < a < 1$,

$$\vartheta(x) \ge a \ln x(\pi(x) - \pi(x^a)).$$

Since $\pi(x) = O(x^\theta)$, we obtain

$$\frac{\vartheta(x)}{x^\theta} \le \frac{\pi(x)}{x^\theta/\ln x} \le \frac{1}{a}\frac{\vartheta(x)}{x^\theta} + O\left(\frac{x^{a\theta}\ln x}{x^\theta}\right),$$

$$\frac{1}{\theta} \le \varliminf_{x\to\infty} \frac{\pi(x)}{x^\theta/\ln x} \le \varlimsup_{x\to\infty} \frac{\pi(x)}{x^\theta/\ln x} \le \frac{1}{a}\frac{1}{\theta}.$$

But since a can be taken arbitrarily close to 1,

$$\lim_{x\to\infty} \frac{\pi(x)}{x^\theta/\ln x} = \frac{1}{\theta}.$$

The theorem is proved.

§2.7. The Hardy-Ramanujan problem

One of the simplest problems on asymptotic formulas in additive number theory is the one concerning an asymptotic formula for the number of "unrestricted partitions" of a natural number.

Suppose n is a natural number. Let $p(n)$ denote the number of solutions in nonnegative integers $x_1, x_2, \ldots$ of the Diophantine equation $n = 1 \cdot x_1 + 2x_2 + 3x_3 + \cdots$ (the natural number n is partitioned in an arbitrary way into natural summands, with 1 occurring x_1 times, 2 occurring x_2 times, and so on).

THEOREM. *As* $n \to \infty$,

$$p(n) = \frac{1}{4\sqrt{3}n} e^{\pi\sqrt{2n/3}} \left(1 + O\left(\frac{1}{n^{1/4-\varepsilon}}\right)\right).$$

PROOF. We will give a simple proof of this theorem due to G. A. Freĭman. It will, however, take up a considerable amount of space, because instead of just presenting it formally we will try to "clarify" why a particular argument is being used.

Consider the product (for $|x| < 1$)

$$F(x) = \prod_{m=1}^{\infty} \frac{1}{1 - x^m},$$

or, using the expansion $1/(1-x^m) = 1 + x^m + x^{2m} + \cdots$ and multiplying out,

$$F(x) = 1 + \sum_{n=1}^{\infty} p(n) x^n.$$

Using Cauchy's formulas for the coefficients of a power series, we obtain

$$p(n) = \frac{1}{2\pi i} \int_C \frac{F(x)}{x^{n+1}}\, dx,$$

where C is a circle of radius less than 1. We make the substitution $x = e^{-u}$, where $u = v + iw$, and put $F(e^{-u}) = f(u)$. For $v > 0$ we have

$$f(u) = \prod_{m=1}^{\infty} \frac{1}{1 - e^{-mu}}$$

and

$$p(n) = \frac{1}{2\pi} \int_{-\pi}^{\pi} f(v + iw) e^{n(v+iw)}\, dw.$$

REMARK. We will now give a rough sketch of the proof, without justifying anything. The symbol $\approx$ will be used to mean "approximately equal." Choose $w_0 > 0$ (where w_0 depends on n and is such that $w_0 \to 0$ as $n \to \infty$):

$$p(n) \approx \frac{1}{2\pi} \int_{-w_0}^{w_0} f(v + iw) e^{n(v+iw)}\, dw.$$

By Taylor's formula with two terms, for $|w| < w_0$ we have

$$\ln f(v + iw) \approx \ln f(v) + iw(\ln f(v))' - \frac{w^2}{2}(\ln f(v))'',$$

$$p(n) \approx \frac{e^{nv} e^{\ln f(v)}}{2\pi} \int_{-w_0}^{w_0} e^{iw((\ln f(v))' + n)} e^{-(w^2/2)(\ln f(v))''}\, dw.$$

Up to this point, v has been a free parameter. We now choose v so that $(\ln f(v))' + n = 0$, i.e., we must choose v to satisfy the equation

$$n = \sum_{m=1}^{\infty} \frac{m}{e^{vm} - 1}$$

(inasmuch as $(\ln f(v))' = -\sum_{m=1}^{\infty} m/(e^{mv} - 1)$). Let us solve this equation approximately. We have

$$\sum_{m=1}^{\infty} \frac{m}{e^{mv} - 1} = \frac{1}{v^2} \sum_{m=1}^{\infty} v \frac{vm}{e^{mv} - 1}.$$

But $\sum_1^\infty v \cdot (vm/(e^{mv} - 1))$ is an integral sum with spacing v for the integral $\int_0^\infty (z/(e^z - 1))\, dz$. Therefore,

$$\sum_{m=1}^{\infty} \frac{m}{e^{vm} - 1} \approx \frac{1}{v^2} \int_0^\infty \frac{z}{e^z - 1}\, dz = \frac{(2\pi)^2}{v^2} \int_0^\infty \frac{x}{e^{2\pi x} - 1}\, dx$$

$$(z = 2\pi x).$$

This integral can be calculated in finite form (see Fikhtengol'ts [49], Chapter XIV, §3, Examples):

$$\int_0^\infty \frac{x}{e^{2\pi x} - 1}\, dx = \frac{1}{(2\pi)^2} \sum_{n=1}^{\infty} \frac{1}{n^2} = \frac{\pi^2}{6} \frac{1}{(2\pi)^2}.$$

Thus,

$$\sum_{m=1}^{\infty} \frac{m}{e^{vm} - 1} \approx \frac{\pi^2}{6} \frac{1}{v^2}.$$

We must take v so that $n \approx (1/v^2) \cdot \pi^2/6$, i.e., $v \approx \pi/\sqrt{6n}$. We choose v to be a root of the equation $(\ln f(v))' + n = 0$. For this choice of v we have

$$\begin{aligned} p(n) &\approx \frac{e^{\ln f(v) + nv}}{2\pi} \int_{-w_0}^{w_0} e^{-(w^2/2)(\ln f(v))''}\, dw \\ &\approx \frac{e^{\ln f(v) + nv}}{2\pi} \int_{-\infty}^{\infty} e^{-(w^2/2)(\ln f(v))''}\, dw = \frac{e^{\ln f(v) + nv}}{\sqrt{2\pi (\ln f(v))''}} \end{aligned}$$

(using the classical integral). As we have seen, $nv \approx \pi\sqrt{n/6}$. Later it will be shown that

$$\ln f(v) \approx \pi \sqrt{\frac{n}{6}} + \frac{1}{4} \ln \frac{1}{24n} \quad \text{and} \quad (\ln f(v))'' \approx \frac{2n\sqrt{6n}}{\pi}.$$

Thus,

$$p(n) \approx \frac{\exp(2\pi\sqrt{\frac{n}{6}} + \frac{1}{4}\ln\frac{1}{24n})}{\sqrt{2\pi \frac{2n\sqrt{6n}}{\pi}}} = \frac{\exp(\pi\sqrt{\frac{2}{3}n})}{2\sqrt[4]{6n24nn^2}} = \frac{\exp(\pi\sqrt{\frac{2}{3}n})}{4\sqrt{3}\, n}.$$

Let us now begin the rigorous argument. We put $w_0 = n^{-3/4+\varepsilon/3}$, $\varepsilon > 0$, and split the integral expression for $p(n)$ into three parts:

$$p(n) = \frac{1}{2\pi}\left(\int_{-\pi}^{-w_0} + \int_{-w_0}^{w_0} + \int_{w_0}^{\pi}\right).$$

LEMMA. *Suppose* $\operatorname{Re} u > 0$ *and* $u \to 0$, *staying within some angle lying in the right half-plane. If* ln *stands for the principal value of the logarithm, then*

$$\ln f(u) = \frac{\pi^2}{6u} + \frac{1}{2}\ln\frac{u}{2\pi} + O(|u|).$$

PROOF. We have

$$\ln f(u) = -\sum_{m=1}^{\infty}\ln(1-e^{-mu}) = \sum_{m=1}^{\infty}\sum_{n=1}^{\infty}\frac{e^{-mnu}}{n} = \sum_{n=1}^{\infty}\frac{1}{n(e^{nu}-1)}.$$

REMARK. The problem of determining the asymptotic behavior of the series $\sum_{n=1}^{\infty} 1/n(e^{nu}-1)$ as $u \to 0$ to within terms of order $O(|u|)$ can be solved in an artificial way. So that the result will not appear to be "unexpected" we will first solve a simpler problem: we assume u is real and determine the asymptotic behavior of the series to within $O(1)$:

$$\sum_{n=1}^{\infty}\frac{1}{n(e^{nu}-1)} = \sum_{n\le 1/u}\frac{1}{n(e^{nu}-1)} + \sum_{n>1/u}\frac{1}{n(e^{nu}-1)}.$$

If $n > 1/u$, then $e^{nu} - 1 \ge e^{nu}/2$ (since $e \ge 2$ and $1/n < u$). Therefore,

$$\sum_{n>1/u}\frac{1}{n(e^{nu}-1)} < u\sum_{n>1/u}\frac{2}{e^{nu}} < 2u\sum_{n=0}^{\infty}\frac{1}{e^{nu}} = \frac{2u}{1-e^{-u}} \le C$$

as $u \to 0$. Thus,

$$\sum_{n=1}^{\infty}\frac{1}{n(e^{nu}-1)} = \sum_{n\le 1/u}\frac{1}{n(e^{nu}-1)} + O(1).$$

In the remaining sum $nu \le 1$, and we use the easily verified relation $1/(e^z-1) = 1/z - \frac{1}{2} + O(z)$ for $|z| \le 1$:

$$\begin{aligned}\sum_{n\le 1/u}\frac{1}{n(e^{nu}-1)} &= \sum_{n\le 1/u}\frac{1}{n}\left(\frac{1}{nu} - \frac{1}{2} + O(nu)\right)\\ &= \frac{1}{u}\sum_{n\le 1/u}\frac{1}{n^2} - \frac{1}{2}\sum_{n\le 1/u}\frac{1}{n} + O\left(\sum_{n\le 1/u} u\right)\\ &= \frac{1}{u}\left(\sum_{n=1}^{\infty}\frac{1}{n^2} - \sum_{n>1/u}\frac{1}{n^2}\right) - \frac{1}{2}\ln\frac{1}{u} + O(1) + O(1)\\ &= \frac{1}{u}\frac{\pi^2}{6} - \frac{1}{2}\ln\frac{1}{u} + O(1),\end{aligned}$$

inasmuch as

$$\frac{1}{u}\sum_{n>1/u}\frac{1}{n^2}=\frac{1}{u}\left(\int_{1/u}^{\infty}\frac{dx}{x^2}+O(u^2)\right)=O(1).$$

Thus,

$$\sum_{n=1}^{\infty}\frac{1}{n(e^{nu}-1)}=\frac{\pi^2}{6}\frac{1}{u}-\frac{1}{2}\ln\frac{1}{u}+O(1).$$

We now return to the proof of the lemma. Suppose, as stated in the hypothesis of the lemma, that u is a complex variable with $\operatorname{Re} u=v>0$ that lies in some neighborhood of zero. To determine the asymptotic behavior of the series $\sum_{n=1}^{\infty}1/n(e^{nu}-1)$ to within terms of order $O(|u|)$ we use the following artificial method. We represent our series in the form

$$\sum_{n=1}^{\infty}\frac{1}{n(e^{nu}-1)}=u\sum_{n=1}^{\infty}\frac{1}{nu(e^{nu}-1)}=u\sum_{n=1}^{\infty}\Phi(nu),$$

where $\Phi(z)=1/z(e^z-1)$. We also represent the expression $\pi^2/6u-\frac{1}{2}\ln(1/u)$ (see the preceding remark) in the form $u\sum_{n=1}^{\infty}\Phi(nu)$ to within terms of order $O(|u|)$. Obviously,

$$\frac{\pi^2}{6}\frac{1}{u}=\frac{1}{u}\sum_{n=1}^{\infty}\frac{1}{n^2}=u\sum_{n=1}^{\infty}\frac{1}{(un)^2}$$

and

$$\frac{1}{2}\ln\frac{1}{u}=-\frac{1}{2}\ln(1-e^{-u})-\frac{1}{2}\ln\frac{u}{1-e^{-u}}.$$

Since $u/(1-e^{-u})=1$ when $u=0$, the function $(1/u)\ln(u/(1-e^{-u}))$ is regular at $u=0$ and is therefore bounded, i.e., $(1/u)\ln(u/(1-e^{-u}))=O(1)$. Consequently,

$$\frac{1}{2}\ln\frac{1}{u}=-\frac{1}{2}\ln(1-e^{-u})+O(|u|),$$

and hence

$$\frac{1}{2}\ln\frac{1}{u}=u\sum_{n=1}^{\infty}\frac{e^{-nu}}{2nu}+O(|u|).$$

Thus,

$$\sum_{n=1}^{\infty}\frac{1}{n(e^{nu}-1)}=\frac{\pi^2}{6}\frac{1}{u}-\frac{1}{2}\ln\frac{1}{u}$$
$$+u\sum_{n=1}^{\infty}\left(\frac{1}{nu(e^{nu}-1)}-\frac{1}{(nu)^2}+\frac{e^{-nu}}{2nu}\right)+O(|u|).$$

Let

$$\varphi(z)=\frac{1}{z(e^z-1)}-\frac{1}{z^2}+\frac{e^{-z}}{2z}.$$

We can view the sum

$$u\sum_{n=1}^{\infty}\left(\frac{1}{nu(e^{nu}-1)}-\frac{1}{(nu)^2}+\frac{1}{2}\frac{e^{-nu}}{nu}\right)$$

as an integral sum with spacing u for the integral

$$\int_L\left(\frac{1}{z(e^z-1)}-\frac{1}{z^2}+\frac{e^{-z}}{2z}\right)dz=\int_L\varphi(z)\,dz,$$

taken along the ray starting at zero and passing through the point with affix u (Figure 4).

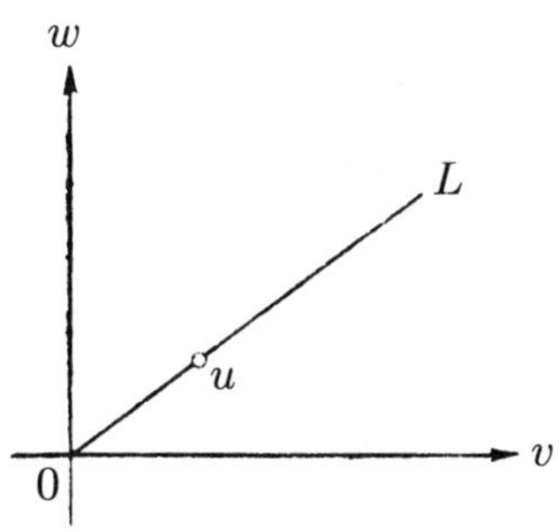

FIGURE 4

This integral is convergent. Its convergence at infinity is obvious. Convergence in a neighborhood of zero follows from the fact that the function $\varphi(z)$ is analytic at $z=0$. Indeed,

$$\begin{aligned}\varphi(z)&=\frac{1}{z(e^z-1)}-\frac{1}{z^2}+\frac{e^{-z}}{2z}\\&=\frac{1}{z^2}-\frac{1}{2z}+\lambda_1(z)-\frac{1}{z^2}+\frac{1}{2z}+\lambda_2(z)\\&=\lambda_1(z)+\lambda_2(z);\end{aligned}$$

$\lambda_1(z)$ and $\lambda_2(z)$ are analytic at $z=0$.

Let us estimate the difference between the integral and the integral sum. We denote the part of the path L from 0 to u by L_1, from u to $2u$ by $L_2,\ldots,$ from $(n-1)u$ to nu by L_n:

$$\begin{aligned}&\int_L\varphi(z)\,dz-u\sum_{n=1}^{\infty}\varphi(nu)\\&\qquad=\int_{L_1}\varphi(z)\,dz-u\varphi(u)+\sum_{n=2}^{\infty}\left(\int_{L_n}\varphi(z)\,dz-\varphi(nu)\right).\end{aligned}$$

Since the length of L_1 is equal to $|u|$ and $\varphi(z)$ is analytic at $z=0$, it follows that

$$\int_{L_1} \varphi(z)\,dz - u\varphi(u) = O(|u|).$$

To estimate the difference $\int_{L_n} \varphi(z)\,dz - \varphi(nu)$ we use the formula obtained in the obvious way by means of integration by parts:

$$\int_{L_n} \varphi(z)\,dz = u\varphi(nu) - \int_{L_n} (z-(n-1)u)\varphi'(z)\,dz.$$

Thus we have

$$\begin{aligned} \left|\int_{L_n} \varphi(z)\,dz - u\varphi(nu)\right| &\le |u| \max_{z\in L_n} |z-(n-1)u|\,|\varphi'(z)| \\ &\le |u^2| \max_{z\in L_n} |\varphi'(z)|. \end{aligned}$$

Consequently,

$$\left|\int_{L} \varphi(z)\,dz - u\sum_{n=1}^{\infty} \varphi(nu)\right| \le O(|u|) + |u|^2 \sum_{n=2}^{\infty} \max_{z\in L_n} |\varphi'(z)|.$$

Since u tends to zero inside some angle, there exists an absolute constant λ such that $\operatorname{Re} u/|u| > \lambda$. We split up the sum in the right-hand side as follows:

$$\sum_{n=2}^{\infty} \max_{z\in L_n} |\varphi'(z)| \le \sum_{2\le n\le 1/\lambda|u|+1} \max_{z\in L_n} |\varphi'(z)| + \sum_{n>1/\lambda|u|+1} \max_{z\in L_n} |\varphi'(z)|.$$

In the first sum, $|z| \le n|u| \le 1/\lambda + |u| \le 1/\lambda + \varepsilon$ (since $u \to 0$). For $|\arg z| \le C_\lambda < \pi$, $|z| \le 1/\lambda + \varepsilon$, the function $\varphi'(z)$ is regular and therefore bounded. Consequently,

$$\sum_{2\le n\le 1/\lambda|u|+1} \max_{z\in L_n} |\varphi'(z)| \le C\frac{1}{|u|},$$

where C is a constant (depending on λ). The second sum can be estimated as follows. Obviously,

$$\begin{aligned} \varphi'(z) &= -\frac{e^z-1+ze^z}{(z(e^z-1))^2} + \frac{2}{z^3} - \frac{e^{-z}}{2z} - \frac{e^{-z}}{2z^2} \\ &= -\frac{1}{z^2(e^z-1)} - \frac{1}{z(e^z-1)} - \frac{1}{z(e^z-1)^2} + \frac{2}{z^3} - \frac{e^{-z}}{2z} - \frac{e^{-z}}{2z^2}. \end{aligned}$$

Therefore,

$$\begin{aligned}
&\sum_{n\geq 1/\lambda|u|+1} \max_{z\in L_n} |\varphi'(z)| \\
&\quad\leq \sum_{n\geq 1/\lambda|u|+1} \max_{z\in L_n} \frac{1}{|z|^2|e^z-1|} + \sum_{n\geq 1/\lambda|u|+1} \max_{z\in L_n} \frac{1}{|z|\,|e^z-1|} \\
&\qquad + \sum_{n\geq 1/\lambda|u|+1} \max_{z\in L_n} \frac{1}{|z|\,|e^z-1|^2} + 2\sum_{n\geq 1/\lambda|u|+1} \max_{z\in L_n} \frac{1}{|z|^3} \\
&\qquad + \frac{1}{2}\sum_{n\geq 1/\lambda|u|+1} \max_{z\in L_n} \frac{1}{|z|\,|e^z|} + \frac{1}{2}\sum_{n\geq 1/\lambda|u|+1} \max_{z\in L_n} \frac{1}{|z|^2|e^z|}.
\end{aligned}$$

Since $1/|z|^2 \leq 1/|z|$ (i.e., $|z| \geq 1$, which is true because $(n-1)|u| \geq 1$), it follows that

$$\begin{aligned}
&\sum_{n\geq 1/\lambda|u|+1} \max_{z\in L_n} |\varphi'(z)| \\
&\quad\ll \sum_{n\geq 1/\lambda|u|+1} \max_{z\in L_n} \frac{1}{|z|\,|e^z-1|} + \sum_{n\geq 1/\lambda|u|+1} \max_{z\in L_n} \frac{1}{|z|\,|e^z-1|^2} \\
&\qquad + \sum_{n\geq 1/\lambda|u|+1} \max_{z\in L_n} \frac{1}{|z|\,|e^z|} + \sum_{n\geq 1/\lambda|u|+1} \max_{z\in L_n} \frac{1}{|z|^3}.
\end{aligned}$$

But

$$\min_{z\in L_n} |z| = (n-1)|u|,$$

$$\min_{z\in L_n} |e^z| = e^{(n-1)\operatorname{Re} u} \geq e^{(n-1)\lambda|u|},$$

$$\min_{z\in L_n} |e^z-1| \geq \min_{z\in L_n} |e^z| - 1 \geq e^{(n-1)\lambda|u|} - 1 \geq \frac{e^{(n-1)\lambda|u|}}{2}$$

(the last inequality holds because $(n-1)\lambda|u| \geq 1$, i.e., $n \geq 1/\lambda|u|+1$). Thus,

$$\begin{aligned}
&\sum_{n\geq 1/\lambda|u|+1} \max_{z\in L_n} |\varphi'(z)| \\
&\quad\ll \frac{1}{|u|}\sum_{n\geq 1/\lambda|u|+1} \frac{e^{-(n-1)\lambda|u|}}{n-1} + \frac{1}{|u|}\sum_{n\geq 1/\lambda|u|+1} \frac{e^{-2(n-1)\lambda|u|}}{n-1} \\
&\qquad + \frac{1}{|u|}\sum_{n\geq 1/\lambda|u|+1} \frac{e^{-(n-1)\lambda|u|}}{n-1} + \frac{1}{|u|^3}\sum_{n\geq 1/\lambda|u|+1} \frac{1}{(n-1)^3} \\
&\quad\ll \frac{1}{|u|}\sum_{n\geq 1/\lambda|u|+1} \frac{e^{-(n-1)\lambda|u|}}{n-1} + \frac{1}{|u|^3}\sum_{n\geq 1/\lambda|u|+1} \frac{1}{(n-1)^3}.
\end{aligned}$$

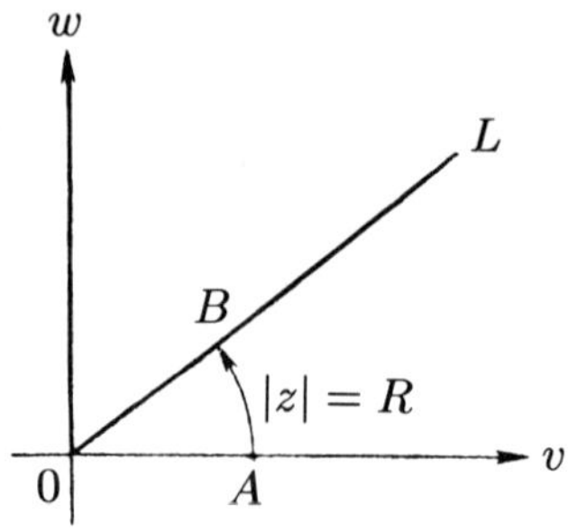

FIGURE 5

But

$$\sum_{n\geq 1/\lambda|u|+1} \frac{e^{-(n-1)\lambda|u|}}{n-1} \ll \sum_{n\geq 1/\lambda|u|+1} e^{-(n-1)\lambda|u|} \leq \frac{1}{1-e^{-\lambda|u|}} = O\left(\frac{1}{|u|}\right).$$

We obtain precisely the same estimate here:

$$\frac{1}{|u|^3} \sum_{n\geq 1/\lambda|u|+1} \frac{1}{(n-1)^3} = \frac{1}{|u|^3} \int_{1/\lambda|u|}^{\infty} \frac{dx}{x^3} + O(1) = O\left(\frac{1}{|u|}\right).$$

Therefore,

$$\sum_{n\geq 1/\lambda|u|+1} \max_{z\in L_n} |\varphi'(z)| \ll \frac{1}{|u|}.$$

It follows that

$$\sum_{n=2}^{\infty} \max_{z\in L_n} |\varphi'(z)| \ll \frac{1}{|u|}$$

and

$$\left|\int_L \varphi(z)\,dz - \sum_{n=1}^{\infty} u\varphi(nu)\right| \ll |u|.$$

Thus, under our assumptions,

$$\sum_{n=1}^{\infty} \frac{1}{n(e^{nu}-1)} = \frac{\pi^2}{6}\frac{1}{u} + \frac{1}{2}\ln u + \int_L \varphi(z)\,dz + O(|u|).$$

Now consider $\int_G \varphi(z)\,dz$, where G is the boundary of sector OAB of radius R (Figure 5).

Since $\varphi(z)$ is regular inside the sector, it follows that $\int_G \varphi(z)\,dz = 0$. But as $R \to \infty$ the integral along the arc of the circle tends to zero. We obtain

$$\int_L \varphi(z)\,dz = \int_0^{\infty} \varphi(x)\,dx.$$

Let

$$I = \int_0^\infty \varphi(x)\,dx = \int_0^\infty \left(\frac{1}{x(e^x-1)} - \frac{1}{x^2} + \frac{e^{-x}}{2x}\right) dx.$$

This integral can be calculated in finite form. As is well known (see Fikhtengol'ts [49], Chapter XIV, §5, Stirling's formula),

$$\ln\Gamma(a) = \int_0^\infty \left((a-1)e^{-x} - \frac{e^{-x}-e^{-ax}}{1-e^{-x}}\right)\frac{dx}{x}, \qquad a > 0.$$

Put $a = 1/2$; we know that $\Gamma(\frac{1}{2}) = \sqrt{\pi}$. We have

$$\begin{aligned}
\frac{\ln\pi}{2} &= -\int_0^\infty \left(\frac{e^{-x}}{2} + \frac{e^{-x}-e^{-x/2}}{1-e^{-x}}\right)\frac{dx}{x} \\
&= -\int_0^\infty \left(\frac{e^{-x}}{2x} + \frac{1-e^{x/2}}{(e^x-1)x}\right) dx, \\
\frac{\ln\pi}{2} &= \int_0^\infty \left(\frac{e^{x/2}}{(e^x-1)x} - \frac{1}{x^2}\right) dx - I.
\end{aligned}$$

Replacing $x/2$ by x in the integral, we obtain

$$\begin{aligned}
I &= \int_0^\infty \left(\frac{e^x}{(e^{2x}-1)x} - \frac{1}{2x^2}\right) dx - \frac{\ln\pi}{2} \\
&= \int_0^\infty \left(\frac{x}{e^x-e^{-x}} - \frac{1}{2}\right)\frac{dx}{x^2} - \frac{\ln\pi}{2}.
\end{aligned}$$

The calculation of this integral can be found in Fikhtengol'ts [49] (Chapter XIII, §1, Examples):

$$I = -\frac{1}{2}\ln 2 - \frac{\ln\pi}{2} = -\frac{\ln 2\pi}{2}.$$

Thus,

$$\sum_{n=1}^\infty \frac{1}{n(e^{nu}-1)} = \frac{\pi^2}{6}\frac{1}{u} + \frac{\ln u}{2} - \frac{\ln 2\pi}{2} + O(|u|).$$

The lemma is proved.

We next take $v = \pi/\sqrt{6n}$.

We will study the integral

$$\frac{1}{2\pi}\int_{-w_0}^{w_0} f(v+iw)e^{n(v+iw)}\,dw.$$

Since $w_0 = n^{-3/4+\varepsilon/3}$ and $v = \pi n^{-1/2}/\sqrt{6}$, it follows that u tends to zero as $n \to \infty$ along a curve tangent to the real axis, and therefore $u = v + iw$ certainly stays within some angle. We can apply the lemma:

$$\begin{aligned}
\ln f(v+iw) &= \frac{\pi^2}{6}\frac{1}{v+iw} + \frac{1}{2}\ln(v+iw) - \frac{\ln 2\pi}{2} + O(|v+iw|) \\
&= \frac{\pi^2}{6}\frac{1}{v+iw} + \frac{1}{2}\ln v + \frac{1}{2}\ln\left(1+i\frac{w}{v}\right) - \frac{\ln 2\pi}{2} + O(|v+iw|).
\end{aligned}$$

For $|w| \le w_0$ and $|w| < v$ we have $|v + iw| = O(v)$. Furthermore,

$$\begin{aligned}\frac{1}{v+iw} &= \frac{1}{v} - \frac{w}{v^2}i - \frac{w^2}{v^3} + \frac{1}{v(1+iw/v)} - \frac{1}{v} + \frac{w}{v^2}i + \frac{w^2}{v^3} \\ &= \frac{1}{v} - \frac{w}{v^2}i - \frac{w^2}{v^3} + \frac{1}{v}\frac{-iw^3/v^3}{1+iw/v} = \frac{1}{v} - \frac{w}{v^2}i - \frac{w^2}{v^3} + O\left(\frac{w_0^3}{v^4}\right).\end{aligned}$$

It is also clear that for $n \ge n_0$

$$\left|\frac{w}{v}\right| \le \frac{\sqrt{6}}{\pi}\frac{1}{n^{1/4-\varepsilon/3}} < \frac{1}{2}.$$

Therefore,

$$\left|\ln\left(1 + i\frac{w}{v}\right)\right| \le \left|\frac{w}{v}\right| + \frac{1}{2}\left|\frac{w}{v}\right|^2 + \frac{1}{3}\left|\frac{w}{v}\right|^3 + \cdots \le \left|\frac{w}{v}\right| \frac{1}{1-|w/v|} = O\left(\frac{w_0}{0}\right).$$

So for $|w| \le |w_0|$

$$\begin{aligned}\ln f(v+iw) &= \frac{\pi^2}{6}\left(\frac{1}{v} - \frac{w}{v^2}i - \frac{w^2}{v^3}\right) + \frac{\ln v}{2} - \frac{\ln 2\pi}{2} + O\left(v + \frac{w_0}{v} + \frac{w_0^3}{v^4}\right) \\ &= \frac{\pi^2}{6}\left(\frac{\sqrt{6n}}{\pi} - \frac{6i}{\pi^2}nw - \frac{(6n)^{3/2}w^2}{\pi^3}\right) + \frac{1}{2\sqrt{6n}} + O(n^{-1/4+\varepsilon}).\end{aligned}$$

Therefore,

$$\ln f(v+iw) = \frac{\pi}{\sqrt{6}}\sqrt{n} - iwn - \frac{n\sqrt{6n}}{\pi}w^2 + \frac{1}{4}\ln\frac{1}{24n} + O(n^{-1/4+\varepsilon}).$$

It follows that

$$\begin{aligned}I_2 &= \int_{-w_0}^{w_0} f(v+iw)e^{n(v+iw)}\,dw \\ &= \frac{1}{2\pi}\exp\left(2\frac{\pi}{\sqrt{6}}\sqrt{n} + \frac{1}{4}\ln\frac{1}{24n} + O(n^{-1/4+\varepsilon})\right) \\ &\quad \times \int_{-w_0}^{w_0} \exp\left(-\frac{n\sqrt{6n}}{\pi}w^2\right)\,dw \\ &= \frac{1}{2\pi}\frac{\exp(\pi\sqrt{\frac{2}{3}n})}{\sqrt[4]{24n}}e^{O(n^{-1/4+\varepsilon})}\int_{-w_0}^{w_0}\exp\left(-\frac{n\sqrt{6n}}{\pi}w^2\right)\,dw \\ &= \frac{\exp(\pi\sqrt{\frac{2}{3}n})}{2\pi\sqrt[4]{24n}}(1 + O(n^{-1/4+\varepsilon}))\int_{-w_0}^{w_0}\exp\left(-\frac{n\sqrt{6n}}{\pi}w^2\right)\,dw.\end{aligned}$$

Making the change of variable

$$w = \sqrt{\frac{\pi}{2}}\frac{1}{\sqrt[4]{6n^3}}z$$

in the integral, we obtain

$$\int_{-w_0}^{w_0} \exp\left(-\frac{n\sqrt{6n}}{\pi}w^2\right) dw = \sqrt{\frac{\pi}{2}} \frac{1}{\sqrt[4]{6n^3}} \int_{-\sqrt{2/\pi}\,\sqrt[4]{6}n^{\varepsilon/3}}^{\sqrt{2/\pi}\,\sqrt[4]{6}n^{\varepsilon/3}} e^{-z^2/2}\,dz$$

$$= \sqrt{\frac{\pi}{2}} \frac{1}{\sqrt[4]{6n^3}} \left(\int_{-\infty}^{\infty} e^{-z^2/2}\,dz + O\left(\int_{\sqrt{2/\pi}\,\sqrt[4]{6}n^{\varepsilon/3}}^{\infty} e^{-z^2/2}\,dz\right)\right).$$

Put $M = \sqrt{2/\pi}\,\sqrt[4]{6}\,n^{\varepsilon/3}$; then $(1/M)e^{-M^2/2}$ is very small as $n \to \infty$. Therefore,

$$I_2 = \frac{1}{2n} \frac{\exp(\pi\sqrt{\frac{2}{3}n})}{\sqrt[4]{144}}(1 + O(n^{-1/4+\varepsilon})) = \frac{\exp(\pi\sqrt{\frac{2}{3}n})}{4\sqrt{3}n}(1 + O(n^{-1/4+\varepsilon})).$$

Thus,

$$p(n) = \frac{\exp(\pi\sqrt{\frac{2}{3}n})}{4\sqrt{3}n}(1 + O(n^{-1/4+\varepsilon})) + \frac{1}{2\pi}\left(\int_{w_0}^{\pi} + \int_{-\pi}^{-w_0}\right).$$

Let us estimate the remaining integrals. It suffices to do one, say from w_0 to π. Let C be a sufficiently large constant and put $w_1 = Cv$:

$$\frac{1}{2\pi}\int_{w_0}^{\pi} = \frac{1}{2\pi}\int_{w_0}^{w_1} + \frac{1}{2\pi}\int_{w_1}^{\pi} = I_1' + I_1''.$$

We will estimate I_1', the integral one the path w_0w_1. On this path, u lies within some angle, so we can apply the lemma:

$$\ln f(u) = \frac{\pi^2}{6}\frac{1}{u} + \frac{1}{2}\ln u + O(1).$$

Therefore,

$$I_1' \ll \int_{w_0}^{w_1} \sqrt{|u|} \exp\left(\frac{\pi^2}{6}\operatorname{Re}\frac{1}{u} + nv\right) dw.$$

But

$$\operatorname{Re}\frac{1}{u} = \frac{v}{v^2 + w^2} \le \frac{v^2}{v^2 + w_0^2}.$$

Moreover, on w_0w_1 we have $|u| \ll v$. Therefore,

$$I_1' \ll \sqrt{v}w_1 \exp\left[\frac{\pi^2}{6}\frac{v}{v^2 + w_0^2} + nv\right]$$

$$\ll v^{3/2}(\exp 2nv)\left(\exp\left[nv - \frac{\pi^2}{6}\frac{v}{v^2 + w_0^2}\right]\right)^{-1}$$

$$\ll n^{-3/4}\left(\exp\pi\sqrt{\tfrac{2}{3}n}\right)\left(\exp\left[nv\left(1 - \frac{1}{1 + w_0^2/v^2}\right)\right]\right)^{-1}.$$

Since

$$1-\frac{1}{1+w_0^2/v^2}=\frac{w_0^2/v^2}{1+w_0^2/v^2}\geq\frac{w_0^2}{v^2},\qquad nv\frac{w_0^2}{v^2}=\frac{\sqrt{6}}{\pi}n^{(2/3)\varepsilon},$$

it follows that

$$I_1'\ll\frac{\exp(\pi\sqrt{\frac{2}{3}n})}{n^{3/4}}\exp\left(-\frac{\sqrt{6}}{\pi}n^{(2/3)\varepsilon}\right)=O\left(\frac{\exp(\pi\sqrt{\frac{2}{3}n})}{\pi^{3/4-\varepsilon}}\right),$$

i.e., the integral I_1' goes into the remainder term. Let us now estimate the integral on the path $w_1\pi$, i.e., the integral I_1''. We have

$$\begin{aligned}|e^{nu}-1|&=\sqrt{(e^{nv}\cos nw-1)^2+e^{2nv}\sin^2 nw}\\&=\sqrt{(e^{nv}-1)^2+4e^{nv}\sin^2(nw/2)}.\end{aligned}$$

Therefore, $|e^{nu}-1|\geq e^{nv}-1$ and (for $n|w|\leq\pi$)

$$|e^{nu}-1|\geq 2\left|\sin\frac{nw}{2}\right|\geq 2\frac{2}{\pi}\frac{n|w|}{2}=\frac{2}{\pi}n|w|.$$

Consequently,

$$\begin{aligned}|\ln f(u)|\leq\sum_{n=1}^{\infty}\frac{1}{n|e^{nu}-1|}&\leq\frac{\pi}{2|w|}+\sum_{n=2}^{\infty}\frac{1}{n(e^{nv}-1)}\\&<\left(\frac{\pi}{2C}+\sum_{n=2}^{\infty}\frac{1}{n^2}\right)\frac{1}{v}=\frac{C_1}{v},\end{aligned}$$

since $|w|\geq Cv$ and $C_1=\pi/2C+\pi^2/6-1$. Therefore,

$$I_1''\ll e^{C_1/v+nv}.$$

For a sufficiently large C,

$$\frac{C_1}{v}+nv<\left(\pi\sqrt{\tfrac{2}{3}}-\delta\right)\sqrt{n},\qquad\delta>0.$$

Therefore I_1'' goes into the remainder.

Finally we obtain

$$p(n)=\frac{1}{4\sqrt{3}n}\exp\left(\pi\sqrt{\tfrac{2}{3}n}\right)(1+O(n^{-1/4+\varepsilon}))$$

as required.

§2.8. Ingham's additive theorem

From Ingham's Tauberian theorem we can prove

THEOREM 1. *Suppose* $0 < \lambda_1 < \lambda_2 < \cdots$ *is a given sequence of real numbers, and*

$$N(u) = Bu^\beta + R(u), \qquad B > 0,\ \beta > 0,$$

where $N(u)$ *is the number of those* λ_ν *not exceeding* u *and where*

$$\int_0^u \frac{R(v)}{v}\,dv = b\ln u + c + o(1)$$

as $u \to \infty$. *For a real number* l, *let* $p(l)$ *be the number of solutions of the equation* $l = r_1\lambda_1 + r_2\lambda_2 + \cdots$ *in integers* $r_\nu \geq 0$. *For a real number* u *and for* $h > 0$, *let* $P(u) = \sum_{l<u} p(l)$, *where the summation extends over the discrete set of numbers* l *for which* $p(l) \neq 0$, *and let*

$$P_h(u) = \frac{P(u) - P(u-h)}{h}.$$

Then as $u \to \infty$

$$P(u) \sim \left(\frac{1-\alpha}{2\pi}\right)^{1/2} e^c M^{-(b+1/2)\alpha} u^{(b+1/2)(1-\alpha)-1/2} e^{(1/\alpha)(Mu)^\alpha},$$

where $\alpha = \beta/(\beta+1)$, $M = (B\beta\Gamma(\beta+1)\varsigma(\beta+1))^{1/\beta}$.

Also,

$$P_h(u) \sim \left(\frac{1-\alpha}{2\pi}\right)^{1/2} e^c M^{-(b-1/2)\alpha} u^{(b-1/2)(1-\alpha)-1/2} e^{(1/\alpha)(Mu)^\alpha},$$

where h *is a positive constant such that* $P_h(u)$ *is a nondecreasing function of* u (*if* h *belongs to the given sequence* λ_ν, *this condition is satisfied*).

Before proving this theorem, we will show how it implies the Hardy-Ramanujan theorem.

Put $h_\nu = \nu$ $(\nu = 1, 2, \ldots)$. If we also put $h = 1$, then $p(u) = P_h(u)$, $N(u) = [u]$, $R(u) = [u] - u$, $B = \beta = 1$, $\alpha = \frac{1}{2}$, $M = \varsigma(2) = \pi^2/6$, and

$$\begin{aligned}
\int_0^u \frac{R(v)}{v}\,dv &= \int_0^u \frac{[v]-v}{v}\,dv = \int_0^u \frac{[v]}{v}\,dv - u \\
&= \sum_{s=1}^{[u]-1} s\ln\frac{s+1}{s} + [u]\ln\frac{u}{[u]} - u = [u]\ln u - \sum_{s=1}^{[u]} \ln s - u \\
&= [u]\ln u - [u]\ln[u] + [u] - \frac{1}{2}\ln[u] - \frac{\ln 2\pi}{2} + o(1) - u \\
&= [u](\ln u - \ln[u]) + [u] - u - \frac{\ln[u]}{2} - \frac{\ln 2\pi}{2} + o(1)
\end{aligned}$$

(we have made use of Stirling's formula). Since

$$\ln u = \ln[u] + \frac{u - [u]}{u} + o\left(\frac{1}{u}\right),$$

it follows that

$$\int_0^u \frac{R(v)}{v}\,dv = -\frac{1}{2}\ln u - \frac{1}{2}\ln 2\pi + o(1).$$

Thus, $b = -\frac{1}{2}$ and $c = -\frac{1}{2}\ln 2\pi$. The theorem yields

$$p(u) \sim \left(\frac{1}{4\pi}\right)^{1/2} e^{-\ln 2\pi/2}\left(\frac{\pi^2}{6}\right)^{1/2} u^{-1}e^{2\sqrt{(\pi^2/6)u}} = \frac{1}{4\sqrt{3}u}e^{\pi\sqrt{(2/3)u}}.$$

PROOF OF INGHAM'S THEOREM. We have

$$\begin{aligned} N(u) = 2\frac{uN(u)}{2u} &\le \int_u^{2u} \frac{N(v)}{v}\,dv \\ &= 2\int_u^{2u} \frac{Bv^\beta}{v}\,dv + 2\int_u^{2u}\frac{R(v)}{v}\,dv = O(u^\beta). \end{aligned}$$

Let $L(u) = \int_0^u (R(v)/v)\,dv$, $S(u) = L(u) - b\ln u - c$, and

$$g(s) = \sum p(l)e^{-ls} = \prod_{\nu=1}^{\infty}(1 - e^{-\lambda_\nu s})^{-1}$$

for $\sigma > 0$. Consider $\ln g(s)$. For $\sigma > 0$ we have

$$\begin{aligned} \ln g(s) &= -\sum_{\nu=1}^{\infty}\ln(1 - e^{-\lambda_\nu s}) = s\int_0^\infty \ln(1 - e^{-us})\,dN(u) \\ &= \int_0^\infty \frac{sN(u)}{e^{us}-1}\,du = \int_0^\infty \frac{s}{e^{us}-1}(Bu^\beta\,du + u\,dL(u)) \\ &= \frac{B}{s^\beta}\int_0^\infty \frac{(us)^\beta d(us)}{e^{us}-1} - \int_0^\infty L(u)\,d\left(\frac{us}{e^{us}-1}\right) \\ &= \frac{B}{s^\beta}\int_0^\infty \frac{z^\beta\,dz}{e^z - 1}\int_0^\infty (b\ln u + c)s\psi'(us)\,du - \int_0^\infty S(u)s\psi'(us)\,du \\ &= \frac{B}{s^\beta}\Gamma(\beta+1)\varsigma(\beta+1) - \int_0^\infty (b\ln u + c)s\psi'(us)\,du \\ &\quad - \int_0^\infty S(u)s\psi'(us)\,du, \end{aligned}$$

where $\psi(z) = z/(e^z - 1)$, $S(u) = o(1)$ as $u \to \infty$, and $S(u) = O(|\ln u|)$ as $u \searrow 0$. Obviously,

$$\begin{aligned} &\int_0^\infty s(b\ln u + c)\psi'(us)\,du \\ &\qquad = \left((b\ln u + c)\frac{us}{e^{us}-1} - b\ln(1 - e^{-us})\right)\Bigg|_{u=0}^{\infty} = b\ln s - c. \end{aligned}$$

Thus,

$$\ln g(s) = \frac{B}{s^\beta}\Gamma(\beta+1)\varsigma(\beta+1) - b\ln s + c - \int_0^\infty S(u)s\psi'(us)\,du.$$

Let us estimate the integral $\int_0^\infty S(u)s\psi'(us)\,du$:

$$\left|\int_0^\infty S(u)s\psi'(us)\,du\right| = O\left(|s|\int_0^{1/\sqrt{\sigma}}(|\ln u|+1)|\psi'(us)|\,du\right) + O\left(|s|\int_{1/\sqrt{\sigma}}^\infty |S(u)|\,|\psi'(us)|\,du\right).$$

If s lies in some arbitrary angle $|t| \le \Delta\sigma$, then:

1) $|s|/\sigma = O(1)$ as $s \to 0$, and

2) $\psi'(us) = O(1)$ as $s \to 0$ if $0 < u < 1/\sqrt{\sigma}$,

since $z/(e^z-1)$ is regular at $z = 0$. Also,

$$\left|\int_0^\infty S(u)s\psi'(us)\,du\right| = O\left(\sigma\int_0^{1/\sqrt{\sigma}}(|\ln u|+1)\,du\right) + O\left(\sigma\int_{1/\sqrt{\sigma}}^\infty |S(u)|\,|\psi'(us)|\,du\right).$$

Since $1/\sqrt{\sigma} \to \infty$ as $\sigma \to 0$, and $S(u) = o(1)$ as $u \to \infty$, it follows that

$$\sigma\int_{1/\sqrt{\sigma}}^\infty |S(u)|\,|\psi'(us)|\,du = o\left(\sigma\int_{1/\sqrt{\sigma}}^\infty |\psi'(us)|\,du\right).$$

Since $\psi'(z)e^{z^2/2}$ remains bounded in any angle $|t| \le \Delta\sigma$, we have

$$\int_0^\infty S(u)s\psi'(us)\,du = O\left(\sigma\left(1+\frac{1}{\sqrt{\sigma}}\ln\frac{1}{\sqrt{\sigma}}+\ln\frac{1}{\sqrt{\sigma}}\right)\right) + o(1) = o(1).$$

Thus in each fixed angle $|t| \le \Delta\sigma$

$$\ln g(s) = \frac{1}{\beta}\left(\frac{M}{s}\right)^\beta - b\ln s + c + o(1),$$

and hence

$$g(s) \sim e^c s^{-b} e^{(1/\beta)(M/s)^\beta}$$

uniformly as $s \to 0$ within each fixed angle. But

$$\int_0^\infty e^{-us}\,dP(u) = g(s),$$

i.e.,

$$\int_0^\infty e^{-us}\,dP(u) \sim e^c s^{-b} e^{(1/\beta)(M/s)^\beta}$$

uniformly as $s \to 0$ within each fixed angle. Also,

$$\int_0^\infty e^{-us}\,dP_h(u) = \frac{1}{h}\left(\int_0^\infty e^{-us}\,dP(u) - \int_0^\infty e^{-(u+h)s}\,dP(u)\right)$$
$$= \frac{1-e^{-hs}}{h}\int_0^\infty e^{-us}\,dP(u).$$

Obviously, $(1-e^{-hs})/h \sim s$ as $s \to 0$ along any path. Therefore,

$$\int_0^\infty e^{-us}\,dP_h(u) \sim e^c s^{-b+1} e^{(1/\beta)(M/s)^\beta}$$

uniformly as $s \to 0$ within each fixed angle. Applying a special case of Ingham's Tauberian theorem, we obtain the assertion of Theorem 1 with $m = (b \pm \frac{1}{2})/\beta$ and $C = M^{1/2-m\beta}e^c$. Moreover, $(1-e^{-hs})g(s)$ has positive coefficients if h belongs to the sequence λ_ν, i.e., $P_h(u)$ is nondecreasing.

THEOREM 2. *Suppose* $0 < \lambda_1 < \lambda_2 < \cdots$ *is a given sequence of real numbers, and* $N(u) = Bu^\beta + R(u)$, *where* $N(u)$ *is the number of those* λ_ν *not exceeding* u *and where*

$$\int_0^u R(v)\,dv = bu + o(u).$$

For a real number l, *let* $p^*(l)$ *be the number of solutions of the equation* $l = r_1\lambda_1 + r_2\lambda_2 + \cdots$ *in integers* r_ν *equal to either* 0 *or* 1. *For a real number* u *and for* $h > 0$, *let* $P^*(u) = \sum_{l<u} p^*(l)$, *where the summation extends over the discrete set of numbers* l *for which* $p^*(l) \neq 0$, *and let*

$$P_h^*(u) = \frac{P^*(u) - P^*(u-h)}{h}.$$

Then as $u \to \infty$

$$P^*(u) \sim \left(\frac{1-\alpha}{2\pi}\right)^{1/2} 2^b (M^*u)^{-\alpha/2} e^{(1/u)(M^*u)^\alpha},$$

where

$$\alpha = \frac{\beta}{\beta+1}, \qquad M^* = \frac{(B\beta\Gamma(\beta+1)\varsigma(\beta+1))^{1/\beta}}{(1-2^{-\beta})^{1/\beta}}.$$

Also,

$$P_h^*(u) \sim \left(\frac{1-\alpha}{2\pi}\right)^{1/2} 2^b M^{*\alpha/2} u^{\alpha/2-1} e^{(1/\alpha)(M^*u)^\alpha},$$

where h *is a positive constant such that* $P_h^*(u)$ *is a nondecreasing function of* u.

REMARK. If the numbers $h, 2h, 2^2h, \ldots$ all belong to the given sequence λ_ν, then $P_h^*(u)$ is a nondecreasing function of u. Obviously,

$$P_h^*(u) = \frac{1}{h}\sum_{u-h\le l<u} p^*(l).$$

The remark will be proved once we show that $p^*(l) \le p^*(l+h)$ for each l such that $p^*(l) \ne 0$. Clearly $l + h = l + 2^s h - 2^{s-1}h - \cdots - h$. Consider the representation $l = \lambda_{i_1} + \cdots + \lambda_{i_m}$. If there is no h among $\lambda_{i_1}, \dots, \lambda_{i_m}$, then

$$l + h = \lambda_{i_1} + \lambda_{i_2} + \cdots + \lambda_{i_m} + h$$

is an admissible representation. Assume that among $\lambda_{i_1}, \dots, \lambda_{i_m}$ there are $h, 2h, \dots, 2^{s-1}h$, but not $2^s h$; then

$$l + h = \lambda_{i_1} + \lambda_{i_2} + \cdots + \lambda_{i_m} + 2^s h - 2^{s-1}h - \cdots - h$$

is an admissible representation. Thus to each representation of l there corresponds a representation of $l + h$.

PROOF OF THE THEOREM. Obviously,

$$\sum_l p^*(l) e^{-ls} = \prod_{\nu=1}^{\infty} (1 + e^{-\lambda_\nu s}) = g^*(s)$$

and

$$g^*(s) \sim 2^b e^{(\beta-1)(M^*/\beta)^s}$$

(where $M^{*\beta} = (1 - 2^{-\beta})M^\beta$) as $s \to 0$ in any angle. The proof is analogous to that of Theorem 1. We have

$$\ln g^*(s) = -\int_0^\infty \ln(1 + e^{-us})\, dN(u) = \int_0^\infty \frac{s}{1 + e^{us}} N(u)\, du.$$

The functions $\ln(1 + e^{-s})$ and $1/(1 + e^s)$ are regular at $s = 0$, and $N(u) = Bu^\beta + R(u)$. We can therefore use $R(u)\, du$ directly without replacing it by $u\, dL(u)$. The proof of Theorem 2 can easily be completed by means of these remarks.

CHAPTER 3

Theory of Functions of a Natural Argument

§3.1. Metric theory of functions of a natural argument

A number of mathematicians of the last century (in particular, N. V. Bugaev) observed the analogy between various number theory operations and certain operations of mathematical analysis. In fact, there are several related analogies here. One of these, which shows up distinctly in problems on congruences with prime power mudulus, led to the formation of the theory of p-adic numbers (see Borevich and Shafarevich [13], Chapter 1).

In this chapter we will mainly have in mind another analogy, whose "sphere of operation" consists of congruences modulo any prime.

The analogue of the concept of interval is the set of natural numbers belonging to an arithmetic progression $Dx + l$, where D is a natural number, $x = 0, 1, 2, \ldots$, and $0 \leq l \leq D - 1$.

We will be concerned with the metric theory of sets of natural numbers.

We first consider the logarithmic density. Suppose $\mathfrak{M}$ is some set of natural numbers. We define the set function $\operatorname{meas}_* \mathfrak{M}$ by the formula

$$\operatorname{meas}_* \mathfrak{M} = \lim_{N\to\infty} \frac{1}{\ln N} \sum_{n\leq N,\, n\in\mathfrak{M}} \frac{1}{n}.$$

If this limit does not exist, then we call the set *nonmeasurable* with respect to logarithmic density. If $\mathfrak{M}$ is a finite set, $\operatorname{meas}_* \mathfrak{M} = 0$. If $\mathfrak{M}$ is the arithmetic progression $Dx+l$, $D > 0$, then $\operatorname{meas}_* \mathfrak{M} = 1/D$. If $\mathfrak{M}$ is the natural sequence, $\operatorname{meas}_* \mathfrak{M} = 1$. The integral with respect to this set function is defined as follows:

$$\int f(n)d\operatorname{meas}_* = \lim_{N\to\infty} \frac{1}{\ln N} \sum_{n\leq N,\, n\in\mathfrak{M}} \frac{f(n)}{n}.$$

It is obvious that the function $\operatorname{meas}_* \mathfrak{M}$ is finitely additive. However, it is not countably additive. We give an example (communicated by Yu. I. Manin) of a partition of the natural sequence into nonoverlapping arithmetic progressions $l_i + D_i x$ such that $\sum 1/D_i < 1$. As the progression $\mathfrak{B}_1$ we take the set of natural numbers divisible by 3 $(3, 6, 9, \ldots)$. The progression $\mathfrak{B}_2$

has difference $3^2 = 9$, and its first term is the smallest natural number not occurring in $\mathfrak{B}_1$, i.e., $\mathfrak{B}_2$ is the progression $1, 10, 19, \ldots$. The progression $\mathfrak{B}_3$ has difference $3^3 = 27$, and its first term is the smallest natural number not occurring in $\mathfrak{B}_1 \cup \mathfrak{B}_2$: $2, 29, 56, \ldots$. We continue the construction of the progressions $\mathfrak{B}_j$ in this way. It is clear that every natural number occurs in some $\mathfrak{B}_j$, i.e., the union of the progressions $\mathfrak{B}_1, \mathfrak{B}_2, \ldots$ is the entire natural sequence. We will prove that when $i \neq j$ the progressions $\mathfrak{B}_i$ and $\mathfrak{B}_j$ do not overlap. Suppose $j > i$ and $l_j + 3^j D_j = l_i + 3^i D_i$. We see that

$$l_j = l_i + 3^i(D_i - 3^{j-i} D_j),$$

i.e., $l_j \in \mathfrak{B}_i$, which contradicts our rule for choosing the initial numbers of the progressions. Thus the natural sequence has been split into nonoverlapping arithmetic progressions. The sum of the reciprocals of the differences of these progressions is

$$\frac{1}{3} + \frac{1}{9} + \frac{1}{27} + \cdots = \frac{1}{2} < 1.$$

Thus the set function $\operatorname{meas}_* \mathfrak{M}$ is not a measure.

Finitely additive functions of sets of natural numbers whose value for the set of all natural numbers is 1 will be called *pseudomeasures*. The logarithmic density is a pseudomeasure.

Another pseudomeasure is the asymptotic density $D(\mathfrak{M})$ of a set $\mathfrak{M}$ of natural numbers:

$$D(\mathfrak{M}) = \lim_{N \to \infty} \frac{1}{N} \sum_{n \leq N,\, n \in \mathfrak{M}} 1$$

(if this limit exists). The density of a finite set is zero. The density of a progression $Dx + l$, $x = 0, 1, \ldots$, is equal to $1/D$. The integral of a function $f(n)$ of a natural argument with respect to this set function is naturally defined to be

$$\lim_{N \to \infty} \frac{1}{N} \sum_{n \leq N} f(n).$$

This limit, if it exists, will be denoted by $M(f)$ and will be called the *mean values* of the function $f(n)$.

The set function $D(\mathfrak{M})$ is not countably additive. This can be shown by the same example used in connection with the pseudomeasure $\operatorname{meas}_* \mathfrak{M}$.

It is easy to prove by means of Abel's transformation that any set that is measurable with respect to asymptotic density is measurable with respect to logarithmic density (and that the numerical values of these pseudomeasures are equal).

Let us give an example of a set that is measurable with respect to logarithmic density, but nonmeasurable with respect to asymptotic density. Consider the sequence

$$2, 3,' 8, 9, 10, 11, 12, 13, 14, 15,' 32, \dots, 63,' \\ \dots,' 2^{2k-1}, \dots, 2^{2k} - 1,' 2^{2k+1}, \dots$$

(the primes following certain numbers are meant to partition the sequence). We denote this set by $\mathfrak{M}$. Suppose P is an increasing parameter. We first consider the case $2^{2s+1} \leq P \leq 2^{2s+2} - 1$. We have

$$\begin{aligned}
\frac{1}{\ln P} \sum_{n \leq P,\, n \in \mathfrak{M}} \frac{1}{n}
&= \frac{1}{\ln P} \left(\sum_{j=0}^{s-1} \sum_{\tau=0}^{2^{2j+1}-1} \frac{1}{2^{2j+1} + \tau} + \sum_{\tau=0}^{P-2^{2s+1}} \frac{1}{2^{2s+1} + \tau} \right) \\
&= \frac{1}{\ln P} \sum_{j=0}^{s-1} \ln \frac{2^{2j+2} - 1}{2^{2j+1}} + O\left(\frac{1}{\ln P}\right) \\
&= \frac{s \ln 2}{\ln P} + O\left(\frac{1}{\ln P}\right) = \frac{1}{2} + O\left(\frac{1}{\ln P}\right).
\end{aligned}$$

If $2^{2s} \leq P \leq 2^{2s+1} - 1$, then

$$\frac{1}{\ln P} \sum_{n \leq P,\, n \in \mathfrak{M}} \frac{1}{n} = \frac{1}{\ln P} \sum_{j=0}^{s-1} \sum_{\tau=0}^{2^{2j+1}} \frac{1}{2^{2j+1} + \tau} = \frac{s \ln 2}{\ln P} = \frac{1}{2} + O\left(\frac{1}{\ln P}\right).$$

We will show that $\mathfrak{M}$ is nonmeasurable with respect to asymptotic density. We have

$$\lim_{k \to \infty} \frac{1}{2^{2k} - 1} \sum_{\substack{n \leq 2^{2k}-1 \\ n \in \mathfrak{M}}} 1 = \lim_{k \to \infty} \frac{2 + 2^3 + \cdots + 2^{2k-1}}{2^{2k-1}} = \frac{2}{3},$$

whereas

$$\lim_{k \to \infty} \frac{1}{2^{2k+1}} \sum_{\substack{n \leq 2^{2k+1} \\ n \in \mathfrak{M}}} 1 = \lim_{k \to \infty} \frac{2(4^k - 1)/3 + 1}{2^{2k+1}} = \frac{1}{3}.$$

We now give an example of a countably additive function defined on the subsets of the natural sequence. We choose $\sigma_0 > 1$. Suppose $\mathfrak{M}$ is some set of natural numbers. We define the set function $\operatorname{meas}_{\sigma_0} \mathfrak{M}$ as follows:

$$\operatorname{meas}_{\sigma_0} \mathfrak{M} = \frac{1}{\varsigma(\sigma_0)} \sum_{n \in \mathfrak{M}} \frac{1}{n^{\sigma_0}},$$

where $\varsigma(s)$ is the Riemann zeta-function. Since the series $\sum_{n \in \mathfrak{M}} 1/n^{\sigma_0}$ are absolutely convergent, this set function is completely (countably) additive.

Thus we have a measure. The measure of the entire set of natural numbers is 1. The measure of a set consisting of a single number is not zero. Any set of natural numbers is measurable.

The integral of a function $f(n)$ of a natural argument is defined by

$$\int f(n) d\,\mathrm{meas}_{\sigma_0} = \frac{1}{\varsigma(\sigma_0)} \sum_{n=1}^{\infty} \frac{f(n)}{n^{\sigma_0}}.$$

§3.2. Wintner's theorem

Any function $f(n)$ of a natural argument can be represented in the form

$$f(n) = \sum_{d|n} \Phi(d);$$

the function $\Phi(d)$ is defined by the Tchebycheff-Möbius inversion formula:

$$\Phi(n) = \sum_{d|n} \mu\left(\frac{n}{d}\right) f(d).$$

We will now state and prove the following theorem of Wintner on summation formulas for arithmetical functions (Wintner [153], p. 180).

THEOREM 1. *Suppose that in the representation* $f(n) = \sum_{d|n} \Phi(d)$ *of the function* $f(n)$ *of a natural argument the function* $\Phi(d)$ *is such that the series* $\sum_1^\infty \Phi(d)/d$ *is absolutely convergent. Then*

$$\lim_{N\to\infty} \frac{1}{N} \sum_{n=1}^{N} f(n) = \sum_{d=1}^{\infty} \frac{\Phi(d)}{d}.$$

PROOF. We have

$$\begin{aligned}\frac{1}{N}\sum_{n\le N} f(n) &= \frac{1}{N}\sum_{n\le N}\sum_{d|n}\Phi(d) = \frac{1}{N}\sum_{d\le N}\Phi(d)\left[\frac{N}{d}\right]\\ &= \sum_{d\le N}\frac{\Phi(d)}{d} + O\left(\frac{1}{N}\sum_{d\le N}|\Phi(d)|\right)\\ &= \sum_{d=1}^{\infty}\frac{\Phi(d)}{d} + O\left(\sum_{d\ge N+1}\frac{|\Phi(d)|}{d}\right) + O\left(\frac{1}{N}\sum_{d\le N}|\Phi(d)|\right),\end{aligned}$$

since $\sum_1^\infty |\Phi(d)|/d$ converges by hypothesis. Note that the convergence of this series implies

$$\lim_{N\to\infty} \sum_{d\le N} |\Phi(d)| = 0.$$

Indeed, let

$$B_k = \sum_{d=1}^{k} \frac{|\Phi(d)|}{d}.$$

Applying Abel's transformation, we see that

$$\frac{1}{N}\sum_{d=1}^{N} |\Phi(d)| = \frac{1}{N}\sum_{d=1}^{N} d\frac{|\Phi(d)|}{d} = \frac{1}{N}\left(NB_N - \sum_{k=1}^{N-1} B_k\right)$$

$$= B_N - \frac{1}{N}\sum_{k=1}^{N-1} B_k.$$

By hypothesis, there is a constant κ such that

$$\lim_{N\to\infty} B_N = \kappa.$$

It is well known that if $\lim_{N\to\infty} B_N = \kappa$, then

$$\lim_{N\to\infty} \frac{1}{N}\sum_{k=1}^{N-1} B_k = \kappa.$$

Consequently,

$$B_N - \frac{1}{N}\sum_{k=1}^{N-1} B_k \to 0.$$

Thus,

$$\frac{1}{N}\sum_{n\le N} f(n) = \sum_{d=1}^{\infty} \frac{\Phi(d)}{d} + o(1) + o(1) = \sum_{d=1}^{\infty} \frac{\Phi(d)}{d} + o(1)$$

(we have used the fact that $\sum_{d\ge N+1} \Phi(d)/d$ is the remainder of a convergent series). The theorem is proved.

REMARK. We will show that Wintner's theorem can be obtained as a corollary of Ikehara's Tauberian theorem. Indeed, if $f(n) = \sum_{d|n} \Phi(d)$, then

$$\sum_{n=1}^{\infty} \frac{f(n)}{n^s} = \zeta(s)\sum_{n=1}^{\infty} \frac{\Phi(n)}{n^s}.$$

The condition that $\sum_1^\infty \Phi(n)/n$ is absolutely convergent guarantees that $F(s) = \sum_1^\infty \Phi(n)/n^s$ converges uniformly to a limit function on any finite interval as $\operatorname{Re} s \to 1$. By Ikehara's theorem,

$$\lim_{N\to\infty} \frac{1}{N}\sum_{n\le N} f(n) = \sum_{n=1}^{\infty} \frac{\Phi(n)}{n}.$$

Wintner's theorem can be applied to functions other than multiplicative ones. Consider, for example, the function $t(n) = \sum_{d|n} 2^{-d}$. By Wintner's theorem,

$$\lim_{N\to\infty} \frac{1}{N} \sum_{n\le N} t(n) = \sum_{d=1}^{\infty} \frac{1}{d2^d} = \ln 2.$$

Functions satisfying the conditions of Wintner's theorem will be said to belong to the *Wintner class* (the class W).

The next theorem will be called the *generalized Wintner theorem.*

THEOREM 2. *Suppose $g(n)$ and $h(n)$ are two (real or complex) functions of a natural argument. Define the function*

$$f(n) = \sum_{d|n} h(d) g\left(\frac{n}{d}\right).$$

Then the following assertions are true:

1) *If g and h are multiplicative, so is f.*

2) *If the limit $M(g) = \lim_{N\to\infty}(1/N)\sum_{n\le N} g(n)$ exists and $\sum_1^\infty |h(n)|/n$ converges, then the limit*

$$M(f) = \lim_{N\to\infty} \frac{1}{N} \sum_{n\le N} f(n)$$

exists and

$$M(f) = M(g) \sum_{n=1}^{\infty} \frac{h(n)}{n}.$$

We prove the first assertion of the theorem. If $(n, m) = 1$, then

$$\begin{aligned} f(nm) &= \sum_{d|nm} h(d) g\left(\frac{nm}{d}\right) = \sum_{d_1|n} \sum_{d_2|m} h(d_1 d_2) g\left(\frac{n}{d_1}\frac{m}{d_2}\right) \\ &= \sum_{d_1|n} \sum_{d_2|m} h(d_1) g\left(\frac{n}{d_1}\right) h(d_2) g\left(\frac{m}{d_2}\right) = f(n) f(m). \end{aligned}$$

Let us prove the second. We have

$$\sum_{n\le x} f(n) = \sum_{n\le x} \sum_{d|n} h(d) g\left(\frac{n}{d}\right) = \sum_{d\le x} h(d) \sum_{m\le x/d} g(m).$$

Suppose $\sum_{n\le x} g(n) = M(g)x + x\eta(x)$, where $\eta(x) \to 0$ as $x \to \infty$. Thus $\eta(x)$ is bounded: $|\eta(x)| \le H$. We have

$$\sum_{n\le x} f(n) = \sum_{d\le x} h(d) \left[M(g)\frac{x}{d} + \frac{x}{d}\eta\left(\frac{x}{d}\right)\right].$$

Therefore,

$$\frac{1}{x}\sum_{n\le x} f(n) = M(g)\sum_{n\le x}\frac{h(n)}{n} + \sum_{n\le x}\frac{h(n)}{n}\eta\left(\frac{x}{n}\right).$$

It suffices to show that

$$\sum_{n\le x}\frac{h(n)}{n}\eta\left(\frac{x}{n}\right) \to 0 \quad \text{as } x\to\infty.$$

Given $\varepsilon > 0$, there exists $x_0 = x_0(\varepsilon) > 1$ such that $|\eta(x)| \le \varepsilon$ for $x \ge x_0$. So for $x \ge x_0$

$$\begin{aligned}\left|\sum_{n\le x}\frac{h(n)}{n}\eta\left(\frac{x}{n}\right)\right| &\le \sum_{n\le x/x_0}\left|\frac{h(n)}{n}\eta\left(\frac{x}{n}\right)\right| + \left|\sum_{x/x_0<n\le x}\frac{h(n)}{n}\eta\left(\frac{x}{n}\right)\right| \\ &\le \varepsilon\sum_{n\le x/x_0}\frac{|h(n)|}{n} + H\sum_{x/x_0<n\le x}\frac{|h(n)|}{n}.\end{aligned}$$

By the Cauchy criterion,

$$\varlimsup_{x\to\infty}\left|\sum_{n\le x}\frac{h(n)}{n}\eta\left(\frac{x}{n}\right)\right| \le \varepsilon\sum_{n=1}^{\infty}\frac{|h(n)|}{n}.$$

The theorem is proved.

THEOREM 1$'$. *Suppose that in the representation $f(n) = \sum_{d|n}\Phi(d)$ of the function $f(n)$ of a natural argument the function $\Phi(d)$ is such that $\sum_1^\infty \Phi(d)/d$ is absolutely convergent. Let a and q be natural numbers such that $(a,q) = 1$. Then*

$$\lim_{N\to\infty}\frac{1}{N}\sum_{m=1}^{N} f(m)e^{2\pi i(a/q)m} = \frac{1}{q}\sum_{t=1}^{\infty}\frac{\Phi(qt)}{t}.$$

PROOF. We have

$$\begin{aligned}\sum_{m=1}^{N} f(m)e^{2\pi i(a/q)m} &= \sum_{m\le N}\left(\sum_{d|m}\Phi(d)\right)e^{2\pi i(a/q)m} \\ &= \sum_{d=1}^{N}\Phi(d)\sum_{l<N/d} e^{2\pi i(ad/q)l} \\ &= \sum_{\substack{d\equiv 0(\operatorname{mod} q)\\ d\le N}}\Phi(d)\left[\frac{N}{d}\right] + O\left(\sum_{d\le N}\frac{|\Phi(d)|}{(ad/q)}\right) \\ &= \sum_{\substack{d\equiv 0(\operatorname{mod} q)\\ d\le N}}\Phi(d)\left[\frac{N}{d}\right] + \sum_{r=1}^{q-1}\frac{1}{(ar/q)}O\left(\sum_{d\le N}|\Phi(d)|\right)\end{aligned}$$

(the symbol (α) denotes the distance from α to the nearest integer). The rest of the proof is carried out as in Theorem 1.

THEOREM 1″. *Suppose the function* $f(n)$ *of a natural argument satisfies the same condition as in Theorem* 1′. *Let* q *be a natural number. Then for any natural number* l, $1 \le l \le q$,

$$\lim_{N\to\infty} \frac{1}{N} \sum_{\substack{m=1\\ m\equiv l \pmod q}}^{N} f(m) = \frac{1}{q}\sum_{d|q} \frac{\varphi(d)}{\varphi(d/(l,d))}\mu\left(\frac{d}{(l,d)}\right)\frac{1}{d}\sum_{t=1}^{\infty}\frac{\Phi(dt)}{t}.$$

To prove this theorem we require the following expression for the Ramanujan sum $S(q,m)$.

LEMMA.

$$S(q,m) = \sum_{\substack{1\le a\le q-1\\ (a,q)=1}} e^{2\pi i(m/q)a} = \frac{\varphi(q)}{\varphi(q/(m,q))}\mu\left(\frac{q}{(m,q)}\right).$$

We prove the lemma. Let

$$f_m^*(x) = \begin{cases} x, & x|m,\\ 0, & x\nmid m.\end{cases}$$

We have

$$\sum_{q'|q} S(q',m) = \sum_{q'|q}\sum_{\substack{1\le t\le q'-1\\ (t,q')=1}} e^{2\pi i(mt/q')} = \sum_{a=1}^{q} e^{2\pi i(am/q)} = \begin{cases} q, & \text{if } q|m,\\ 0, & \text{if } q\nmid m.\end{cases}$$

Thus,

$$\sum_{q'|q} S(q',m) = f_m^*(q).$$

Therefore, by the inversion formula,

$$S(q,m) = \sum_{q'|q}\mu\left(\frac{q}{q'}\right) f_m^*(q') = \sum_{q'|(q,m)}\mu\left(\frac{q}{q'}\right)q'.$$

In particular, $S(q,1)=\mu(q)$. Put $(m,q)=d$. It is clear that

$$\sum_{\substack{1\le a\le q-1\\ (a,q)=1}} e^{2\pi i(m/q)a} = \sum_{\substack{1\le a\le q-1\\ (a,q)=1}} e^{2\pi i(m'/q')a},$$

where $m'=m/d$ and $q'=q/d$; hence $(m',q')=1$. The congruence $m'x\equiv l \pmod{q'}$, where $(q',m')=1$, has one solution: $x=a_1$, $(a_1,q')=1$. This solution yields the following solutions modulo $q=q'd$:

$$a_1, a_1+q', \dots, a_1+(d-1)q'. \tag{1}$$

We will show this set contains exactly $\varphi(q)/\varphi(q')$ numbers that are relatively prime to d. Suppose $p_1,\dots,p_s$ are the primes occurring in q, and $p_1,\dots,p_t$

are those among them that occur in q'. The numbers in the set (1) range over $d/p_1 \cdots p_t$ complete residue systems modulo $p_1 \cdots p_t$; the number of them that are relatively prime to $p_1 \cdots p_t$ is

$$\varphi(p_1 p_2 \cdots p_t)\frac{d}{p_1 \cdots p_t} = \frac{\varphi(q)}{\varphi(q')}.$$

Thus,

$$\sum_{\substack{1 \le a \le q-1 \\ (a,q)=1}} e^{2\pi i (m'/q')a} = \frac{\varphi(q)}{\varphi(q')} \sum_{(a,q')=1} e^{2\pi i (a/q')}$$

$$= \frac{\varphi(q)}{\varphi(q')} S(q', 1) = \frac{\varphi(q)}{\varphi(q')} \mu(q').$$

Consequently,

$$S(a,q) = \frac{\varphi(q)}{\varphi(q')}\mu(q'),$$

and the lemma is proved.

We now turn to the proof of Theorem 1″. We have

$$\frac{1}{N} \sum_{\substack{m \equiv l (\operatorname{mod} q) \\ m \le N}} f(m) = \frac{1}{N} \sum_{m=1}^{N} f(m) \frac{1}{q} \sum_{s=0}^{q-1} e^{2\pi i ((m-l)/q)s}$$

$$= \frac{1}{q} \sum_{s=0}^{q-1} e^{-2\pi i (s/q) l} \frac{1}{N} \sum_{m=1}^{N} f(m) e^{2\pi i (s/q) m}.$$

Put $q/(s,q) = d_s$ and $s/(s,q) = t_s$. Then

$$\frac{1}{N} \sum_{\substack{m \equiv l (\operatorname{mod} q) \\ m \le N}} f(m) = \frac{1}{q} \sum_{s=0}^{q-1} e^{-2\pi i (t_s/d_s) l} \frac{1}{N} \sum_{m=1}^{N} f(m) e^{2\pi i (t_s/d_s) m}.$$

Therefore,

$$\lim_{N \to \infty} \frac{1}{N} \sum_{\substack{m \equiv l (\operatorname{mod} q) \\ m \le N}} f(m) = \frac{1}{q} \sum_{s=0}^{q-1} e^{-2\pi i (t_s/d_s) l} \frac{1}{d_s} \sum_{t=1}^{\infty} \frac{\Phi(d_s t)}{t},$$

i.e.,

$$\lim_{N \to \infty} \frac{1}{N} \sum_{\substack{m \equiv l (\operatorname{mod} q) \\ m \le N}} f(m) = \frac{1}{q} \sum_{d \mid q} \left(\sum_{(d,t)=1} e^{-2\pi i (t/d) l} \right) \frac{1}{d} \sum_{t=1}^{\infty} \frac{\Phi(dt)}{t}$$

$$= \frac{1}{q} \sum_{d \mid q} \frac{\varphi(d)}{\varphi(d/(l,d))} \mu\left(\frac{d}{(l,d)}\right) \frac{1}{d} \sum_{t=1}^{\infty} \frac{\Phi(dt)}{t}.$$

Theorem 1″ is proved.

§3.3. Almost periodic functions of a natural argument

In the Introduction we touched upon the analogies between number theory and analysis. We now consider these in more detail.

The analogue of the uniformly continuous functions are the functions $f(n)$ of a natural argument that possess the following property: for any $\varepsilon > 0$ there exists a natural number $N(\varepsilon)$ such that $|f(a) - f(b)| < \varepsilon$ for $N \geq N(\varepsilon)$ and $a \equiv b \pmod{N!}$. We will call such functions *polyadically continuous.*

An example of a polyadically continuous function is

$$f(n) = \frac{\varphi(n)\sigma(n)}{n^2}.$$

Indeed, if the canonical decomposition of n has the form $p_1^{\alpha_1} \cdots p_s^{\alpha_s}$, then

$$\begin{aligned} f(n) = \frac{\varphi(n)\sigma(n)}{n^2} &= \prod_{i=1}^{s} \left(1 - \frac{1}{p_i}\right)\left(1 + \frac{1}{p_i} + \cdots + \frac{1}{p_i^{\alpha_i}}\right) \\ &= \prod_{i=1}^{s} \left(1 - \frac{1}{p_i^{\alpha_i+1}}\right). \end{aligned}$$

Given $\varepsilon > 0$, we put $k = [3/\varepsilon] + 1$. There exists a natural number $N_0(\varepsilon)$ such that if $N \geq N_0(\varepsilon)$ and $a \equiv b \pmod{N!}$ then the primes less than k in the canonical decompositions of a and b are the same. Let

$$a = p_1^{\alpha_1} \cdots p_s^{\alpha_s}, \qquad b = q_1^{\beta_1} \cdots q_r^{\beta_r};$$

then

$$\begin{aligned} &\left| \frac{\varphi(a)\sigma(a)}{a^2} - \frac{\varphi(b)\sigma(b)}{b^2} \right| \\ &\qquad = \left| \prod_{p_i | a} \left(1 - \frac{1}{p_i^{\alpha_i+1}}\right) - \prod_{q_i | b} \left(1 - \frac{1}{q_i^{\beta_i+1}}\right) \right| \leq \sum_{n=k}^{\infty} \frac{2}{n^2} \leq \frac{3}{k} \leq \varepsilon, \end{aligned}$$

as required.

We have spoken of the analogue of the concept of a measure on the set of natural numbers; in particular, we introduced a pseudomeasure—the asymptotic density

$$\operatorname{meas} \mathfrak{M} = \lim_{N \to \infty} \frac{1}{N} \sum_{n \leq N,\, n \in \mathfrak{M}} 1.$$

Consider any function $f(n)$ of a natural argument. The mean value of this function over the numbers $1, \ldots, N$, i.e., the expression $N^{-1} \sum_1^n f(n)$, can be viewed as an analogue of an integral sum for $f(n)$ in which the "interval of integration" is partitioned into intervals $1+Nx,\ 2+Nx,\dots,N+Nx$ of the same length. A theorem on the existence of the limit

$$\lim_{N\to\infty}\frac{f(1)+f(2)+\cdots+f(N)}{N}$$

for functions $f(n)$ of some class becomes an analogue of the existence theorem for the Riemann integral.

By analogy with the proof of the existence theorem for the integral of a continuous function we can prove

THEOREM 1. *For polyadically continuous functions $f(n)$ the limit*

$$\lim_{N\to\infty}\frac{f(1)+f(2)+\cdots+f(N)}{N}$$

exists.

A class of functions containing the polyadically continuous functions comprises the functions that can be well approximated by periodic ones. We will say that a function $f(n)$ of a natural argument can be *well approximated by periodic functions* if for any $\varepsilon>0$ there exists a periodic function $f(n,A)$ with natural period A such that

$$\overline{\lim_{N\to\infty}}\frac{1}{N}\sum_{n\le N}|f(n)-f(n,A)|\le\varepsilon.$$

The class of functions that can be well approximated by periodic functions is a variant of the Besicovitch class of almost periodic functions. We will denote by B^1 the class of functions that can be well approximated by periodic functions.

It is clear that any periodic function $f(n)$ with integral period (we denote this period by A) belongs to the class B^1; we can take $f(n,A)=f(n)$.

We will show that the class B^1 contains the class of functions satisfying the condition of Wintner's theorem (we will, as above, call this class the Wintner class and denote it by W). Namely, we have the following

LEMMA. *Represent a function $f(n)$ of a natural argument in the form $f(n)=\sum_{d|n}\Phi(d)$. Suppose $\Phi(d)$ is such that $\sum_1^\infty\Phi(d)/d$ is absolutely convergent. Then $f(n)$ belongs to the class B^1.*

PROOF. Let $p_1,\dots,p_l$ be the first l primes of the natural sequence. Let $\mathfrak{M}_l$ denote the set of natural numbers n having the form

$$n=p_1^{\alpha_1}\cdots p_l^{\alpha_l},\qquad 0\le\alpha_i\le l.$$

We see that $\mathfrak{M}_l$ contains the number 1. Suppose $\varepsilon > 0$ is given. Since $\sum_1^\infty |\Phi(d)|/d$ converges, there exists l_0 such that for $l \geq l_0$

$$\sum_{d \notin \mathfrak{M}_l} \frac{|\Phi(d)|}{d} < \varepsilon.$$

We define a function $f_\varepsilon(n)$ by

$$f_\varepsilon(n) = \sum_{d \in \mathfrak{M}_l,\, d|n} \Phi(d).$$

It is clear that $f_\varepsilon(n)$ is periodic with period n_0, where $n_0 = p_1^l \cdots p_l^l$. We have

$$\begin{aligned} \frac{1}{N} \sum_{n=1}^{N} |f(n) - f_\varepsilon(n)| &\leq \frac{1}{N} \sum_{n \leq N} \sum_{d|n,\, d \notin \mathfrak{M}_l} |\Phi(d)| \\ &\leq \frac{1}{N} \sum_{d \notin \mathfrak{M}_l,\, d \leq N} |\Phi(d)| \left[\frac{N}{d}\right] \leq \sum_{d \notin \mathfrak{M}_l} \frac{|\Phi(d)|}{d} \leq \varepsilon. \end{aligned}$$

Therefore,

$$\overline{\lim_{N \to \infty}} \frac{1}{N} \sum_{n=1}^{N} |f(n) - f_\varepsilon(n)| \leq \varepsilon,$$

as required.

Thus any function of class W is also of class B^1. We will now construct examples of functions that belong to B^1 but not to the Wintner class, i.e., we will show that B^1 is larger than W.

THEOREM 2. ([1]) *A character of the second kind with respect to any modulus does not belong to* W.

PROOF. Let $\chi^{(2)}(n)$ be a character of the second kind modulo m. Let us write

$$\chi^{(2)}(n) = \sum_{d|n} \Phi(d).$$

Then, using Möbius inversion and the multiplicativity of $\chi^{(2)}(n)$, we obtain

$$\Phi(n) = \prod_{p|n} (\chi^{(2)}(p^{\alpha_p(n)}) - \chi^{(2)}(p^{\alpha_p(n)-1}))$$

($\Phi(1) = 1$). Noting also that

$$\chi_0(n)|\Phi(n)| = \chi_0(n) \prod_{p|n} (1 - \chi^{(2)}(p))$$

([1])This example is taken from Mustafin [103].

($\chi_0(n)$ is the principal character), we see that

$$\sum_{n\le x}\frac{|\Phi(n)|}{n}\ge\sum_{n\le x}\frac{\chi_0(n)|\Phi(n)|}{n}$$
$$>\sum_{\substack{\chi^{(2)}(p)=-1\\ p\le x}}\frac{1}{p}=\frac{k}{\varphi(m)}\ln\ln x+O(1)$$

(the theorem on the distribution of primes in the arithmetic progression $mq+a$), where k is the number of classes in a reduced residue system modulo m for which $\chi^{(2)}(p)$ assumes the value -1. It follows from a property of quadratic characters that $k>0$ (see Venkov [145], Chapter I, §9). The theorem is proved.

Since $\chi^{(2)}(n)$ is a periodic function, it belongs to the class B^1.

Consider the function $r(n)$. We will show that for any bounded function $f(n)$ we have

$$\varliminf_{x\to\infty}\frac{1}{x}\sum_{n\le x}|r(n)-f(n)|\ge\pi,$$

which implies that $r(n)$ is not of class B^1. Let the function $b(n)$ be defined as follows:

$$b(n)=\begin{cases}1 & \text{if } n \text{ is representable as a sum of two squares,}\\ 0 & \text{if } n \text{ is not representable as a sum of two squares.}\end{cases}$$

Suppose $|f(n)|\le M$. We have

$$\frac{1}{x}\sum_{n\le x}|r(n)-f(n)|\ge\frac{1}{x}\sum_{n\le x}b(n)|r(n)-f(n)|$$
$$\ge\frac{1}{x}\Big(\sum_{n\le x}r(n)-\sum_{n\le x}b(n)|f(n)|\Big)$$
$$\ge\frac{1}{x}\sum_{n\le x}r(n)-\frac{M}{x}\sum_{n\le x}b(n).$$

Since $\sum_{n\le x}r(n)$ is the number of lattice points in a disc of radius $\sqrt{x}$, it follows that

$$\frac{1}{x}\sum_{n\le x}r(n)=\pi+o(1).$$

In the Introduction we mentioned the theorem of Landau [90]

$$\sum_{n\le x}b(n)\sim A\frac{x}{\sqrt{\ln x}},$$

which implies that

$$\sum_{n\leq x} b(n) = O\left(\frac{x}{\sqrt{\ln x}}\right).$$

This yields

$$\frac{1}{x}\sum_{n\leq x} |r(n) - f(n)| \geq \pi + o(1) + O\left(\frac{1}{\sqrt{\ln x}}\right),$$

which implies our assertion.

Erdős and Wintner [42] showed that the Möbius function $\mu(n)$ is not of class B^1. The basis of their argument is the following inequality of Davenport [24]: for any $h > 0$

$$\sum_{n\leq x} \mu(n)e^{2\pi i n\theta} = O\left(\frac{x}{\ln^h x}\right)$$

uniformly in θ.

Let

$$M(x, A) = \sum_{m+Aj\leq x} \mu(m + Aj).$$

It is known (see Gel′fond and Linnik [60], Chapter 3, §5, Lemma 4) that the asymptotic law of the distribution of primes in an arithmetic progression can be deduced from the following assertion, which we will call the weak form of this law.

LEMMA. *If* $d_m = (m, A) = 1$, *then* $M(x, A) = o(x)$.

It is easy to see that Davenport's inequality is a generalization and strengthening of this assertion.

Using the weak form of the asymptotic law of the distribution of primes in an arithmetic progression, we will prove

THEOREM 3. *For any natural number* A *and any periodic function* $f(n, A)$ *of a natural argument with period* A,

$$\varliminf_{n\to\infty} \frac{1}{N}\sum_{n=1}^{N} |\mu(n) - f(n, A)| \geq \frac{3}{\pi^2}.$$

PROOF. It is a simple matter to verify that

$$\frac{\mu^2(n) + \mu(n)}{2} = \begin{cases} 0, & \mu(n) = 0, \\ 0, & \mu(n) = -1, \\ 1, & \mu(n) = 1, \end{cases}$$

$$\frac{\mu^2(n) - \mu(n)}{2} = \begin{cases} 0, & \mu(n) = 0, \\ 1, & \mu(n) = -1, \\ 0, & \mu(n) = 1. \end{cases}$$

Suppose A is a natural number and $f(n, A)$ is a periodic function of a natural argument with period A. Let

$$f(1, A) = a_1, \quad f(2, A) = a_2, \quad \ldots, \quad f(A, A) = a_A.$$

We have

$$\frac{1}{N}\sum_{n=1}^{N} |\mu(n) - f(n, A)| \geq \frac{1}{N}\sum_{n=1}^{A[N/A]} |\mu(n) - f(n, A)|$$
$$= \frac{1}{N}\sum_{m=1}^{A}\sum_{j=0}^{[N/A]-1} |\mu(m + Aj) - a_m|.$$

Let us denote the real part of a_m by $\operatorname{Re} a_m$. If $\operatorname{Re} a_m > 0$, then

$$|\mu(m + Aj) - a_m| \geq \frac{\mu^2(m + Aj) - \mu(m + Aj)}{2}.$$

If $\operatorname{Re} a_m < 0$, then

$$|\mu(m + Aj) - a_m| \geq \frac{\mu^2(m + Aj) + \mu(m + Aj)}{2}$$

and

$$\frac{1}{N}\sum_{n=1}^{N} |f(n) - f(n, A)|$$
$$\geq \frac{1}{2N}\sum_{m=1}^{A}\sum_{j=0}^{[N/A]-1} \mu^2(m + Aj) + \frac{1}{2N}\sum_{m=1}^{A}\left(\pm_m \sum_{j=0}^{[N/A]} \mu(m + Aj)\right)$$
$$= \frac{1}{2N}\sum_{n=1}^{A[N/A]} \mu^2(n) + \frac{1}{2N}\sum_{m=1}^{A}\left(\pm_m \sum_{j=0}^{[N/A]-1} \mu(m + Aj)\right);$$

the symbol $\pm_m$ stands for either $+$ or $-$, in accordance with the sign of $\operatorname{Re} a_m$.

We have spoken of the weak form of the asymptotic law: If $d_m = (m, A) = 1$, then

$$M(x, A) = \sum_{m+Aj \leq x} \mu(m + Aj) = o(x).$$

It is easy to see that the restriction $d_m = 1$ is inessential. If $\mu(d_m) = 0$, then, for each j, $\mu(m + Aj) = 0$ and the estimate

$$\sum_{m+Aj \leq x} \mu(m + Aj) = o(x)$$

automatically holds. If $\mu(d_m) \neq 0$, then

$$\sum_{m+Aj\le x} \mu(m+Aj) = \sum_{j_0=1}^{d_m} \sum_{m+Aj_0+Ad_m t\le x} \mu(m+Aj_0+Ad_m t)$$
$$= \sum_{\substack{j_0=1\\((m+Aj_0)/d_m,\, d_m)=1}}^{d_m} \sum_{m+Aj_0+Ad_m t\le x} \mu\left(\frac{m+Aj_0}{d_m}+At\right) = o(x).$$

Thus,

$$\frac{1}{N}\sum_{n=1}^{N} |\mu(n) - f(n,A)| \ge \frac{1}{2N}\sum_{n=1}^{A[N/A]} \mu^2(n) + o(1)$$
$$= \frac{1}{2N}\left(\frac{6}{\pi^2}N + O(\sqrt{N})\right) + o(1) = \frac{3}{\pi^2} + o(1).$$

This proves the theorem.

The theory of functions of class B^1 consists of generalizations of properties of periodic functions. We will prove several theorems on summation formulas and the distribution of values of functions of class B^1.

THEOREM 4. *For any function $f(n)$ of class B^1 the limit*

$$\lim_{N\to\infty} \frac{1}{N}\sum_{n=1}^{N} f(n)$$

exists.

PROOF. We establish the existence of the limit by means of Cauchy's convergence criterion. Given $\varepsilon > 0$, we can find a periodic function $f_\varepsilon(n)$ such that

$$\overline{\lim_{N\to\infty}} \frac{1}{N}\sum_{n=1}^{N} |f(n) - f_\varepsilon(n)| \le \varepsilon.$$

Take natural numbers N_1 and N_2; we have

$$\frac{1}{N_1}\sum_{n=1}^{N_1} f(n) = \frac{1}{N_1}\sum_{n=1}^{N_1} f_\varepsilon(n) + \frac{\theta_1}{N_1}\sum_{n=1}^{N_1} |f(n) - f_\varepsilon(n)|, \qquad |\theta_1| \le 1;$$
$$\frac{1}{N_2}\sum_{n=1}^{N_2} f(n) = \frac{1}{N_2}\sum_{n=1}^{N_2} f_\varepsilon(n) + \frac{\theta_2}{N_2}\sum_{n=1}^{N_2} |f(n) - f_\varepsilon(n)|, \qquad |\theta_2| \le 1.$$

Therefore,

$$\left|\frac{1}{N_1}\sum_{n=1}^{N_1} f(n) - \frac{1}{N_2}\sum_{n=1}^{N_2} f(n)\right| \le \left|\frac{1}{N_1}\sum_{n=1}^{N_1} f_\varepsilon(n) - \frac{1}{N_2}\sum_{n=1}^{N_2} f_\varepsilon(n)\right|$$
$$+\frac{1}{N_1}\sum_{n=1}^{N_1} |f(n) - f_\varepsilon(n)| + \frac{1}{N_2}\sum_{n=1}^{N_2} |f(n) - f_\varepsilon(n)|.$$

Since $\lim_{N\to\infty}(1/N)\sum_{n\le N} f_\varepsilon(n)$ exists (the function $f_\varepsilon(n)$ is periodic), it follows from Cauchy's criterion that for $N_1 \ge N_0(\varepsilon)$ and $N_2 \ge N_0(\varepsilon)$

$$\left|\frac{1}{N_1}\sum_{n=1}^{N_1} f_\varepsilon(n) - \frac{1}{N_2}\sum_{n=1}^{N_2} f_\varepsilon(n)\right| \le \varepsilon.$$

Furthermore, since

$$\overline{\lim_{N\to\infty}}\, \frac{1}{N}\sum_{n\le N} |f(n) - f_\varepsilon(n)| \le \varepsilon,$$

it follows that for $N_1 \ge \overline{N}_0(\varepsilon)$ and $N_2 \ge \overline{N}_0(\varepsilon)$ we have

$$\frac{1}{N_1}\sum_{n=1}^{N_1} |f(n) - f_\varepsilon(n)| \le 2\varepsilon, \qquad \frac{1}{N_2}\sum_{n=1}^{N_2} |f(n) - f_\varepsilon(n)| \le 2\varepsilon.$$

Thus for $N_1 \ge \max(N_0(\varepsilon), \overline{N}_0(\varepsilon))$, $N_2 \ge \max(N_0(\varepsilon), \overline{N}_0(\varepsilon))$, we have

$$\left|\frac{1}{N_1}\sum_{n=1}^{N_1} f(n) - \frac{1}{N_2}\sum_{n=1}^{N_2} f(n)\right| \le \varepsilon + 2\varepsilon + 2\varepsilon = 5\varepsilon.$$

By Cauchy's criterion, $\lim_{N\to\infty}(1/N)\sum_{n\le N} f(n)$ exists. The theorem is proved.

The theory of mean values of arithmetical functions is the starting point for subsequent constructions. We will now establish existence theorems for limits of the form

$$\lim_{N\to\infty} \frac{1}{N}\sum_{n\le N} F(f(n)).$$

We will prove the following fundamental theorem.([2])

THEOREM 5a. *Suppose $f(n)$ is a function of class B^1. Assume it is bounded and its values belong to a finite interval $[A, B]$. Suppose $F(x)$ is a continuous function on $[A, B]$. Then the limit*

$$\lim_{N\to\infty} \frac{1}{N}\sum_{n\le N} F(f(n))$$

exists.

For the proof we will need the following lemma.

([2])This result was obtained in a different way by Novoselov [111], §§26 and 19.

LEMMA. *Suppose the function $f(n)$ is of class B^1. Suppose also that a function $F(x)$ is defined on $(-\infty, \infty)$ and satisfies a first-order Lipschitz condition*

$$|F(x_1) - F(x_2)| \leq C|x_1 - x_2|.$$

Then the function $F(f(n))$ is also of class B^1.

PROOF. Consider, along with $F(f(n))$, the function $F(f_{\varepsilon/C}(n))$, where C is the constant in the Lipschitz condition and $f_{\varepsilon/C}(n)$ is the notation used in the proof of Theorem 4. Obviously $F(f_{\varepsilon/C}(n))$ is a periodic function. Also,

$$\overline{\lim_{N\to\infty}} \frac{1}{N} \sum_{n\leq N} |F(f(n)) - F(f_{\varepsilon/C}(n))| \leq \overline{\lim_{N\to\infty}} \frac{C}{N} \sum_{n\leq N} |f(n) - f_{\varepsilon/C}(n)| \leq C\frac{\varepsilon}{C} = \varepsilon.$$

The lemma is proved.

PROOF OF THE THEOREM. Assume $F(x)$ is a continuous function on $[A, B]$ satisfying the first-order Lipschitz condition

$$|F(x_1) - F(x_2)| \leq C|x_1 - x_2|.$$

We extend the definition of $F(x)$ to $(-\infty, A)$ and (B, ∞):

$$\begin{aligned} F(x) &= F(A) + \arctan(x - A), \qquad -\infty < x \leq A; \\ F(x) &= F(B) + \arctan(x - B), \qquad B \leq x < \infty. \end{aligned}$$

Clearly the extended function is continuous on $(-\infty, \infty)$ and satisfies the Lipschitz condition

$$|F(x_1) - F(x_2)| \leq \max(1, C)|x_1 - x_2|.$$

The function $F(f(n))$ is of class B^1 (by the lemma). Thus by Theorem 4 the limit

$$\lim_{N\to\infty} \frac{1}{N} \sum_{n\leq N} F(f(n))$$

exists. Note that the extended definition of the function has no significance for the sum $\sum_{n\leq N} F(f(n))$, since $A \leq f(n) \leq B$.

Let us assume first that $F(x)$ is only continuous on $[A, B]$. Then it is uniformly continuous there, i.e., for any $\varepsilon > 0$ we can find $N_0 = N_0(\varepsilon)$ such that if we partition $[A, B]$ into parts of length $(B - A)/N_0(\varepsilon)$, then for two points x' and x'' in the same part

$$|F(x') - F(x'')| \leq \varepsilon.$$

We replace the curve $y = F(x)$ by the polygonal line joining the points

$$\left(A + k\frac{B-A}{N_0},\ F\left(A + k\frac{B-A}{N_0}\right)\right), \qquad k = 0, 1, \dots, N_0.$$

Suppose the equation of this polygonal line is given by the function $y = F_\varepsilon(x)$. Clearly $|F(x) - F_\varepsilon(x)| \le \varepsilon$; $F_\varepsilon(x)$ satisfies a Lipschitz condition (with a constant depending on ε). Consequently, the limit

$$\lim_{N\to\infty} \frac{1}{N} \sum_{n\le N} F_\varepsilon(f(n))$$

exists. Also,

$$\overline{\lim_{N\to\infty}} \frac{1}{N} \sum_{n\le N} |F(f(n)) - F_\varepsilon(f(n))| < \varepsilon.$$

Repeating the argument used in the proof of Theorem 4, we obtain the desired result.

THEOREM 5b. *Represent a function $f(n)$ of a natural argument in the form $f(n) = \sum_{d|n} \Phi(d)$; assume $\sum \Phi(d)/d$ is absolutely convergent. Assume also that the values of $f(n)$ lie in a finite interval $[A, B]$. Suppose $F(x)$ is a continuous function defined on $[A, B]$. Then the limit*

$$\lim_{N\to\infty} \frac{1}{N} \sum_{n\le N} F(f(n))$$

exists.

Theorem 5b is a direct consequence of Theorem 5a. Theorems 5a and 5b will be called *Novoselov's theorems.*

The following result will be proved by a method suggested by Novoselov (see Tyan (Tian) [141], Theorem 5).

THEOREM 5c. *Let*

$$t_3(n) = \sum_{d|n} \frac{1}{d^2+1}.$$

The values $t_3(n)$ lie in the interval $[1/2, \alpha]$, where

$$\alpha = \sum_{d=1}^{\infty} \frac{1}{d^2+1}.$$

Suppose a function $F(x)$ satisfies a first-order Lipschitz condition on $[1/2, \alpha]$. Then there exists a number $A(F)$ such that

$$\frac{1}{N} \sum_{n\le N} F(t_3(n)) = A(F) + O\left(\frac{1}{\ln^2 N}\right).$$

PROOF. Let T_N denote the least common multiple of the numbers $1, 2, \dots, [(1-\beta)\ln N]$, where β is fixed, $0 < \beta < 1$. Since

$$\ln T_N = \sum_{n=1}^{[(1-\beta)\ln N]} \Lambda(n),$$

it follows from the asymptotic law of the distribution of primes that

$$T_N \leq N^{(1-\beta)(1+o(1))}.$$

We divide N by T_N with a remainder:

$$N = R_N + \theta T_N, \qquad 0 \leq \theta < 1.$$

We next let $R_N(m)$ be the positive remainder of the division of the natural number m by R_N, $1 \leq R_N(m) \leq R_N$. Putting $f(m) = F(t_3(m))$ and $M = \max_{0\leq x\leq 1} |F(x)|$, we have

$$\begin{aligned}
&\left|\frac{1}{R_N}\sum_{m=1}^{R_N} f(m) - \frac{1}{N}\sum_{m=1}^{N} f(m)\right| \\
&\quad= \left|\left(1-\frac{R_N}{N}\right)\frac{1}{R_N}\sum_{m=1}^{R_N} f(m) - \frac{1}{N}\sum_{m=R_N+1}^{N} f(m)\right| \\
&\quad\leq \left(1-\frac{R_N}{N}\right)M + \frac{1}{N}(N-R_N)M \leq \frac{M}{N^{\beta(1+o(1))}}.
\end{aligned}$$

Since the function $f(R_N(m))$ is periodic with period R_N, it follows that

$$\lim_{T\to\infty}\frac{1}{T}\sum_{m=1}^{T} f(R_N(m)) = \frac{1}{R_N}\sum_{m=1}^{R_N} f(m).$$

By assumption, $|F(x_1) - F(x_2)| < C|x_1 - x_2|$. Therefore,

$$\begin{aligned}
&\left|\frac{1}{R_N}\sum_{m=1}^{R_N} f(m) - \lim_{T\to\infty}\frac{1}{T}\sum_{m=1}^{T} f(m)\right| \\
&\quad= \left|\lim_{T\to\infty}\frac{1}{T}\sum_{m=1}^{T}(f(R_N(m)) - f(m))\right| \\
&\quad\leq \overline{\lim_{T\to\infty}}\frac{C}{T}\sum_{m=1}^{T}|t_3(R_N(m)) - t_3(m)| \\
&\quad\leq \overline{\lim_{T\to\infty}}\frac{C}{T}\sum_{m=1}^{T}\left|\sum_{d\mid R_N(m)}\frac{1}{d^2+1} - \sum_{d\mid m}\frac{1}{d^2+1}\right|.
\end{aligned}$$

Since R_N is divisible by T_N, which in turn is divisible by all natural numbers n such that $1 \le n \le (1-\beta)\ln N$, it follows that the divisors of m and $R_N(m)$ that do not exceed $(1-\beta)\ln N$ are the same. We have

$$\begin{aligned}
&\left|\frac{1}{R_N}\sum_{m=1}^{R_N} f(m) - \lim_{T\to\infty}\frac{1}{T}\sum_{m=1}^{T} f(m)\right| \\
&\quad\le \varlimsup_{T\to\infty}\frac{C}{T}\sum_{m=1}^{T}\left|\sum_{d|R_N(m)}\frac{1}{d^2+1}-\sum_{d|m}\frac{1}{d^2+1}\right| \\
&\quad\le \varlimsup_{T\to\infty}\frac{C}{T}\sum_{m=1}^{T}\left|\sum_{\substack{d|R_N(m)\\ d>(1-\beta)\ln N}}\frac{1}{d^2+1}-\sum_{\substack{d|m\\ d>(1-\beta)\ln N}}\frac{1}{d^2+1}\right| \\
&\quad\le \varlimsup_{T\to\infty}\frac{C}{T}\sum_{m=1}^{T}\sum_{\substack{d|R_N(m)\\ d>(1-\beta)\ln N}}\frac{1}{d^2+1} \\
&\qquad+ \varlimsup_{T\to\infty}\frac{C}{T}\sum_{m=1}^{T}\sum_{\substack{d|m\\ d>(1-\beta)\ln N}}\frac{1}{d^2+1}.
\end{aligned}$$

Also,

$$\begin{aligned}
&\left|\frac{1}{R_N}\sum_{m=1}^{R_N} f(m) - \lim_{T\to\infty}\frac{1}{T}\sum_{m=1}^{T} f(m)\right| \\
&\quad\le \varlimsup_{T\to\infty}\frac{C}{T}\sum_{d>(1-\beta)\ln N}^{T}\frac{1}{d^2+1}\sum_{\substack{1\le m\le T\\ m=dx+R_N y\\ 1\le dx\le R_N}} 1 \\
&\qquad+ \varlimsup_{T\to\infty}\frac{C}{T}\sum_{d>(1-\beta)\ln N}^{T}\frac{1}{d^2+1}\left[\frac{T}{d}\right] \\
&\quad\le \varlimsup_{T\to\infty}\frac{C}{T}\sum_{d>(1-\beta)\ln N}^{T}\frac{1}{d^2+1}\sum_{x=1}^{[R_N/d]}\left(\left[\frac{T-dx}{R_N}\right]+1\right) \\
&\qquad+ C\sum_{d>(1-\beta)\ln N}\frac{1}{d^3} \\
&\quad= O\left(\varlimsup_{T\to\infty}\frac{1}{T}\sum_{d>(1-\beta)\ln N}^{T}\frac{1}{d^2+1}\sum_{x=1}^{[R_N/d]}\frac{T-dx}{R_N}\right)+O\left(\frac{1}{\ln^2 N}\right)
\end{aligned}$$

(continued)

(*continued*)

$$= O\left(\overline{\lim_{T\to\infty}}\,\frac{1}{T}\sum_{d>(1-\beta)\ln N}^{T}\frac{1}{(d^2+1)R_N}\,\frac{T+T-d[R_N/d]}{2}\left[\frac{R_N}{d}\right]\right)$$
$$+O\left(\frac{1}{\ln^2 N}\right)$$
$$= O\left(\overline{\lim_{T\to\infty}}\sum_{d>(1-\beta)\ln N}^{T}\frac{1}{d(d^2+1)}\right)+O\left(\frac{1}{\ln^2 N}\right)=O\left(\frac{1}{\ln^2 N}\right),$$

as required.

Suppose $f(n)$ is a real-valued function of a natural argument. Let λ be a fixed real number. We denote by $P_N(f(n) < \lambda)$ the number of natural numbers among $1, \dots, N$ for which $f(n) < \lambda$; we also put

$$\nu_N(f(n) < \lambda) = P_N(f(n) < \lambda)/N.$$

THEOREM 6a. *Suppose $f(n)$ is a function of a natural argument of class B^1. Then for all real λ, except perhaps for a countable set, the limit*

$$\lim_{N\to\infty} \nu_N(f(n) < \lambda) = V(\lambda)$$

exists.

For the proof of this theorem (suggested by Yu. V. Prokhorov in a supplement to the Russian translation of Kac [77]) we need the following

LEMMA. *Given a sequence $\alpha_1, \alpha_2, \dots$ of real numbers situated in some finite interval $[A, B]$, assume that for any continuous function $F(x)$ on $[A, B]$ satisfying a first-order Lipschitz condition the limit*

$$\lim_{N\to\infty}\frac{1}{N}\sum_{n\le N} F(\alpha_n)$$

exists. Then for all numbers λ belonging to $[A, B]$, except perhaps for a countable set, the limit

$$\lim_{N\to\infty} \nu_N(\alpha_n < \lambda) = V(\lambda)$$

exists.

PROOF. Suppose λ is any number such that $A \le \lambda \le B$. We define on $[A, B]$ the functions

$$F_{\lambda,\varepsilon}(x) = \begin{cases} 1, & A \le x \le \lambda, \\ \text{the linear function equal to 1 when } x = \lambda \text{ and} & \\ \text{equal to 0 when } x = \lambda+\varepsilon, & \lambda \le x \le \lambda+\varepsilon, \\ 0, & \lambda+\varepsilon \le x \le B; \end{cases}$$

$$F_{\lambda,-\varepsilon}(x)=\begin{cases}1, & A\le x\le\lambda-\varepsilon,\\ \text{the linear function equal to 1 when } x=\lambda-\varepsilon \text{ and} \\ \text{equal to 0 when } x=\lambda,\ \lambda-\varepsilon\le x\le\lambda, \\ 0, & \lambda\le x\le B.\end{cases}$$

These functions are continuous and satisfy a Lipschitz condition. Therefore the limits

$$\lim_{N\to\infty}\frac{1}{N}\sum_{n\le N}F_{\lambda,\varepsilon}(\alpha_n)=W_\varepsilon(\lambda),$$

$$\lim_{N\to\infty}\frac{1}{N}\sum_{n\le N}F_{\lambda,-\varepsilon}(\alpha_n)=W_{-\varepsilon}(\lambda)$$

exist. Since $W_\varepsilon(\lambda)\ge W_{-\varepsilon}(\lambda)$ and, for $\varepsilon_1>\varepsilon_2$,

$$W_{\varepsilon_1}(\lambda)\ge W_{\varepsilon_2}(\lambda),\qquad W_{-\varepsilon_1}(\lambda)\le W_{-\varepsilon_2}(\lambda),$$

it follows that the limits

$$\lim_{\varepsilon\to0}W_\varepsilon(\lambda)=W_+(\lambda),\qquad \lim_{\varepsilon\to0}W_{-\varepsilon}(\lambda)=W_-(\lambda)$$

exist. It is obvious that

$$W_+(\lambda)\ge\overline{\lim_{N\to\infty}}\,\nu_N(\alpha_n<\lambda)\ge\underline{\lim_{N\to\infty}}\,\nu_N(\alpha_n<\lambda)\ge W_-(\lambda). \tag{1}$$

It is also clear that if $\lambda_1>\lambda_2$, then $W_-(\lambda_1)\ge W_+(\lambda_2)$. Consequently, there can be at most a countable set of numbers λ such that $W_+(\lambda)$ is strictly greater than $W_-(\lambda)$. Indeed, suppose λ is such that $W_+(\lambda)>W_-(\lambda)$. Choose a rational number $p(\lambda)$ in the interval $(W_-(\lambda),W_+(\lambda))$. Since $W_-(\lambda_1)\ge W_+(\lambda_2)$ for $\lambda_1>\lambda_2$, the points $p(\lambda_1)$ and $p(\lambda_2)$ are distinct. Thus the set of points at which $W_+(\lambda)>W_-(\lambda)$ is not larger than the set of rational numbers, i.e., it is a countable set. In view of (1), there is at most a countable set of points λ such that

$$\overline{\lim_{N\to\infty}}\,\nu_N(\alpha_n<\lambda)\ne\underline{\lim_{N\to\infty}}\,\nu_N(\alpha_n<\lambda).$$

For all points outside this countable set we put $V(\lambda)=W_+(\lambda)=W_-(\lambda)$. The lemma is proved.

Suppose $f(n)$ is a function of class B^1.

1) Assume that $f(n)$ is bounded and its values belong to the interval $[A,B]$. Then for any continuous function $F(x)$ on $[A,B]$ the limit

$$\lim_{N\to\infty}\frac{1}{N}\sum_{n\le N}F(f(n))$$

exists. According to the lemma, the limit

$$\lim_{N\to\infty} \nu_N(f(N) < \lambda)$$

exists for all λ, except perhaps for a countable set of values.

2) Assume the function $f(n)$ is unbounded. Consider the auxiliary function $F(n) = \arctan f(n)$. The function $F(n)$ is of class B^1 and is obviously bounded. Indeed, suppose $f(n)$ is a periodic function such that

$$\overline{\lim_{N\to\infty}} \frac{1}{N} \sum_{n\le N} |f(n) - f_\varepsilon(n)| \le \varepsilon.$$

The function $F_\varepsilon(n) = \arctan f_\varepsilon(n)$ is periodic. Since $0 < (\arctan x)' \le 1$, it follows that

$$|F(n) - F_\varepsilon(n)| \le |f(n) - f_\varepsilon(n)|,$$

and hence

$$\overline{\lim_{N\to\infty}} \frac{1}{N} \sum_{n\le N} |F(n) - F_\varepsilon(n)| \le \varepsilon.$$

We have seen that for all Λ in the interval $(-\pi/2, \pi/2)$, except perhaps for a countable set, the limit

$$\lim_{N\to\infty} \nu_N(\arctan f(n) < \Lambda)$$

exists. Put $\lambda = \tan\Lambda$. Under the transformation $\lambda = \tan\Lambda$ the interval $(-\pi/2, \pi/2)$ is mapped onto the whole line $(-\infty, \infty)$. Thus the limit

$$\lim_{N\to\infty} \nu_N(\tan \arctan f(n) < \lambda) = \lim_{N\to\infty} \nu_N(f(n) < \lambda)$$

exists for all λ, except perhaps for a countable set of values.

The theorem is proved.

THEOREM 6b. *Represent a function $f(n)$ of a natural argument in the form $f(n) = \sum_{d|n} \Phi(d)$, and assume $f(n)$ is such that $\sum_1^\infty \Phi(d)/d$ is absolutely convergent. Then for all λ, except perhaps for a countable set, the limit*

$$\lim_{N\to\infty} \nu_N(f(n) < \lambda)$$

exists.

PROOF. The condition of the theorem shows that the function $f(n)$ is of class B^1. It remains to apply Theorem 6a.

§3.4. Independent functions of a natural argument

In §3.1 we introduced the measure $\text{meas}_{\sigma_0} \mathfrak{M}$ and certain pseudomeasures, namely the logarithmic density $\text{meas}_* \mathfrak{M}$ and the asymptotic density $D(\mathfrak{M})$.

We are now able to construct analogues of the theory of independent functions for functions of a natural argument. In studying independent functions we will employ the concepts and methods of probability theory.

When dealing with $\operatorname{meas}_{\sigma_0} \mathfrak{M}$, there is no need to talk about analogies, because the set function $\operatorname{meas}_{\sigma_0} \mathfrak{M}$ is a measure and we have strong foundations for applying the theorems of probability theory.

Suppose p and q are primes. We will prove that the set $\mathfrak{M}_p$ of numbers divisible by p and the set $\mathfrak{M}_q$ of numbers divisible by q are independent. We have

$$\operatorname{meas}_{\sigma_0} \mathfrak{M}_p = \frac{1}{\varsigma(\sigma_0)} \frac{1}{p^{\sigma_0}} \varsigma(\sigma_0) = \frac{1}{p^{\sigma_0}}$$

and

$$\operatorname{meas}_{\sigma_0} \mathfrak{M}_q = \frac{1}{\varsigma(\sigma_0)} \frac{1}{q^{\sigma_0}} \varsigma(\sigma_0) = \frac{1}{q^{\sigma_0}},$$

and also

$$\operatorname{meas}_{\sigma_0}(n : p|n, q|n) = \frac{1}{\varsigma(\sigma_0)} \frac{1}{p^{\sigma_0}} \frac{1}{q^{\sigma_0}}.$$

Hence

$$\operatorname{meas}_{\sigma_0}(n : p|n, q|n) = \operatorname{meas}_{\sigma_0}(n : p|n) \operatorname{meas}_{\sigma_0}(n : q|n).$$

An analogous assertion holds for an arbitrary number of primes.

Suppose t is a fixed real number. We define, for each prime p, the function

$$f_p(n) = e^{-it\alpha_p(n) \ln p}.$$

For two different primes p and q the functions $f_p(n)$ and $f_q(n)$ are independent. Let δ_p be any value that the function $f_p(n)$ can assume, and δ_q any value that $f_q(n)$ can assume. By definition,

$$\begin{aligned} \operatorname{meas}_{\sigma_0}(n : f_p(n) = \delta_p) &= \frac{1}{\varsigma(\sigma_0)} \sum_{\alpha_p(m) = \alpha_p(n)} \frac{1}{m^{\sigma_0}} \\ &= \frac{1}{\varsigma(\sigma_0)} \frac{1}{p^{\sigma_0 \alpha_p(n)}} \sum_{(m,p)=1} \frac{1}{m^{\sigma_0}} = \left(1 - \frac{1}{p^{\sigma_0}}\right) \frac{1}{p^{\sigma_0 \alpha_p(n)}}. \end{aligned}$$

Also,

$$
\begin{aligned}
\operatorname{meas}_{\sigma_0}&(n : f_p(n) = \delta_p, f_q(n) = \delta_q)\\
&= \frac{1}{\varsigma(\sigma_0)} \sum_{\substack{\alpha_p(m)=\alpha_p(n)\\ \alpha_q(m)=\alpha_q(n)}} \frac{1}{m^{\sigma_0}}\\
&= \frac{1}{\varsigma(\sigma_0)} \frac{1}{p^{\sigma_0 \alpha_p(n)}} \frac{1}{q^{\sigma_0 \alpha_p(n)}} \sum_{(m,p)=(n,q)=1} \frac{1}{m^{\sigma_0}}\\
&= \frac{1}{\varsigma(\sigma_0)} \frac{1}{p^{\sigma_0 \alpha_p(n)}} \frac{1}{q^{\sigma_0 \alpha_p(n)}} \left(1 - \frac{1}{p^{\sigma_0}}\right)\left(1 - \frac{1}{q^{\sigma_0}}\right) \varsigma(\sigma_0)\\
&= \left(1 - \frac{1}{p^{\sigma_0}}\right) \frac{1}{p^{\sigma_0 \alpha_p(n)}} \left(1 - \frac{1}{q^{\sigma_0}}\right) \frac{1}{q^{\sigma_0 \alpha_p(n)}}\\
&= \operatorname{meas}_{\sigma_0}(n : f_p(n) = \delta_p) \operatorname{meas}_{\sigma_0}(n : f_q(n) = \delta_q),
\end{aligned}
$$

as required.

We calculate the integral

$$
\begin{aligned}
\int f_p(n)\, d\operatorname{meas}_{\sigma_0} &= \frac{1}{\varsigma(\sigma_0)} \sum_{n=1}^{\infty} \frac{f_p(n)}{n^{\sigma_0}} = \frac{1}{\varsigma(\sigma_0)} \sum_{n=1}^{\infty} \frac{e^{-it\alpha_p(n)\ln p}}{n^{\sigma_0}}\\
&= \frac{1}{\varsigma(\sigma_0)} \left(1 + \frac{e^{-it\ln p}}{p^{\sigma_0}} + \frac{e^{-2it\ln p}}{p^{2\sigma_0}} + \cdots\right) \sum_{(n,p)=1} \frac{1}{n^{\sigma_0}}\\
&= \frac{1 - 1/p^{\sigma_0}}{1 - 1/p^{\sigma_0 + it}}.
\end{aligned}
$$

Consider the function $f(n) = e^{-it\ln n}$. We have

$$\int f(n)\, d\operatorname{meas}_{\sigma_0} = \frac{\varsigma(\sigma_0 + it)}{\varsigma(\sigma_0)}.$$

It is clear that

$$f(n) = \prod_p e^{-it\ln p\alpha_p(n)} = \prod_p f_p(n).$$

In probability theory there is a theorem to the effect that the expectation of a product of independent random variables is equal to the product of their expectations (see Gnedenko [61], §28, Theorem 3). The proof depends essentially on the countable additivity of a measure. Applying this theorem, we obtain a proof of a classical identity of Euler in the form

$$\frac{\sum_{n=1}^{\infty} 1/n^{\sigma_0+it}}{\sum_{n=1}^{\infty} 1/n^{\sigma_0}} = \frac{\prod_p (1 - 1/p^{\sigma_0})}{\prod_p (1 - 1/p^{\sigma_0+it})}.$$

We give an example of an application of the central limit theorem of probability theory. We define a function of a natural argument, $f(n) = \ln n$,

$n = 1, 2, \ldots$. Since the natural sequence is a measure space, we can view $f(n)$ as a random variable ξ. It is easy to calculate the expectation and variance of this variable:

$$a = -\frac{\varsigma'(\sigma_0)}{\varsigma(\sigma_0)}, \qquad b^2 = \left(\frac{\varsigma'(\sigma)}{\varsigma(\sigma)}\right)'\Bigg|_{\sigma=\sigma_0}.$$

Since for $\sigma > 1$

$$\left(\frac{\varsigma'(\sigma)}{\varsigma(\sigma)}\right)' = \sum_{n=2}^{\infty} \frac{\Lambda(n)\ln n}{n^\sigma},$$

it follows that $b \neq 0$. We add n times the independent variables ξ and put $\varsigma_n = \xi + \cdots + \xi$ (n times). As usual, we denote by $\tau_n(N)$ the number of solutions in natural numbers of the equation $N = x_1 \cdots x_n$. The variable ς_n assumes the value $\ln N$, $N = 1, 2, \ldots$, with probability $\tau_n(N)/\varsigma^n(\sigma_0)N^{\sigma_0}$. The mean and variance of this law are respectively equal to

$$-n\frac{\varsigma'(\sigma_0)}{\varsigma(\sigma_0)} \quad \text{and} \quad n\left(\frac{\varsigma'(\sigma)}{\varsigma(\sigma)}\right)'\Bigg|_{\sigma=\sigma_0}.$$

The central limit theorem yields the following result.

THEOREM. *Suppose $\sigma_0 > 1$ is any fixed number. Let*

$$b^2(\sigma_0) = \left(\frac{\varsigma'(\sigma)}{\varsigma(\sigma)}\right)'\Bigg|_{\sigma=\sigma_0}.$$

Suppose α and β are any real numbers. Then as $n \to \infty$

$$\sum_{\substack{-n\varsigma'(\sigma_0)/\varsigma(\sigma_0)+\alpha\sqrt{b^2(\sigma_0)n} \\ \leq N \leq -n\varsigma'(\sigma_0)/\varsigma(\sigma_0)+\beta\sqrt{b^2(\sigma_0)n}}} \frac{\tau_n(N)}{N^{\sigma_0}} \sim \varsigma^n(\sigma_0)\frac{1}{\sqrt{2\pi}}\int_\alpha^\beta e^{-t^2/2}\,dt.$$

If we now turn to probabilistic theorems involving the asymptotic density, we encounter the difficulty that the density $D(\mathfrak{M})$ is not a countably additive set function. This difficulty can be overcome by additional arguments. We give an example taken from Kac [78], p. 13.

THEOREM. *Let $\omega(n)$ denote the total number of primes occurring in the canonical decomposition of n. Then*

$$\begin{aligned} \rho_k &= D(\omega(n) - \nu(n) = k) \\ &= \frac{1}{2\pi}\int_0^{2\pi} \left(\prod_p \left(1 - \frac{1}{p}\right)\left(1 + \frac{1}{p - e^{i\xi}}\right)\right) e^{-i\xi k}\,d\xi, \end{aligned}$$

where the product in the integrand extends over all primes.

Before proving this theorem, we give its probabilistic interpretation.

Suppose $p_1, \dots, p_r$ is a fixed finite set of primes and $k_1, \dots, k_r$ is a fixed set of natural numbers. It is known (Vinogradov [148], Chapter II, Problem 19a) that the number of natural numbers not exceeding $N/p_1^{k_1} \cdots p_r^{k_r}$ and relatively prime to $p_1 \cdots p_r$ is equal to

$$\sum_{d|p_1\cdots p_r} \mu(d) \left[\frac{N}{p_1^{k_1} \cdots p_r^{k_r} d}\right].$$

For fixed $p_1, \dots, p_r$ and $k_1, \dots, k_r$ and for $N \to \infty$ this expression is asymptotically equivalent to

$$\frac{N}{p_1^{k_1} \cdots p_r^{k_r}} \left(1 - \frac{1}{p_1}\right) \dots \left(1 - \frac{1}{p_r}\right).$$

Thus,

$$D(\alpha_{p_1}(n) = k_1, \dots, \alpha_{p_r}(n) = k_r) = \frac{1}{p_1^{k_1} \cdots p_r^{k_r}} \left(1 - \frac{1}{p_1}\right) \dots \left(1 - \frac{1}{p_r}\right).$$

This formula can be interpreted as the following assertion: For any finite set of primes $p_1, \dots, p_r$ and any set of nonnegative integers $k_1, \dots, k_r$

$$D(\alpha_{p_1}(n) = k_1, \dots, \alpha_{p_r}(n) = k_r) = D(\alpha_{p_1}(n) = k_1) \cdots D(\alpha_{p_r}(n) = k_r).$$

We see that the functions $\alpha_{p_i}(n)$ behave analogously to independent random variables, i.e., the functions $\alpha_{p_i}(n)$ are independent.

We introduce the functions

$$\beta_{p_i}(n) = \begin{cases} \alpha_{p_i}(n) - 1, & \text{if } \alpha_{p_i}(n) \geq 1, \\ 0, & \text{if } \alpha_{p_i}(n) = 0. \end{cases}$$

It is easy to see that the independence of the functions $\alpha_{p_i}(n)$ implies that of the functions $\beta_{p_i}(n)$.

Note that

$$\omega(n) - \nu(n) = \sum_p \beta_p(n), \tag{1}$$

where the sum extends over all primes.

In carrying out the analogy, it is natural to call the mean

$$M(f) = \lim_{N\to\infty} \frac{1}{N} \sum_{n=1}^{N} f(n)$$

the *expectation* of the function $f(n)$, if this limit exists.

Let us calculate the mean value of the function $e^{i\xi\beta_p(n)}$. To do this we define a function $h_p(n)$ as follows:

$$h_p(n) = \begin{cases} 1, & n = 1, \\ 0, & n \neq p^\alpha, \alpha > 0, \\ 0, & n = p, \\ e^{i\xi(\alpha-1)} - e^{i\xi(\alpha-2)}, & n = p^\alpha, \alpha \geq 2. \end{cases}$$

Obviously $h_p(n)$ is a multiplicative function. It is easy to verify that

$$\sum_{d|n} h_p(d) = e^{i\xi\beta_p(n)}.$$

As in the proof of Wintner's theorem,

$$\begin{aligned} \frac{1}{N}\sum_{n\leq N} e^{i\xi\beta_p(n)} &= \frac{1}{N}\sum_{n\leq N}\sum_{d|n} h_p(d) = \frac{1}{N}\sum_{n\leq N} h_p(n)\left[\frac{N}{n}\right] \\ &= \sum_{n\leq N}\frac{h_p(n)}{n} + O\Big(\frac{1}{N}\sum_{n\leq N}|h_p(n)|\Big) \\ &= \sum_{n=1}^{\infty}\frac{h_p(n)}{n} + O\Big(\sum_{n>N}\frac{|h_p(n)|}{n}\Big) + O\Big(\frac{1}{N}\sum_{n\leq N}|h_p(n)|\Big) \\ &= \sum_{n=1}^{\infty}\frac{h_p(n)}{n} + O\Big(\sum_{\alpha>\ln N/\ln p}\frac{1}{p^\alpha}\Big) + O\left(\frac{\ln N}{N}\right) \\ &= \sum_{n=1}^{\infty}\frac{h_p(n)}{n} + O\left(\frac{\ln N}{N}\right). \end{aligned}$$

Thus,

$$\begin{aligned} M(e^{i\xi\beta_p(n)}) &= \sum_{n=1}^{\infty}\frac{h_p(n)}{n} = 1 + \frac{e^{i\xi}-1}{p^2} + \frac{e^{2i\xi}-e^{i\xi}}{p^3} + \dots \\ &= \left(1-\frac{1}{p}\right)\left(1+\frac{1}{p-e^{i\xi}}\right). \end{aligned}$$

Consequently,

$$\begin{aligned} F(\xi) &= M(e^{i\xi(\omega(n)-\nu(n))}) = M(e^{i\xi\sum_p \beta_p(n)}) \\ &= M\Big(\prod_p e^{i\xi\beta_p(n)}\Big) = \prod_p M\left(e^{i\xi\beta_p(n)}\right). \end{aligned}$$

If the density were countably additive, then by the theorem that the expectation of a product of random variables is equal to the product of their expectations we could write down immediately on the basis of (1) that

$$M(e^{i\xi(\omega(n)-\nu(n))}) = \prod_p M(e^{i\xi\beta_p(n)}).$$

Let us give a rigorous proof of the theorem. From the classical formula

$$\frac{1}{2\pi}\int_0^{2\pi} e^{i\xi(\omega(n)-\nu(n))}e^{-i\xi k}\,d\xi = \begin{cases} 1, & \text{if } \omega(n)-\nu(n)=k, \\ 0, & \text{if } \omega(n)-\nu(n)\neq k, \end{cases}$$

we obtain

$$\frac{1}{N}\sum_{\substack{\omega(n)-\nu(n)=k\\ n\le N}} 1 = \frac{1}{2\pi}\int_0^{2\pi}\left(\frac{1}{N}\sum_{n\le N} e^{i\xi(\omega(n)-\nu(n))}\right)e^{-i\xi k}\,d\xi.$$

To study the trigonometric sum

$$\frac{1}{N}\sum_{n\le N} e^{i\xi(\omega(n)-\nu(n))}$$

we introduce a function $h(n)$, which we define as follows:

$$h(n) = \begin{cases} 1, & \text{if } n=1, \\ 0, & \text{if } n \text{ contains at least one prime to the first power}, \\ \prod_{p|n}(e^{i\xi(\alpha_i-1)} - e^{i\xi(\alpha_i-2)}), & \text{if } n = p_1^{\alpha_1}\cdots p_s^{\alpha_s},\ \alpha_1,\dots,\alpha_s\ge 2; \end{cases}$$

the function $h(n)$ is multiplicative. If $n = p_1^{\gamma_1}\cdots p_l^{\gamma_l}$, then

$$\sum_{d|n} h(d) = \prod_{p|n}(1+h(p)+\cdots+h(p^\gamma)) = e^{i\xi(\omega(n)-\nu(n))}.$$

Thus,

$$\begin{aligned}\frac{1}{N}\sum_{n\le N} e^{i\xi(\omega(n)-\nu(n))} &= \frac{1}{N}\sum_{n\le N} h(n)\left[\frac{N}{n}\right] \\ &= \sum_{n\le N}\frac{h(n)}{n} + O\left(\frac{1}{N}\sum_{n\le N}|h(n)|\right).\end{aligned}$$

For numbers n whose canonical decomposition contains no primes to the first power we have $|h(n)|\le 2^{\nu(n)}$. Therefore,

$$\begin{aligned}\frac{1}{N}\sum_{n\le N}|h(n)| &= O\left(\frac{1}{N}\sum_{l^2\le N} 2^{\nu(l)}\sum_{n\le N/l^2,\, n|l} 1\right) \\ &= O\left(\frac{1}{N}\sum_{l\le\sqrt{N}} 2^{\nu(l)}\tau(l)\right) = O\left(\frac{1}{N}\sum_{l\le\sqrt{N}}\tau^2(l)\right) \\ &= O\left(\frac{N^{2\varepsilon}}{N}\sum_{l\le\sqrt{N}} 1\right) = O\left(\frac{N^{2\varepsilon}\sqrt{N}}{N}\right) = o(1).\end{aligned}$$

We next show that $\sum_1^\infty h(n)/n$ is absolutely convergent for any ξ:

$$\sum_{n=1}^{N} \frac{|h(n)|}{n} \leq \sum_{l^2 \leq N} \frac{2^{\nu(l)}}{l^2} \sum_{u|l, u \leq N/l^2} \frac{1}{u} + O\left(\sum_{l \leq \sqrt{N}} \frac{\tau^2(l)}{l^2} \right).$$

Let

$$\lim_{N\to\infty} \frac{1}{N} \sum_{n=1}^{N} e^{i\xi(\omega(n)-\nu(n))} = F(\xi).$$

By Lebesgue's theorem on the integration of a bounded sequence of functions, the limit

$$\lim_{N\to\infty} \frac{1}{2\pi} \int_0^{2\pi} \left(\frac{1}{N} \sum_{n\leq N} e^{i\xi(\omega(n)-\nu(n))} \right) e^{-i\xi k}\, d\xi = \frac{1}{2\pi} \int_0^{2\pi} F(\xi) e^{-i\xi k}\, d\xi$$

exists. Since $F(\xi) = \sum_1^\infty h(n)/n$ is absolutely convergent and the function $h(n)/n$ is multiplicative, $F(\xi)$ can be expressed as an infinite product:

$$\begin{aligned}
F(\xi) &= \prod_p \left(1 + \frac{h(p)}{p} + \frac{h(p^2)}{p^2} + \cdots \right) \\
&= \prod_p \left(1 + \frac{e^{i\xi} - 1}{p^2} + \frac{e^{2i\xi} - e^{i\xi}}{p^3} + \frac{e^{3i\xi} - e^{2i\xi}}{p^4} + \cdots \right) \\
&= \prod_p \left(1 - \frac{1}{p^2} + \frac{e^{i\xi}}{p^2} \left(1 - \frac{1}{p} \right) + \frac{e^{2i\xi}}{p^3} \left(1 - \frac{1}{p} \right) + \cdots \right) \\
&= \prod_p \left(1 - \frac{1}{p} \right) \left(1 + \frac{1}{p(1 - e^{i\xi}/p)} \right) \\
&= \prod_p \left(1 - \frac{1}{p} \right) \left(1 + \frac{1}{p - e^{i\xi}} \right),
\end{aligned}$$

as required.

REMARK. It follows from the theorem that ρ_k is the coefficient of z^k in the expansion (which converges for $|z| < 1$) of the function

$$\prod_p \left(1 - \frac{1}{p}\right)\left(1 + \frac{1}{p - z}\right)$$

into a power series:

$$\prod_p \left(1 - \frac{1}{p}\right)\left(1 + \frac{1}{p - z}\right) = \rho_0 + \rho_1 z + \cdots + \rho_k z^k + \cdots .$$

In particular,

$$\rho_0 = \prod_p \left(1 - \frac{1}{p}\right)\left(1 + \frac{1}{p}\right) = \frac{6}{\pi^2}.$$

This is a well-known result on the density of square-free numbers.

The problem of investigating the asymptotic behavior of the quantity

$$\sum_{n \le x, \omega(n) - \nu(n) = k} 1$$

is called *Rényi's problem*, because Alfréd Rényi proved the first results in this direction (see Rényi [123]).

Delange [32] and Kátai [80] studied the remainder term in Rényi's problem.

The additive function $g(n) = \omega(n) - \nu(n)$ has the property that it is equal to zero at prime numbers. Kubilius established ([87], Theorem 4.8) an asymptotic law for such functions. Levin and Faĭnleĭb ([94], Theorem 2.2.2) proved the following result.

THEOREM. *Suppose $g(n)$ is an additive arithmetical function that assumes only integral values and is such that $g(p) = 0$ for all primes p. Then for any integer k*

$$\frac{1}{N} \sum_{n \le N, g(n) = k} 1 = \rho_k + O(\exp(-(\ln N)^{3/8 - \varepsilon})),$$

where the ρ_k are defined by the equality

$$\sum_{k=-\infty}^{\infty} \rho_k e^{i\xi k} = \prod_p \left(1 - \frac{1}{p}\right)\left(1 + \frac{1}{p} + \sum_{r=2}^{\infty} \frac{e^{i\xi g(p^r)}}{p^r}\right).$$

Theorem 9 of Kátai [81] provides a broad generalization of Rényi's law; it is a multidimensional local theorem for additive functions of values of an integer-valued polynomial.

§3.5. Polyadic analysis and its applications

The analogy between the theory of functions of a natural argument and the theory of functions of a real variable, of which we spoke in the Introduction, acquires a more specific character if we extend the domain of definition of the functions under consideration not only to the ring of integers, but also to some ring containing the ring of integers as a subring.

Let S denote the ring of integers. Consider the system of sets Σ^* consisting of the ideals (m). It is easy to verify the following properties.

1) The only element common to all sets in the system Σ^* is 0.

2) The intersection of two sets (m) and (n) in Σ^* is again a set in Σ^* (and is the ideal generated by the least common multiple of m and n).

3) For any set (m) in Σ^* there exists a set (n) in Σ^* such that (m) contains all differences $nx - ny$ (we must take n to be a multiple of m).

4) Suppose (m) is any set in Σ^* and mx_0 is an element of (m). Then for any (n) in Σ^* we have the set-theoretic inclusion $(n)mx_0 \subset (m)$.

5) For any set (m) in Σ^* there exists (n) in Σ^* such that $(n) \subset (m)$. We can take as n any multiple of m.

By Theorem 9 in Chapter III of Pontryagin's book [114], the system Σ^* can be taken as a complete system of neighborhoods of zero in the additive group Z of integers, and addition is continuous in this topology. Recall that the sets of the form $a + (m)$ comprise a complete system of neighborhoods in Z.

We will prove that multiplication is continuous in this topology, i.e., for any integers a and b and any neighborhood W of ab there exist a neighborhood U of a and a neighborhood V of b such that $UV \subset W$. Suppose W is $ab + (l)$. It is clear that as U we can take $a + (l)$ and as V we can take $b + (l)$.

Thus we have turned S into a topological ring, which we denote by S_τ.

The above topology is not discrete: the sequence $a + 1!, a + 2!, \ldots$ does not contain a but converges to it.

THEOREM 1. *The function*

$$\rho(x, y) = \sum_{m=1}^{\infty} \frac{1}{2^m} \left(\!\left(\frac{x-y}{m} \right)\!\right),$$

where $(\!(t)\!)$ *denotes the distance from* t *to the nearest integer, is a metric on* S_τ.

PROOF. Note first that if $\rho(x, y) = 0$, then $(\!((x-y)/m)\!) = 0$ for any m, i.e., for any m, $m | x - y$, hence $x = y$. Note also that the function $(\!(t)\!)$ satisfies the inequality

$$(\!(y - z)\!) \le (\!(y - x)\!) + (\!(x - z)\!),$$

or

$$(\!(\alpha)\!) \le (\!(\alpha + \beta)\!) + (\!(\beta)\!).$$

To prove this we must analyze each of the following cases separately:

1) $\alpha \le \frac{1}{2}$, $\beta \le \frac{1}{2}$, $\alpha + \beta \le \frac{1}{2}$;

2) $\alpha \leq \frac{1}{2}$, $\beta \leq \frac{1}{2}$, $\alpha + \beta \geq \frac{1}{2}$;
3) $\alpha \leq \frac{1}{2}$, $\beta \geq \frac{1}{2}$, $\alpha + \beta \leq 1$;
4) $\alpha \geq \frac{1}{2}$, $\beta \leq \frac{1}{2}$, $\alpha + \beta \leq 1$;
5) $\alpha \geq \frac{1}{2}$, $\beta \geq \frac{1}{2}$, $\alpha + \beta \leq \frac{3}{2}$;
6) $\alpha \geq \frac{1}{2}$, $\beta \geq \frac{1}{2}$, $\alpha + \beta \geq \frac{3}{2}$.

In view of this observation, the function $\rho(x, y)$ satisfies the triangle inequality

$$\rho(y, z) \leq \rho(x, y) + \rho(x, z).$$

Suppose an integer a is fixed. Given $\varepsilon > 0$, we can find m_0 such that

$$\sum_{m=m_0+1}^{\infty} \frac{1}{2^m} \left(\frac{a}{m}\right) < \varepsilon.$$

Consequently, $\rho(x, a) < \varepsilon$ for all $x \in a + (m_0!)$. Also, since

$$\min_{x \notin a+(m)} \left(\frac{x-a}{m}\right) = \frac{1}{m},$$

it follows that $\rho(x, a) \geq 1/2^m m$ if $x \notin a + (m)$. Thus if $\rho(x, a) < 1/2^m m$, then $x = a + bm$; in particular, if $\rho(x, a) < 1/2^{n!} n!$, then $x \equiv a \pmod{n!}$. The theorem is proved.

Note that the metrization of S_τ effected by the function $\rho(x, y)$ is not unique. Novoselov [107] gives as a metric the function

$$\overline{\rho}(x, y) = \sum_{m=1}^{\infty} \frac{\varphi_m(x, y)}{2^m},$$

where

$$\varphi_m(x, y) = \begin{cases} 0, & x \equiv y \pmod{m}, \\ 1, & x \not\equiv y \pmod{m}. \end{cases}$$

THEOREM 2. *The ring S_τ is totally bounded.*

PROOF. We have $\lim_{n \to \infty} n! = 0$.(³) Consequently, for $\varepsilon > 0$ there exists m such that $\rho(m!, 0) \leq \varepsilon$. The set $\{0, 1, \dots, m! - 1\}$ forms a finite ε-net. Indeed, if $x \equiv i \pmod{m!}$, then $\rho(x, i) \leq \varepsilon$.

It follows from this theorem that any infinite sequence of integers contains a convergent (in the sense of the topology of S_τ) subsequence; this fact can also be easily proved directly.

We give examples of functions of an integral argument that are continuous in the topology of S_τ. First of all, any periodic function $f(n)$ with integral period k is continuous. Indeed, suppose $k | s!$; then, if $n \equiv m \pmod{s!}$, for any

(³)The limit is to be understood in the sense of the topology of S_τ, i.e., $\lim_{n\to\infty} a_n = b$, where $a_n \in S_\tau$ and $b \in S_\tau$, means that for each natural number k there exists N such that $a_n - b \equiv 0 \pmod{k!}$ for $n > N$.

$\varepsilon > 0$ we have $|f(n) - f(m)| < \varepsilon$. As an example of a polyadically continuous function we have mentioned

$$f(n) = \sigma(n)\varphi(n)/n^2.$$

We extend this function to all integers n by putting $f(0) = 0$ and $f(-n) = f(n)$. It is easy to see that the function so defined is continuous in the topology of S_τ; we must, of course, consider the prime factorizations of n and $-n$ to be the same.

We call an infinite sequence of integers $a_1, a_2, \ldots$ a *null sequence* if $\lim_{n\to\infty} a_n = 0$ (the limit is understood in the sense of the topology of S_τ).

We call an infinite sequence of integers $a_1, a_2, \ldots$ a *fundamental sequence* if for each natural number k there exists a natural number N such that $a_n \equiv a_m \pmod{k!}$ for any natural numbers n and m greater than N.

It is obvious that any null sequence is fundamental.

THEOREM 3. *The space S_τ is not complete.*

PROOF. The sequence $1!$, $1! + 2!$, $1! + 2! + 3!, \ldots$ is fundamental but has no limit in S_τ.

It is easy to prove that the sum, difference, and product (elementwise) of two fundamental sequences are fundamental. Thus the fundamental sequences form a ring.

We say that a fundamental sequence $a_1, a_2, \ldots$ is *equivalent* to a fundamental sequence $b_1, b_2, \ldots$ if their difference is a null sequence. This notion of equivalence is reflexive, symmetric, and transitive. Therefore the following definition is meaningful.

DEFINITION. By a *polyadic number* we mean an equivalence class of fundamental sequences of integers.

Constructions of polyadic numbers were suggested by Prüfer [119], von Neumann [106], and Novoselov [107]. An account of the theory of polyadic numbers can be found in the book of Hewitt and Ross [71] (§10), where the set of polyadic numbers is denoted by Λ_a ($a = 2, 3, \ldots$).

It is easy to prove that if a sequence A_1 is equivalent to a sequence A_2, and a sequence B_1 is equivalent to a sequence B_2, then the sequences $A_1 \pm B_1$ and A_1B_1 are equivalent to $A_2 \pm B_2$ and A_2B_2 respectively. This observation enables us to define addition, subtraction, and multiplication on the set of polyadic numbers, i.e., to turn the set of polyadic numbers into a ring. We denote the ring of polyadic numbers by $\mathfrak{S}$.

The ring $\mathfrak{S}$ contains a part isomorphic to the ring S; an embedding of S into $\mathfrak{S}$ can be effected by associating to an element $a \in S$ the class of fundamental sequences containing the sequence $a, a, a, \ldots$.

The ring S_τ is a metric space. The completion of the metric space S_τ by the method described, for example, in Aleksandrov [1] (§6 of Chapter VII) leads to the construction of a topological space $\mathfrak{S}_\tau$. Suppose an element $x \in \mathfrak{S}_\tau$ is defined by a fundamental sequence $\{x_k\}$, and an element $y \in \mathfrak{S}_\tau$ by a fundamental sequence $\{y_k\}$. If we define the distance between x and y by

$$\rho(x, y) = \lim_{k\to\infty} \rho(x_k, y_k),$$

we obtain a metric on $\mathfrak{S}_\tau$.

THEOREM 4. *The ring $\mathfrak{S}_\tau$ is totally bounded.*

This follows from the fact that if a metric space is totally bounded, then so is its completion.

THEOREM 5. *The system of sets $\alpha + (n)$, where $\alpha \in \mathfrak{S}$, $n \in S$, $n \geq 1$, and (n) is the ideal of $\mathfrak{S}$ generated by n, is a topological basis of $\mathfrak{S}_\tau$.*

The proof can be found in Novoselov [107].

LEMMA. *Each polyadic number can be defined by a sequence of rational integers $x_0, x_1, \ldots$ such that*

$$x_n \equiv x_{n+1} \pmod{(n+1)!}.$$

PROOF. Consider a polyadic number defined by a fundamental sequence $a_1, a_2, \ldots$. Let n_1 be the smallest number such that $m, n \geq n_1$ implies $a_m \equiv a_n \pmod{2!}$. Put $\overline{n}_1 = n_1$.

Let n_2 be the smallest number such that $m, n \geq n_2$ implies $a_m \equiv a_n \pmod{3!}$. Put $\overline{n}_2 = \max(n_2, \overline{n}_1 + 1)$.

Let n_3 be the smallest number such that $m, n \geq n_3$ implies $a_m \equiv a_n \pmod{4!}$. Put $\overline{n}_3 = \max(n_3, \overline{n}_2 + 1)$, and so on. We construct the sequence

$$x_k = a_{\overline{n}_k}, \qquad k = 1, 2, \ldots.$$

Suppose

$$\overline{n}_k \leq l < \overline{n}_{k+1}, \qquad a_l - x_l = a_l - a_{\overline{n}_k}.$$

By the definition of $\overline{n}_k$, we have $a_l - x_l \equiv 0 \pmod{(k+1)!}$. Thus the sequence $\{x_k\}$ is equivalent to the original. Also, by construction, $x_k \equiv x_{k+1} \pmod{(k+1)!}$.

By a *canonical sequence* we mean a sequence of positive integers of the form

$$a_1,\ a_1 + a_2 2!,\ a_1 + a_2 2! + a_3 3!, \ldots,$$

where $0 \leq a_j \leq j$.

It is obvious that a canonical sequence is fundamental and defines a polyadic number.

LEMMA. *Each polyadic number can be defined by a canonical sequence.*

A polyadic number is given by a sequence of rational integers $x_1, x_2, \ldots$, where $x_n \equiv x_{n+1} \pmod{(n+1)!}$. There exists a unique number a_1 such that $a_1 \equiv x_1 \pmod 2$ and $a_1 = 0$ or 1. Put $y_1 = a_1$. Clearly $y_1 \equiv x_1 \pmod 2$.

Next, there exists a unique number a_2 such that

$$a_2 \equiv (x_2 - a_1)/2 \pmod 3, \qquad 0 \le a_2 \le 2.$$

Put $y_2 = a_1 + 2a_2$. Obviously $y_2 \equiv x_2 \pmod{3!}$, and hence $y_2 \equiv x_3 \pmod{3!}$.

Suppose we have constructed

$$y_n = a_1 + 2!a_2 + \cdots + n!a_n, \qquad 0 \le a_n \le n,$$
$$y_n \equiv x_n \pmod{(n+1)!}.$$

Since $y_n \equiv x_{n+1} \pmod{(n+1)!}$, it follows that $(x_{n+1} - y_n)/(n+1)!$ is an integer. We define a_{n+1} from the conditions

$$a_{n+1} \equiv \frac{x_{n+1} - y_n}{(n+1)!} \pmod{(n+2)}, \qquad 0 \le a_{n+1} \le n+1,$$

and we put $y_{n+1} = y_n + (n+1)!a_{n+1}$. It is clear that

$$y_{n+1} \equiv x_{n+1} \pmod{(n+2)!}.$$

The sequence $y_1, y_2, \ldots$ is equivalent to the sequence $x_1, x_2, \ldots$ and is canonical. The lemma is proved.

Applying the algorithm described in this lemma to the sequence $\{-1\}$, we see that the sequence

$$\{1! + 2 \cdot 2! + 3 \cdot 3! + \cdots + nn!\}$$

is equivalent to the sequence $\{-1\}$, i.e.,

$$-1 = \sum_{n=1}^{\infty} n \cdot n!\,.$$

It is easy to prove that two different canonical sequences define two different polyadic numbers.

The arithmetic in the ring of polyadic numbers was investigated by Novoselov [107].

The study of power series on the ring of polyadic numbers has significance for number theory. This aspect of polyadic analysis is discussed in Novoselov [100], and applications to number theory are given in Novoselov [109].

We consider here the theory of measure and integral in the polyadic domain. On the additive group of the ring $\mathfrak{S}_\tau$, as on a compact group, there exists a translation-invariant measure, i.e., a Haar measure. It is possible to introduce a measure on $\mathfrak{S}_\tau$ by means of the Haar construction, but, following the example of Sprindzhuk ([136], Part II, Chapter 1, §2), we introduce a measure on the ring of polyadic numbers by means of the canonical representation of polyadic numbers. The canonical representation yields a mapping

$$\mathfrak{S}_\tau \leftrightarrow (a_1, a_2, \dots),$$

where the a_n are integers such that $0 \le a_n \le n$. We will introduce a measure on the set of sequences. The mapping will carry this measure over to $\mathfrak{S}_\tau$.

Suppose s is any natural number. The set M of sequences $a_1, a_2, \dots$ $(0 \le a_n \le n)$ in which the first s symbols are fixed will be called an *elementary cylindrical set*. By a *cylindrical set* we mean a finite set-theoretic sum of elementary cylindrical sets. It is easy to prove that the cylindrical sets form an algebra.

Note that the elementary cylindrical set obtained by fixing a finite number of elements $a_1, \dots, a_s$ is a progression in $\mathfrak{S}_\tau$:

$$a_1 + a_2 2! + \dots + a_s s! + ((s+1)!).$$

The algebra of cylindrical sets contains all progressions $b + (n)$, where n is a natural number and b is an integer, $0 \le b \le n-1$. Indeed, let $s+1$ be the smallest number such that $n | (s+1)!$. The progression $b + (n)$ splits into $(s+1)!/n$ progressions modulo $(s+1)!$.

The measure of the cylindrical set determined by fixing the elements $a_1, \dots, a_s$ is defined to be $1/(s+1)!$. The measure of a cylindrical set is defined to be the sum of the measures of nonoverlapping elementary cylindrical sets whose union is the given cylindrical set. It is easy to see that this definition is unambiguous. Under this definition, the measure of the progression $b + (n)$, where n is a natural number, is $1/n$.

By a well-known theorem of measure theory, we can uniquely extend this measure to the smallest σ-algebra containing the algebra of cylindrical sets. The measure of a measurable set E will be denoted by μE.

The proof of the following result is analogous to that of Lemma 4 of Sprindzhuk [136], Part II, Chapter 1, §3.

LEMMA. *Suppose E is a measurable set of points in $\mathfrak{S}_\tau$ and $\omega_0 \in \mathfrak{S}_\tau$. Let $E_0 = \omega_0 + E$. Then E_0 is measurable, and $\mu E_0 = \mu E$.*

The measure μ is called the *Haar-Lebesgue measure* on $\mathfrak{S}_\tau$. The following result is proved in Novoselov [107].

LEMMA. *Suppose $(n_i, n_j) = 1$. Then*

$$\mu\left(\bigcup_{k=1}^{m} \alpha_k + (n_k)\right) = \sum_{k=1}^{m} \frac{1}{n_k} \prod_{i=1}^{k-1} \left(1 - \frac{1}{n_i}\right) = 1 - \prod_{k=1}^{m} \left(1 - \frac{1}{n_k}\right).$$

Novoselov [108] gives applications of the theory of measure on $\mathfrak{S}_T$ to the calculation of densities of certain subsets of the natural sequence. We will not dwell on these connections here. We will restrict ourselves to giving elementary proofs of certain theorems used by Novoselov to illustrate his theory.

THEOREM 6. *Suppose a sequence of natural numbers*

$$m_1 < m_2 < \cdots \tag{1}$$

is such that the smallest prime divisor of m_n tends to infinity as $n \to \infty$. Then the density of the set of numbers in this sequence is zero.

PROOF. We number the primes in increasing order: $p_1 = 2$, $p_2, \ldots$. Let $s(k)$ denote the smallest number in the sequence (1) such that no $m_i > s(k)$ is divisible by $p_1, \ldots, p_k$. By assumption, $s(k) \to \infty$ as $k \to \infty$. If $N \geq 1$, we have

$$\sum_{m_i \leq N} 1 \leq s(k) + \sum_{\substack{l \in \mathfrak{M}_{p_1 \cdots p_k} \\ s(k) \leq l \leq N}} 1,$$

where $\mathfrak{M}_{p_1 \cdots p_k}$ is the set of numbers relatively prime to $p_1 \cdots p_k$. Therefore

$$\sum_{m_i \leq N} 1 \leq s(k) + \sum_{\substack{l \in \mathfrak{M}_{p_1 \cdots p_k} \\ l \leq N}} 1 = s(k) + \sum_{d | p_1 \cdots p_k} \mu(d) \left[\frac{N}{d}\right]$$

$$= N\left(1 - \frac{1}{p_1}\right) \cdots \left(1 - \frac{1}{p_k}\right) + s(k) + 2^k.$$

It follows that for any k

$$\overline{\lim_{N \to \infty}} \frac{1}{N} \sum_{m_i \leq N} 1 = \left(1 - \frac{1}{p_1}\right) \ldots \left(1 - \frac{1}{p_k}\right).$$

But since

$$\left(1 - \frac{1}{p_1}\right) \ldots \left(1 - \frac{1}{p_k}\right) \sim \frac{e^{-\gamma}}{\ln p_k} \to 0$$

as $k \to \infty$, we have

$$\lim_{N \to \infty} \frac{1}{N} \sum_{m_i \leq N} 1 = 0.$$

It follows from this theorem, by the way, that

$$\pi(x) = o(x), \tag{2}$$

since the sequence of primes satisfies the conditions of the problem. Our argument is related to the classical proof of (2) given by Ingham ([74], Chapter I, §3, Theorem 2) and Prachar ([118], Chapter I, §2, Theorem 2.3).

THEOREM 7. *The density of the set of numbers made up of two prime factors is equal to zero.*

Let $p_1, \dots, p_r$ be the first r primes, arranged in increasing order. Let $M(x, p_i)$, $i = 1, \dots, r$, be the number of natural numbers not exceeding x that are divisible by p_i and relatively prime to $p_1 \cdots p_r / p_i$. It is obvious that

$$M(x,p_i) = \sum_{\substack{n \equiv 0 \pmod{p_i} \\ n \le x}} \sum_{d \mid \left(n, \frac{p_1 \cdots p_r}{p_i}\right)} \mu(d)$$

$$= \sum_{\substack{d \mid \frac{p_1 \cdots p_r}{p_i} \\ d \le x}} \mu(d) \left[\frac{x}{dp_i}\right] = \frac{x}{p_i} \prod_{\substack{j=1 \\ j \ne i}}^{r} \left(1 - \frac{1}{p_j}\right) + \theta 2^{r-1}$$

Let $\pi_2(x)$ denote the number of natural numbers not exceeding x and made up of two prime factors. Suppose u is a positive number and $p_1, \dots, p_r$ are the primes not exceeding u. It is obvious that for $x > u$

$$\pi_2(x) \le \pi_2(u) + \sum_{i=1}^{r} M(x, p_i).$$

Therefore,

$$\pi_2(x) \le \pi_2(u) + x \sum_{i=1}^{r} \frac{1}{p_i} \prod_{j \ne i} \left(1 - \frac{1}{p_j}\right) + r \cdot 2^{r-1}$$

$$\le u + r2^{r-1} + x \prod_{p_i \le u} \left(1 - \frac{1}{p_i}\right) \sum_{p_i \le u} \frac{1}{p_i - 1}$$

$$\le u + r2^{r-1} + xO\left(\frac{\ln \ln u}{\ln u}\right)$$

with an absolute constant implied by the symbol O. Consequently,

$$\overline{\lim_{x \to \infty}} \frac{\pi_2(x)}{x} \le C \frac{\ln \ln u}{\ln u}.$$

But u can be arbitrarily large. Thus,

$$\lim_{x \to \infty} \frac{\pi_2(x)}{x} = 0.$$

In a similar way we can prove

THEOREM 8. *Let $\pi_k(x)$ denote the number of natural numbers not exceeding x and made up of exactly k prime factors. Then $\pi_k(x) = o(x)$.*

Theorems 7 and 8, as we know, can be strengthened by the generalized asymptotic law of the distribution of primes proved by Landau:

$$\pi_k(x) \sim \frac{x (\ln \ln x)^{k-1}}{(k-1)! \ln x}.$$

Integration on $\mathfrak{S}_\tau$ can be constructed according to the classical model by means of the Haar–Lebesgue measure.

Novoselov [111] applied the theory of measure and integral on $\mathfrak{S}_\tau$ to questions concerning the distribution of values of arithmetical functions. The main feature here is continuation of a function of a natural argument to the ring $\mathfrak{S}_\tau$, where the continuation is such that the function with extended domain of definition belongs to some "nice" class. Let us be more specific. Novoselov [108] proved that the sequence of natural numbers $1, 2, \ldots$ is uniformly distributed in $\mathfrak{S}_\tau$; this means that

$$\lim_{N\to\infty} \frac{1}{N} \sum_{n=1}^{N} f(n) = \int f(x)\, d\mu$$

for any continuous function $f(x)$ on $\mathfrak{S}_\tau$. Therefore, if we can prove that some function of a natural argument can be extended to $\mathfrak{S}_\tau$ so that it becomes continuous on $\mathfrak{S}_\tau$, then this implies the existence for $f(x)$ of the limit

$$\lim_{N\to\infty} \frac{1}{N} \sum_{n=1}^{N} f(n).$$

A polyadically continuous function $f(n)$ of a natural argument can always be continued to $\mathfrak{S}_\tau$. We put $f(0) = 0$. Suppose α is defined by the canonical series

$$\alpha = \sum_{j=1}^{\infty} a_j j!, \qquad 0 \le a_j \le j.$$

By the property of polyadic continuity, the sequence $f(\sum_1^n a_j j!)$ satisfies the Cauchy criterion; hence the limit

$$\lim_{n\to\infty} f\left(\sum_{j=1}^{n} a_j j!\right)$$

exists; then $f(\alpha)$ is defined to be the value of this limit. It is easy to show that the function so defined is continuous on $\mathfrak{S}_\tau$. Thus we see that for any polyadically continuous function $f(n)$ of a natural argument the limit

$$\lim_{N\to\infty} \frac{1}{N} \sum_{n\le N} f(n)$$

exists.

The theory of continuous functions on $\mathfrak{S}_\tau$ is identical to the theory of almost periodic functions on $\mathfrak{S}_\tau$, since, as is well known, any continuous function on a compact group is almost periodic (see, for example, Levitan [97], Part I, Chapter VI, §8, Theorem 6.8.1).

In [111] Novoselov introduced and studied certain classes of functions on $\mathfrak{S}_\tau$. We denote by L_H the class of continuous functions on $\mathfrak{S}_\tau$, by L_C the class of continuous almost everywhere (with respect to the measure μ) and bounded functions, and by L_* the class of continuous almost everywhere functions on $\mathfrak{S}_\tau$. The symbol $\Rightarrow$ will denote convergence in $\mathfrak{S}_\tau$. Suppose $\{N_k\}$ is an arbitrary but fixed sequence of natural numbers that converges to zero in $\mathfrak{S}_\tau$ and satisfies the conditions $N_{k+1} > N_k$ and $N_{k+1}/N_k \to 1$. Letting $s(n) = 1^2 + 2^2 + \cdots + n^2$, we can put $N_k = (k - s(n) + n + 2)n!$ if $s(n) \leq k < s(n+1)$. We denote by $R_K(x)$ the smallest positive residue of the polyadic number x modulo N_K. Let $\mathfrak{H}^0$ be the class of functions $f(x)$ such that $f(R_K(x))$ converges to $f(x)$ with respect to the measure μ as $k \to \infty$. Novoselov [111] showed that the definition of the class $\mathfrak{H}^0$ does not essentially depend on the choice of the specific sequence N_k. Let $\mathfrak{H}^r$, $r > 0$, denote the class of functions $f(x)$ for which

$$\lim_{k\to\infty} \left(\int |f(R_K(x)) - f(x)|^r \, d\mu \right)^{1/r} = 0.$$

The definition of $\mathfrak{H}^r$ also does not essentially depend on the choice of N_k. Note that $r_1 \geq r_2$ implies $\mathfrak{H}^{r_1} \subset \mathfrak{H}^{r_2}$. The class $\mathfrak{H}_C$ is defined to be the set of bounded functions in $\mathfrak{H}^0$. The relations among these classes are illustrated in Figure 6; arrows denote inclusions.

$$\begin{array}{ccccccc}
 & & & & \mathfrak{H}_C & \longrightarrow & \mathfrak{H}^r \\
 & & & \nearrow & & & \\
L_H & \longrightarrow & L_C & & & & \downarrow \\
 & & & \searrow & & & \\
 & & & & L_* & \longrightarrow & \mathfrak{H}^0
\end{array}$$

FIGURE 6

We state some properties of the above classes.

THEOREM 9. *Every function in $\mathfrak{H}^0$ is measurable and is finite almost everywhere. If the real-valued function $f(x) \in \mathfrak{H}^0$, then the sequence of distribution functions*

$$F_N(\lambda) = \frac{1}{N} \sum_{f(m)<\lambda, m\leq N} 1$$

is weakly convergent to some limit distribution function.

THEOREM 10. *If the real-valued function $f(x)$ belongs to $\mathfrak{H}^0$, then for any real t*

$$\lim_{N\to\infty} \frac{1}{N} \sum_{n\leq N} e^{itf(n)} = \int e^{itf(n)} \, d\mu.$$

This is Proposition 26 of [111].

THEOREM 11. *Suppose $f(x) \in \mathfrak{H}^r$, $r > 0$. Let M denote the closure of the numerical sequence $f(m)$, $m = 1, 2, \ldots$. Suppose the function $g(z)$ is defined for $z \in M$ and satisfies a first-order Lipschitz condition on M. Then $g(f(x)) \in \mathfrak{H}^r$.*

This is Proposition 24 of [111].

THEOREM 12. *If $f(x) \in \mathfrak{H}^1$, then for any natural number m and for $0 \le l \le m-1$*

$$\lim_{N\to\infty} \frac{1}{N} \sum_{\substack{n\equiv l \pmod m \\ n\le N}} f(n) = \int_{l+(m)} f(x)\, d\mu.$$

See also [111], §2.

Novoselov gives a criterion (and methods) for extending functions defined for natural numbers m to the ring $\mathfrak{S}_\tau$. For brevity, instead of saying "the function $f(m)$ of a natural argument can be extended to $\mathfrak{S}_\tau$ so that the extended function is of class..." we will simply say "$f(m)$ is of class...".

THEOREM 13. *Suppose that $f_d(m)$, $1 \le d < \infty$, is a sequence of periodic functions with natural periods k_d and that the series*

$$\sum_{d=1}^{\infty} \frac{1}{k_d} \sum_{m=1}^{k_d} |f_d(m)|$$

converges. Then the function $f(m) = \sum_{k_d \le m} f_d(m)$ is of class $\mathfrak{H}^1$, and

$$\lim_{N\to\infty} \frac{1}{N} \sum_{m\le N} f(m) e^{2\pi i(a/n)m} = \sum_{k_d \equiv 0 \pmod n} \frac{1}{k_d} \sum_{m=1}^{k_d} f_d(m) e^{2\pi i(a/n)m}.$$

See [111], §4.

A special case of this theorem is

THEOREM 14. *Functions of a natural argument that belong to the Wintner class W also belong to $\mathfrak{H}^1$.*

THEOREM 15. *Suppose a function $f(n)$ is represented in the form $f(n) = \sum_{d|n} \Phi(d)$, and the function $\Phi(d)$ is such that the series*

$$\sum_{|\Phi(d)|\ge 1} \frac{1}{d}, \qquad \sum_{|\Phi(d)|<1} \frac{|\Phi(d)|}{d}$$

converge. Then $f(n) \in \mathfrak{H}^0$.

This is Proposition 44 of [111].

In [111] Novoselov interpreted a number of propositions of the theory of multiplicative functions in terms of polyadic analysis. We will not discuss this matter here.

Finally, one can find in [111] some details on the history of polyadic analysis and the necessary bibliographic references.

Polyadic analysis enables us to comprehend from a general point of view many results pertaining to the distribution of values of arithmetical functions, and not only multiplicative ones. It enables us to discover new concealed facts; we mention, as an example, a theorem on the continuity of a function:

$$\Phi(x) = \lim_{N\to\infty} \frac{1}{N} \sum_{\substack{n\le N \\ \sum_{d|n} 1/2d \le x}} 1$$

(see [111], §5, Example 1).

§3.6. Axer's theorem

The following result, which is called Axer's theorem (see Hardy [66], p. 378), is a strengthening of Wintner's theorem.

THEOREM. *Suppose a function $f(n)$ of a natural argument is represented in the form $f(n) = \sum_{d|n} \Phi(d)$, and it is known that* (1) *$\sum_1^\infty \Phi(d)/d$ converges, and* (2) *$\sum_{d\le N} \mid \Phi(d) \mid = O(N)$. Then*

$$\lim_{N\to\infty} \frac{1}{N} \sum_{n=1}^{N} f(n) = \sum_{d=1}^{\infty} \frac{\Phi(d)}{d}.$$

PROOF. We first show that convergence of $\sum_1^\infty \Phi(d)/d$ implies $\sum_{d\le N} \Phi(d) = o(N)$.

Let $\sum_1^\infty \Phi(d)/d = A$ and $\sum_1^N \Phi(d)/d = S_N$. By hypothesis, $S_N = A + o(1)$ as $N \to \infty$. But

$$\sum_{d=1}^{N} \Phi(d) = \sum_{d=1}^{N} \frac{\Phi(d)}{d} d = S_1 + \sum_{d=2}^{N} (S_d - S_{d-1}) d$$
$$= NS_N - S_1 - S_2 - \cdots - S_{N-1}.$$

Therefore,

$$\sum_{d=1}^{N} \Phi(d) = N(A + o(1)) - NA + o(N) = o(N).$$

Suppose the conditions of the theorem are satisfied. We have

$$\begin{aligned}\frac{1}{N}\sum_{n\le N}\sum_{d|N}\Phi(d) &= \frac{1}{N}\sum_{d\le N}\Phi(d)\left[\frac{N}{d}\right]\\ &= \sum_{d\le N}\frac{\Phi(d)}{d} - \frac{1}{N}\sum_{d\le N}\Phi(d)\left\{\frac{N}{d}\right\}\\ &= \sum_{d=1}^{\infty}\frac{\Phi(d)}{d} + o(1) - \frac{1}{N}\sum_{d\le N}\Phi(d)\left\{\frac{N}{d}\right\}.\end{aligned}$$

Everything will be proved once we show that

$$\frac{1}{N}\sum_{d\le N}\Phi(d)\left\{\frac{N}{d}\right\}\to 0.$$

Take any δ such that $0<\delta<1$. We have

$$\begin{aligned}&\left|\frac{1}{N}\sum_{d\le N}\Phi(d)\left\{\frac{N}{d}\right\}\right|\\ &\quad\le \left|\frac{1}{N}\sum_{d\le N\delta}\Phi(d)\left\{\frac{N}{d}\right\}\right| + \left|\frac{1}{N}\sum_{N\delta<d\le N}\Phi(d)\left\{\frac{N}{d}\right\}\right|\\ &\quad\le \frac{1}{N}\sum_{d\le N\delta}|\Phi(d)| + \frac{1}{N}\left|\sum_{n=[\delta N]+1}^{N}(A_n-A_{n-1})\left\{\frac{N}{n}\right\}\right|,\end{aligned}$$

where $A_0=0$ and $A_n=\sum_{d\le n}\Phi(d)$. For an integer N

$$\begin{aligned}\left|\frac{1}{N}\sum_{d\le N}\Phi(d)\left\{\frac{N}{d}\right\}\right| &\le \frac{1}{N}\sum_{d\le N\delta}|\Phi(d)|\\ &\quad+\frac{1}{N}\left|-A_{[\delta N]}\left\{\frac{N}{[\delta N]+1}\right\}+\sum_{n=[\delta N]+1}^{N-1}A_n\left(\left\{\frac{N}{n}\right\}-\left\{\frac{N}{n+1}\right\}\right)\right|\\ &\le \frac{1}{N}\sum_{d\le N\delta}|\Phi(d)| + \frac{1}{N}|A_{[\delta N]}|\\ &\quad+\frac{1}{N}\max_{[\delta N]+1\le n\le N-1}|A_n|\sum_{n=[\delta N]+1}^{N-1}\left|\left\{\frac{N}{n}\right\}-\left\{\frac{N}{n+1}\right\}\right|.\end{aligned}$$

In the interval $[\delta N]+1\le n\le N-1$ there are at most $1/\delta+2$ values of n such that $[N/n]\ne[N/(n+1)]$. For these n we use the estimate

$$|\{N/n\}-\{N/(n+1)\}|\le 2.$$

For those n for which $[N/n]=[N/(n+1)]$ we use the estimate

$$\left|\left\{\frac{N}{n}\right\}-\left\{\frac{N}{n+1}\right\}\right| = \frac{N}{n(n+1)} \le \frac{N}{(\delta N)^2},$$

and there are at most $N(1-\delta)$ such values of n in the interval $[\delta N]+1 \le n \le N-1$. Thus,

$$\left|\frac{1}{N}\sum_{d\le N}\Phi(d)\left\{\frac{N}{d}\right\}\right| \le \frac{1}{N}\sum_{d\le N\delta}|\Phi(d)| + \frac{1}{N}|A_{[\delta N]}| + \frac{1}{N}\max_{[\delta N]+1\le n\le N-1}|A_n|\left(\frac{2}{\delta}+4+\frac{1-\delta}{\delta^2}\right).$$

For $\varepsilon > 0$ we can find a δ such that for $N \ge N_0$

$$\frac{1}{N}\sum_{d\le N\delta}|\Phi(d)| \le \frac{\varepsilon}{4}.$$

Furthermore, for a fixed δ we can find $N_1 > N_0$ such that

$$\frac{1}{N}|A_{[\delta N]}| \le \frac{\varepsilon}{4},$$

$$\frac{1}{N}\max_{[\delta N]+1\le n\le N-1}|A_n|\left(\frac{2}{\delta}+4+\frac{1-\delta}{\delta^2}\right) \le \frac{\varepsilon}{4}.$$

Thus for $N \ge N_1$

$$\left|\frac{1}{N}\sum_{d\le N}\Phi(d)\left\{\frac{N}{d}\right\}\right| < \varepsilon,$$

i.e.,

$$\overline{\lim_{N\to\infty}}\frac{1}{N}\left|\sum_{d\le N}\Phi(d)\left\{\frac{N}{d}\right\}\right|\varepsilon.$$

But ε can be arbitrarily small. The theorem is proved.

We will show that Wintner's theorem is a consequence of Axer's theorem. Namely, we can make the following assertion.

REMARK. If $\sum_1^\infty \Phi(d)/d$ is absolutely convergent, then

$$\lim_{N\to\infty}\frac{1}{N}\sum_{d\le N}|\Phi(d)| = 0.$$

Let $B_k = \sum_1^k |\Phi(d)|/d$. Using Abel's transformation, we obtain

$$\frac{1}{N}\sum_{d=1}^{N}|\Phi(d)| = \frac{1}{N}\sum_{d=1}^{N} d\frac{|\Phi(d)|}{d} = B_N - \frac{1}{N}\sum_{k=1}^{N-1}B_k.$$

By hypothesis, there exists a constant κ such that $\lim_{N\to\infty} B_N = \kappa$. But by the theorem on the limit of an arithmetic mean,

$$\lim_{N\to\infty}\frac{1}{N}\sum_{k=1}^{N-1}B_k = \kappa,$$

and hence $B_N - (1/N)\sum_{k=1}^{N-1} B_k \to 0$ as $N \to \infty$. Consequently,

$$\frac{1}{N}\sum_{d=1}^{N} |\Phi(d)| = o(1),$$

as required.

The method used in the proof of Axer's theorem can be applied to functions other than multiplicative ones. Consider, for example, the function

$$P_\alpha(n) = \sum_{d|n} \sin 2\pi d\alpha,$$

where α is not an integer. Here $\Phi(d) = \sin 2\pi\, d\alpha$. It is obvious that

$$\sum_{d\le N} |\Phi(d)| = O(N),$$

and the convergence of $\sum_{d=1}^{\infty} d^{-1} \sin 2\pi\, d\alpha$ is a well-known fact of the theory of Fourier series (Fikhtengol′ts [50], Chapter XIX, §2, Examples):

$$\sum_{d=1}^{\infty} \frac{\sin 2\pi d\alpha}{d} = -\pi\left(\{\alpha\} - \frac{1}{2}\right).$$

We obtain

$$\lim_{N\to\infty} \frac{1}{N}\sum_{n\le N} P_\alpha(n) = -\pi\left(\{\alpha\} - \frac{1}{2}\right).$$

For generalizations of Axer's theorem we refer the reader to Atkinson and Cherwell [3].

CHAPTER 4

Theory of Multiplicative Functions

§4.1. Upper bounds of multiplicative functions

The following theorem is due to Drozdova and Freĭman [33].

THEOREM 1. *Suppose the canonical decomposition of a number N is $p_1^{\alpha_1}\cdots p_r^{\alpha_r}$, and let there be given a function $l(\alpha)$, defined for natural numbers α and assuming positive values, such that*

$$\ln\frac{l(s+1)}{l(s)} = O\left(\frac{1}{s}\right). \tag{1}$$

Let $f(N) = l(\alpha_1)\cdots l(\alpha_r)$. Then

$$f(N) \le \exp\left(\sup_{m\ge 1}\frac{\ln l(m)}{m}\frac{\ln N}{\ln\ln N} + O\left(\frac{\ln N}{(\ln\ln N)^2}\ln\ln\ln N\right)\right). \tag{2}$$

We consider a more general situation. Suppose G is a free multiplicative semigroup consisting of natural numbers. If 1 does not belong to G, we include it. Let $\omega_1, \omega_2, \dots$ be a basis of G. Suppose we are given a positive function $l(\alpha)$ of a natural argument satisfying (2). We define a function on G as follows: if $N = \pi_1^{\alpha_1}\cdots\pi_s^{\alpha_s}$ is the canonical decomposition of N, then $f(N) = l(\alpha_1)\cdots l(\alpha_s)$.

THEOREM 2. *Arrange the elements of the basis of G in increasing order: $\omega_1 < \omega_2 < \cdots$. Assume that, as $s\to\infty$, $\ln\omega_s = \ln s + O(\ln\ln s)$. Then*

$$f(N) \le \exp\left(\sup_{m\ge 1}\frac{\ln l(m)}{m}\frac{\ln N}{\ln\ln N} + O\left(\frac{\ln N}{(\ln\ln N)^2}\ln\ln\ln N\right)\right).$$

PROOF. Suppose the canonical decomposition of N is $\pi_1^{\alpha_1}\cdots\pi_r^{\alpha_r}$. We may assume without loss of generality that $\alpha_1 \ge \alpha_2 \ge \cdots \ge \alpha_r$. We define numbers k_s for $s = 1, \dots, \alpha_1$ as follows:

$$k_s = \begin{cases} i, & \text{if } \alpha_{i+1} < s < \alpha_i, \\ t, & \text{if } s = \alpha_i = \alpha_{i+1} = \cdots = \alpha_t > \alpha_{t+1}. \end{cases}$$

It is clear that k_s is a nonincreasing function of the index s. We can write $f(N)$ in the form

$$\begin{aligned} f(N) &= l(1)^{k_1-k_2} l(2)^{k_2-k_3} \cdots l(\alpha_1-1)^{k_{\alpha_1-1}-k_{\alpha_1}} l(\alpha_1)^{k_{\alpha_1}} \\ &= l(1)^{k_1} \left(\frac{l(2)}{l(1)}\right)^{k_2} \cdots \left(\frac{l(\alpha_1)}{l(\alpha_1-1)}\right)^{k_{\alpha_1}} \\ &= \exp\left(k_1 \ln l(1) + k_2 \ln \frac{l(2)}{l(1)} + \cdots + k_{\alpha_1} \ln \frac{l(\alpha_1)}{l(\alpha_1-1)}\right) = e^{\Phi(N)}. \end{aligned}$$

Let $s_1 = [(\ln \ln N)^2]$. We denote by s' the largest s for which

$$k_s \geq (\ln N)/(\ln \ln N)^2.$$

Finally, we put $s_2 = \min(s_1, s')$.

We split the sum $\Phi(N)$ into three parts:

$$\Phi(N) = \sum_{s=1}^{s_2} + \sum_{s=s_2+1}^{s_1} + \sum_{s=s_1+1}^{\alpha_1}$$

(if $s_1 \leq s'$, then $s_1 = s_2$ and the middle sum is empty). Let $M_k = \omega_1 \cdots \omega_k$. By the hypothesis of the theorem, $\ln \omega_k = \ln k + O(\ln \ln k)$. Hence

$$\ln M_k = k \ln k + O(k \ln \ln k)$$

and therefore

$$\ln k = \ln \ln M_k + O(\ln \ln \ln M_k),$$

which implies

$$k = \frac{\ln M_k}{\ln \ln M_k} + O\left(\frac{\ln M_k}{(\ln \ln M_k)^2} \ln \ln \ln M_k\right). \tag{3}$$

By a theorem on rearrangements (see Hardy, Littlewood, and Pólya [69], §10.2), the quantity $\sum_1^r \alpha_j \ln \pi_j$ attains its minimum when along with the inequalities $\alpha_1 \geq \cdots \geq \alpha_r$ assumed at the outset we have $\pi_1 < \cdots < \pi_r$. Therefore,

$$N = \pi_1^{\alpha_1} \cdots \pi_r^{\alpha_r} \geq \omega_1^{\alpha_1} \cdots \omega_r^{\alpha_r}.$$

It follows that

$$N \geq (\omega_1 \cdots \omega_{k_1})(\omega_1 \cdots \omega_{k_2}) \cdots (\omega_1 \cdots \omega_{k\alpha_1}) \geq (\omega_1 \cdots \omega_{k_s})^s$$

for any $s = 1, \dots, \alpha_1$. This together with (3) implies that for $s \le s_1$

$$\begin{aligned} k_s &\le \frac{\ln N^{1/s}}{\ln \ln N^{1/s}}\left(1 + O\left(\frac{\ln \ln \ln N}{\ln \ln N^{1/s}}\right)\right) \\ &= \frac{\ln N}{s \ln(\ln N/s)}\left(1 + O\left(\frac{\ln \ln \ln N}{\ln(\ln N/(\ln \ln N)^2)}\right)\right) \\ &= \frac{\ln N}{s(\ln \ln N - \ln s)}\left(1 + O\left(\frac{\ln \ln \ln N}{\ln \ln N}\right)\right) \\ &= \frac{\ln N}{s \ln \ln N(1 + O(\frac{\ln \ln \ln N}{\ln \ln N}))} = \frac{\ln N}{s \ln \ln N}\left(1 + O\left(\frac{\ln \ln \ln N}{\ln \ln N}\right)\right); \end{aligned}$$

hence for $s \le s_1$ and a sufficiently large N

$$k_s \le \frac{2}{s}\,\frac{\ln N}{\ln \ln N}. \tag{4}$$

For $s \le s_2$, in view of the inequality

$$\frac{\ln N}{(\ln \ln N)^2} \le k_s = O\left(\frac{\ln N}{s \ln \ln N}\right) = O\left(\frac{\ln N}{\ln \ln N}\right),$$

we obtain

$$\ln k_s = \ln \ln N + O(\ln \ln \ln N). \tag{5}$$

From $N \ge (\omega_1 \cdots \omega_{k_1})(\omega_1 \cdots \omega_{k_2}) \cdots (\omega_1 \cdots \omega_{k_{\alpha_1}})$ we have

$$\ln N \ge \sum_{j=1}^{\alpha_1} \ln M_{k_j} = \sum_{j=1}^{\alpha_1} (k_j \ln k_j + O(k_j \ln \ln k_j)).$$

This yields the rough estimate

$$\sum_{j=1}^{\alpha_1} k_j = O(\ln N), \tag{6}$$

but for $s \le s_2$, in view of (5), we have a more precise estimate:

$$\begin{aligned} \ln N &\ge \sum_{s=1}^{s_2} k_s(\ln \ln N + O(\ln \ln \ln N)) + O\left(\ln \ln \ln N \sum_{s=1}^{s_2} k_s\right) \\ &= \left(\sum_{s=1}^{s_2} k_s\right)(\ln \ln N + O(\ln \ln \ln N)). \end{aligned}$$

It follows that

$$k_1 + k_2 + \cdots + k_{s_2} \le \frac{\ln N}{\ln \ln N} + O\left(\frac{\ln N}{(\ln \ln N)^2} \ln \ln \ln N\right). \tag{7}$$

We now estimate the sum $\sum_{s=1}^{s_2}$. We have

$$\begin{aligned}\sum_{s=1}^{s_2} &= k_1 \ln l(1) + \sum_{s=2}^{s_2} k_s \ln \frac{l(s)}{l(s-1)} \\ &= (k_1 - k_2)\ln l(1) + (k_2 - k_3)\ln l(2) \\ &\qquad + \cdots + (k_{s_2-1} - k_{s_2})\ln l(s_2 - 1) + k_{s_2} \ln l(s_2) \\ &= (k_1 - k_2)\ln l(1) + (2k_2 - 2k_3)\frac{\ln l(2)}{2} \\ &\qquad + \cdots + ((s_2-1)k_{s_2-1} - (s_2-1)k_{s_2})\frac{\ln l(s_2-1)}{s_2-1} + s_2 k_{s_2}\frac{\ln l(s_2)}{s_2} \\ &\le (k_1 + k_2 + \cdots + k_{s_2}) \sup_{m\ge 1} \frac{\ln l(m)}{m} \\ &\le \sup_{m\ge 1}\frac{\ln l(m)}{m}\frac{\ln N}{\ln\ln N} + O\left(\frac{\ln N}{(\ln\ln N)^2}\ln\ln\ln N\right).\end{aligned}$$

Let us estimate the sum $\sum_{s_2+1}^{s_1}$. Since $k_s \le (\ln N)/(\ln\ln N)^2$ for $s > s_2 - s'$, we obtain

$$\sum_{s=s_2+1}^{s_1} k_s \ln\frac{l(s)}{l(s-1)} = O\left(\sum_{s=s_2+1}^{s_1}\frac{\ln N}{(\ln\ln N)^2}\frac{1}{s}\right) = O\left(\frac{\ln N\ln\ln\ln N}{(\ln\ln N)^2}\right).$$

Now consider $\sum_{s=s_1+1}^{\alpha_1}$. We have

$$\sum_{s=s_1+1}^{\alpha_1} k_s \ln\frac{l(s)}{l(s-1)} = O\left(\sum_{s=s_1+1}^{\alpha_1}\frac{k_s}{s}\right) = O\left(\frac{1}{(\ln\ln N)^2}\sum_{s=s_1+1}^{\alpha_1} k_s\right).$$

Therefore, in view of (7),

$$\sum_{s=s_1+1}^{\alpha_1} k_s \ln\frac{l(s)}{l(s-1)} = O\left(\frac{\ln N}{(\ln\ln N)^2}\right).$$

Thus,

$$\Phi(N) \le \sup_{m\ge 1}\frac{\ln l(m)}{m}\frac{\ln N}{\ln\ln N} + O\left(\frac{\ln N}{(\ln\ln N)^2}\ln\ln\ln N\right),$$

and hence

$$f(N) \le \exp\left[\sup_{m\ge 1}\frac{\ln l(m)}{m}\frac{\ln N}{\ln\ln N} + O\left(\frac{\ln N}{(\ln\ln N)^2}\ln\ln\ln N\right)\right].$$

Theorem 2 is proved.

The Drozdova–Freĭman theorem follows from the above result: we take as G the whole multiplicative semigroup of natural numbers. By a theorem of Tchebycheff,

$$C_1 x/\ln x < \pi(x) < C_2 x/\ln x,$$

and hence for the nth prime q_n we have

$$C_1'n \ln n \le q_n \le C_2'n \ln n.$$

Thus, $\ln q_n = \ln n + O(\ln \ln n)$, and the conditions of Theorem 2 are satisfied.

We will use the Drozdova–Freĭman theorem to obtain an upper bound for the function $\nu(N)$. Suppose $N = p_1^{\alpha_1} \cdots p_r^{\alpha_r}$; then

$$2^{\nu(N)} = \prod_{p_i | N} l(\alpha_i),$$

where $l(\alpha) \equiv 2$. By the Drozdova–Freĭman theorem,

$$2^{\nu(N)} < 2^{\frac{\ln N}{\ln \ln N} + O(\frac{\ln N}{(\ln \ln N)^2} \ln \ln \ln N)},$$

i.e.,

$$\nu(N) < \frac{\ln N}{\ln \ln N} + O\left(\frac{\ln N}{(\ln \ln N)^2} \ln \ln \ln N\right).$$

We will show that the bound for $\nu(N)$ cannot be improved. Take the sequence of numbers N_j defined by $N_j = 2 \cdot 3 \cdot 5 \cdots q_j$ (q_j is the jth prime). Then

$$2^{\nu(N_j)} = 2^{\pi(q_j)} = 2^{q_j / \ln q_j + O(q_j/(\ln q_j)^2)}.$$

But

$$\ln N_j = \sum_{p \le q_j} \ln p.$$

By the asymptotic law of the distribution of primes,

$$\ln N_j = q_j + O(q_j/(\ln q_j)^2).$$

Consequently, for the chosen sequence of numbers N_j

$$2^{\nu(N_j)} = 2^{\ln N_j / \ln \ln N_j + O(\ln N_j/(\ln \ln N_j)^2)}.$$

We give an example in which our generalization of the Drozdova–Freĭman theorem is used.

We will find an upper bound for the function $r(N)$. We represent N in the form $N = N_1 N_2$, where N_1 is composed only of primes of the form $4k+1$ and N_2 contains no such primes. Obviously $r(N) \le r(N_1)$ and $N_1 \le N$. To estimate $r(N_1)$ we consider the multiplicative semigroup G of natural numbers that are composed only of primes of the form $4k+1$. If $N_1 = \pi_1^{\alpha_1} \cdots \pi_r^{\alpha_r}$, then

$$r(N_1) = 4(\alpha_1 + 1) \cdots (\alpha_r + 1).$$

In our case, $l(\alpha) = \alpha + 1$. By Theorem 2,

$$\begin{aligned} r(N) \le r(N_1) &\le 2^{\frac{\ln N_1}{\ln \ln N_1} + O(\frac{\ln N_1}{(\ln \ln N_1)^2} \ln \ln \ln N_1)} \\ &\le 2^{\frac{\ln N}{\ln \ln N} + O(\frac{\ln N}{(\ln \ln N)^2} \ln \ln \ln N)}. \end{aligned}$$

This too is an accessible bound.

§4.2. Summation of values of Euler's function

A summation formula for Euler's function $\varphi(n)$ was considered in the last century by Mertens ([100], pp. 290–291). We will prove the following

THEOREM. *As* $N \to \infty$

$$\sum_{m=1}^{N} \varphi(m) = \frac{3}{\pi^2} N^2 + O(N \ln N).$$

PROOF. It is known that $\varphi(m) = m \sum_{d|m} \mu(d)/d$. Therefore,

$$\begin{aligned}\sum_{m\le N} \varphi(m) &= \sum_{m\le N} m \sum_{d|m} \frac{\mu(d)}{d} = \sum_{d\le N} \frac{\mu(d)}{d} \sum_{k\le N/d} kd \\ &= \sum_{d\le N} \mu(d) \sum_{k\le N/d} k = \frac{N^2}{2} \sum_{d\le N} \frac{\mu(d)}{d^2} + O\left(\frac{N^2}{N} \sum_{d\le N} \frac{1}{d}\right) \\ &= \frac{N^2}{2} \sum_{d=1}^{\infty} \frac{\mu(d)}{d^2} + O\left(N^2 \sum_{d\ge N} \frac{1}{d^2}\right) + O(N \ln N) \\ &= \frac{6}{\pi^2} \frac{N^2}{2} + O(N \ln N),\end{aligned}$$

as required.

REMARK. Using estimates of trigonometric sums, Walfisz [151] (Chapter IV) and Saltykov [129] proved the following nontrivial result: For any $\varepsilon > 0$

$$\sum_{m=1}^{N} \varphi(m) = \frac{3}{\pi^2} N^2 + O((N \ln N)^{2/3} (\ln \ln N)^{1+\varepsilon}).$$

We state a theorem of Erdős [37] on sums of the function $\varphi(n)/n$ over an interval of the natural sequence.

THEOREM. *Suppose* $g(n)/\ln \ln \ln n \to \infty$. *Then*

$$\sum_{m=n}^{n+g(n)} \frac{\varphi(m)}{m} = \frac{6}{\pi^2} g(n)(1 + o(1)).$$

We present a result on a summation function for $\varphi(n)$ due to Pillai and Chowla [113].

THEOREM.

$$\sum_{n\le x} \varphi(n) - \frac{3x^2}{\pi^2} = \Omega(x \ln \ln \ln x).$$

PROOF. Let

$$E(x) = \sum_{n\le x} \varphi(n) - \frac{3}{\pi^2} x^2.$$

Let $P(a,b)$ denote the product of all primes between a and b. We denote by x_0 the smallest positive solution of the system of congruences

$$\begin{aligned} x &\equiv 0 \pmod 2,\\ x+1 &\equiv 0 \pmod{P(2,2^3)},\\ x+3 &\equiv 0 \pmod{P(2^3,2^{3^2})},\\ &\dots\\ x+2k-1 &\equiv 0 \pmod{P(2^{3^{k-1}},2^{3^k})}. \end{aligned}$$

It is obvious that x_0 is an even number. Since the moduli in the system of congruences are relatively prime, it follows that x_0 is less than the product of the moduli, i.e., $x_0 \le P(1,2^{3^k})$. Hence

$$\ln x_0 \le \sum_{p\le 2^{3^k}} \ln p,$$

but since $\sum_{p\le u} \ln p \le Cu$, it follows that $C_1 \ln\ln\ln x_0 \le k$. Furthermore, x_0+2k+1 is divisible by some prime $\bar{p} > 2^{3^{k-1}}$; hence $x_0+2k+1 > 2^{3^{k-1}}$. This implies $k \le C_2 \ln\ln x_0$. By Mertens' formula,

$$\prod_{p\le T}\left(1-\frac{1}{p}\right) \sim \frac{e^{-\gamma}}{\ln T}.$$

Thus,

$$\prod_{T\le p\le T^3}\left(1-\frac{1}{p}\right) = \frac{1}{3}+o(1)$$

and for $T \ge 2^{3^f}$, where f is an absolute constant,

$$\prod_{T\le p\le T^3}\left(1-\frac{1}{p}\right) < \frac{1}{2}. \tag{1}$$

Now

$$E(x_0+2k)-E(x_0) = \left(\sum_{x_0+1}^{x_0+2k}\varphi(n)\right) - \frac{12}{\pi^2}k^2 - \frac{12}{\pi^2}x_0k.$$

Suppose t is an even number. Since $\varphi(n)/n \le 1/2$ for even numbers, $\varphi(x_0+t)/(x_0+t) \le 1/2$. If t is an odd number in the set $2f+1,\ 2f+3,\dots,$ $2k-1$ (for $k \ge f+1$), then, since $P(2^{3^s},2^{3^{s+1}}) \mid x_0+t$ for $s=(t-1)/2$, we have, in view of (1), $\varphi(x_0+t)/(x_0+t) \le 1/2$. For $t=1,\dots,2f$ we use the inequality $\varphi(x_0+t)/(x_0+t) < 1$. We continue the calculation:

$$E(x_0+2k)-E(x_0) = -\frac{12}{\pi^2}k^2 - \frac{12}{\pi^2}x_0k + \sum_{x_0+1}^{x_0+2f}\varphi(n) + \sum_{x_0+2f+1}^{x_0+2k}\varphi(n).$$

From this and the estimates for $\varphi(x_0+t)/(x_0+t)$ we obtain

$$E(x_0+2k)-E(x_0) < -\frac{12}{\pi^2}k^2-\frac{12}{\pi^2}x_0k \\ +f(2x_0+2f+1)+(k-f)(2x_0+2k+2f+1).$$

Since $f=O(1)$, it follows that

$$E(x_0+2k)-E(x_0)=(1-12/\pi^2)x_0k+O(x_0)+O(k^2).$$

But $12/\pi^2-1>0$ and $k=O(\ln\ln\ln x_0)$, from which it follows that

$$|E(x_0+2k)-E(x_0)| \geq C_3x_0k+O(x_0)+O((\ln\ln\ln x_0)^2) \\ \geq C_4x_0\ln\ln\ln x_0+O(x_0),$$

where C_3 and C_4 are positive constants. Thus,

$$\max(|E(x_0+2k)|,|E(x_0)|) \geq \tfrac{1}{2}C_4x_0\ln\ln\ln x_0.$$

If

$$\max(|E(x_0+2k)|,|E(x_0)|)=|E(x_0)|,$$

the theorem is proved. Suppose

$$\max(|E(x_0+2k)|,|E(x_0)|)=|E(x_0+2k)|.$$

Putting $x_0+2k=\bar{x}_0$, we obtain

$$E(\bar{x}_0) \geq \tfrac{1}{2}C_4(\bar{x}_0-2k)\ln\ln\ln(\bar{x}_0-2k) > C_5\bar{x}_0\ln\ln\ln\bar{x}_0.$$

Thus the theorem is proved in this case also.

THEOREM. *Suppose k is an integer, $k\geq 2$, b is an integer, and $(k,b)=c$. Then*

$$\sum_{\substack{m\equiv b \pmod k\\ m\leq N}}\varphi(m)=\frac{3}{\pi^2}\frac{k}{\varphi_2(k)}\prod_{p\,|\,c}\left(1-\frac{1}{p}\right)N^2+O(N\ln N),$$

where

$$\varphi_2(k)=k^2\prod_{p\,|\,k}\left(1-\frac{1}{p^2}\right).$$

The constant implied by the symbol O is absolute.

PROOF. We have

$$\begin{aligned}\sum_{\substack{m\equiv b \pmod k\\ m\le N}} \varphi(m) &= \sum_{\substack{m\equiv b \pmod k\\ m\le N}} m \sum_{d\,|\,m} \frac{\mu(d)}{d}\\
&= \sum_{d\le N} \frac{\mu(d)}{d} \sum_{\substack{dx\equiv b \pmod k\\ x\le N/d}} dx = \sum_{d\le N} \mu(d) \sum_{\substack{dx\equiv b \pmod k\\ x\le N/d}} x\\
&= \sum_{\tau\,|\,k} \sum_{\substack{(d,k)=\tau\\ d\le N}} \mu(d) \sum_{\substack{dx\equiv b \pmod k\\ x\le N/d}} x.\end{aligned}$$

The inner sum is nonempty only when $\tau\,|\,c$, since $dx - ky = b$ implies $\tau = (d,k)\,|\,b$. Thus,

$$\begin{aligned}\sum_{\substack{m\equiv b \pmod k\\ m\le N}} \varphi(m) &= \sum_{\tau\,|\,c} \sum_{\substack{(d,k)=\tau\\ d\le N}} \mu(d) \sum_{\substack{dx/\tau\equiv b/\tau \pmod{k/\tau}\\ x\le N/d}} x\\
&= \sum_{\tau\,|\,c} \sum_{\substack{(d,k)=\tau\\ d\le N}} \mu(d) \sum_{\substack{x\equiv x_0 \pmod{k/\tau}\\ x\le N/d}} x\\
&= \sum_{\tau\,|\,c} \sum_{\substack{(d,k)=\tau\\ d\le N}} \mu(d)\frac{1}{2k/\tau}\left(\frac{N}{d}\right)^2 + O\Bigg(\sum_{\tau\,|\,c} \sum_{\substack{(d,k)=\tau\\ d\le N}} \frac{1}{k/\tau}\frac{N}{d}\Bigg)\\
&= \frac{N^2}{2k}\sum_{\tau\,|\,c}\tau \sum_{\substack{(d,k)=\tau\\ d\le N}} \frac{\mu(d)}{d^2} + O\Bigg(\frac{N}{k}\sum_{\tau\,|\,c}\tau \sum_{l\le N/\tau}\frac{1}{l\tau}\Bigg)\\
&= \frac{N^2}{2k}\sum_{\tau\,|\,c}\tau \sum_{\substack{(d,k)=\tau\\ d\le N}} \frac{\mu(d)}{d^2} + O\Bigg(\frac{N}{k}\sum_{\tau\,|\,c}\ln N\Bigg)\\
&= \frac{N^2}{2k}\sum_{\tau\,|\,c}\frac{\tau}{\tau^2} \sum_{(l,k/\tau)=1} \frac{\mu((l,\tau))}{l^2} + O(N\ln N)\\
&= \frac{N^2}{2k}\sum_{\tau\,|\,c}\frac{\mu(\tau)}{\tau} \sum_{\substack{l\le N/\tau,\ (l,\tau)=1,\\ (l,k/\tau)=1}} \frac{\mu(l)}{l^2} + O(N\ln N).\end{aligned}$$

The two conditions $(l,\tau)=1$ and $(l,k/\tau)=1$ can be replaced by the one condition $(l,k)=1$:

$$\sum_{\substack{m\le N\\ m\equiv b \pmod k}} \varphi(m) = \frac{N^2}{2k}\sum_{\tau\,|\,c}\frac{\mu(\tau)}{\tau} \sum_{\substack{l\le N/\tau\\ (l,k)=1}} \frac{\mu(l)}{l^2} + O(N\ln N).$$

It is clear that

$$\sum_{\substack{l\le N/\tau\\(l,k)=1}} \frac{\mu(l)}{l^2} = \sum_{\substack{l=1\\(l,k)=1}}^{\infty} \frac{\mu(l)}{l^2} + O\Bigg(\sum_{l=[N/\tau]+1}^{\infty} \frac{1}{l^2}\Bigg) = \sum_{\substack{l=1\\(l,k)=1}}^{\infty} \frac{\mu(l)}{l^2} + O\left(\frac{\tau}{N}\right).$$

Therefore,

$$\begin{aligned}\sum_{\substack{m\le N\\ m\equiv b \pmod k}} \varphi(m) &= \frac{N^2}{2k}\sum_{\tau\,|\,c}\frac{\mu(\tau)}{\tau}\frac{\prod_p(1-1/p^2)}{\prod_{p\,|\,k}(1-1/p^2)} \\ &\quad + O\Bigg(\frac{N}{k}\sum_{\tau\,|\,c}|\mu(\tau)|\Bigg) + O(N\ln N) \\ &= \frac{N^2}{2k\prod_{p\,|\,k}(1-1/p^2)}\frac{6}{\pi^2}\prod_{p\,|\,c}\left(1-\frac{1}{p}\right) + O(N\ln N) \\ &= \frac{3}{\pi^2}\frac{k}{\varphi_2(k)}\prod_{p\,|\,c}\left(1-\frac{1}{p}\right)N^2 + O(N\ln N),\end{aligned}$$

as required.

Romanov [127] studied the series $\sum_2^\infty(\varphi(n)/n)x^n$ (this series obviously converges for $|x|<1$). He established the identity

$$\sum_{n=2}^{\infty}\frac{\varphi(n)}{n}x^n = \frac{6}{\pi^2}\sum_{n=1}^{\infty}\frac{\mu(n)}{n^2\prod_{p\,|\,n}(1-1/p^2)}Q_n(x),$$

where

$$Q_n(x) = \sum_{\rho(n)}\frac{\rho^2x^2}{1-\rho x},$$

the summation extending over all primitive nth roots $\rho(n)$ of unity.

Let us briefly consider some nonlinear problems connected with Euler's function. Let

$$S_N = \sum_{n\le N}\frac{\varphi(n^2+1)}{n^2+1}.$$

THEOREM. *As* $N\to\infty$

$$S_N = \frac{1}{2}\prod_{p\equiv 1 \pmod 4}\left(1-\frac{2}{p^2}\right)N + O(\ln^2 N).$$

PROOF. We have

$$S_N = \sum_{n\le N} \frac{\varphi(n^2+1)}{n^2+1} = \sum_{n\le N}\sum_{d\,|\,n^2+1} \frac{\mu(d)}{d} = \sum_{d\le N^2+1} \frac{\mu(d)}{d} \sum_{\substack{n^2+1\equiv 0 \pmod d\\ n\le N}} 1.$$

Let $L(d)$ denote the number of solutions of the congruence $x^2+1\equiv 0 \pmod d$ with $1\le x\le d$; then

$$\begin{aligned} S_N &= \sum_{d\le N^2+1} \frac{\mu(d)}{d} L(d)\left(\left[\frac{N}{d}\right]+\theta\right) \\ &= N\sum_{d\le N^2+1} \frac{\mu(d)L(d)}{d^2} + O\left(\sum_{d\le N^2+1} \frac{L(d)}{d}\right). \end{aligned}$$

As is well known (see Davenport [25], Chapter II, §6). $L(d)$ is a multiplicative function and (see Venkov [145], Chapter I, §9)

$$\begin{aligned} L(2)&=1, & L(2^\alpha)&=0 & &\text{if } \alpha\ge 2, \\ L(p^\alpha)&=2, & p&\equiv 1 \pmod 4 & &\text{for any } \alpha, \\ L(p^\alpha)&=0, & p&\equiv 3 \pmod 4 & &\text{for any } \alpha. \end{aligned}$$

It follows that $L(d)\le \tau(d)$. Therefore,

$$\sum_{d\le N^2+1} \frac{L(d)}{d} \le \sum_{d\le N^2+1} \frac{\tau(d)}{d} = O(\ln^2(N^2+1)) = O(\ln^2 N),$$

and we have

$$S_N = N\sum_{d\le N^2+1} \frac{\mu(d)L(d)}{d^2} + O(\ln^2 N).$$

Also,

$$S_N = N\sum_{d=1}^{\infty} \frac{\mu(d)L(d)}{d^2} + O\left(N\sum_{d>N^2+1} \frac{\tau(d)}{d^2}\right) + O(\ln^2 N).$$

Therefore,

$$\begin{aligned}\sum_{d>N^2+1}\frac{\tau(d)}{d^2}&=\sum_{n>N^2+1}\frac{\sum_{d\,|\,n}1}{n^2}\\&=\sum_{n>N^2+1}\frac{\sum_{d\le N^2+1,\ d\,|\,n}1}{n^2}+\sum_{n>N^2+1}\frac{\sum_{d>N^2+1,\ d\,|\,n}1}{n^2}\\&=\sum_{d\le N^2+1}\frac{1}{d^2}\sum_{k\ge(N^2+1)/d}\frac{1}{k^2}+\sum_{d>N^2+1}\frac{1}{d^2}\sum_{k\ge(N^2+1)/d}\frac{1}{k^2}\\&=O\Bigl(\frac{1}{N^2}\sum_{d\le N^2+1}\frac{1}{d}\Bigr)+O\Bigl(\sum_{d\ge N^2+1}\frac{1}{d^2}\Bigr)+O\Bigl(\frac{\ln N}{N^2}\Bigr).\end{aligned}$$

Thus,

$$S_N=N\sum_{d=1}^{\infty}\frac{\mu(d)}{d^2}+O(\ln^2 N).$$

In view of the expressions for $L(p^\alpha)$,

$$\sum_{d=1}^{\infty}\frac{\mu(d)L(d)}{d^2}=\prod_p\Bigl(1-\frac{L(p)}{p^2}\Bigr)=\frac{1}{2}\prod_{p\equiv1\ (\mathrm{mod}\,4)}\Bigl(1-\frac{2}{p^2}\Bigr),$$

as required.

Schwarz [134] proved a general theorem from which the following result can be obtained.

THEOREM. *Suppose $f(n)$ is a polynomial with integral coefficients and a nonzero discriminant; assume that the greatest common divisor of the coefficients is 1. Suppose also that $f(n)>0$ if $n>0$. Let $L(d)$ denote the number of solutions of the congruence $f(n)\equiv 0 \pmod d$. Then as $N\to\infty$*

$$\sum_{n\le N}\frac{\varphi(f(n))}{f(n)}=N\prod_{p\ge2}\Bigl(1-\frac{L(p)}{p^2}\Bigr)+O(\ln^c N),$$

where $c>0$ is a constant.

We will calculate the mean value of the function $\ln(\varphi(n)/n)$. We have

$$\begin{aligned}\frac{1}{N}\sum_{n\le N}\ln\frac{\varphi(n)}{n}&=\frac{1}{N}\sum_{n\le N}\sum_{p\,|\,n}\ln\Bigl(1-\frac{1}{p}\Bigr)\\&=\frac{1}{N}\sum_{p\le N}\ln\Bigl(1-\frac{1}{p}\Bigr)\Bigl[\frac{N}{p}\Bigr]\\&=\sum_{p\le N}\frac{\ln(1-1/p)}{p}+O\Bigl(\frac{1}{N}\sum_{p\le N}\ln\frac{1}{1-1/p}\Bigr).\end{aligned}$$

By Mertens' formula,

$$-\sum_{p\le N}\ln\Big(1-\frac{1}{p}\Big)=-\ln\prod_{p\le N}\Big(1-\frac{1}{p}\Big)=O(\ln\ln N).$$

The series $\sum_p(\ln(1-1/p))/p$ is majorized by a convergent series and therefore converges itself. Thus,

$$\frac{1}{N}\sum_{n\le N}\ln\frac{\varphi(n)}{n}=A+O\Big(\sum_{p\ge N}\frac{1}{p^2}\Big)+O\Big(\frac{\ln\ln N}{N}\Big),$$

where $A=\sum_p(\ln(1-1/p))/p$. Since $p_n\ge n\ln n$, it follows that

$$\sum_{p\ge N}\frac{1}{p^2}\le\sum_{n\ge\pi(N)}\frac{1}{n^2\ln^2 n}\le\sum_{n\ge cN/\ln N}\frac{1}{n^2\ln^2 n}$$
$$=O\Big(\frac{1}{\ln^2 N}\sum_{n\ge cN/\ln N}\frac{1}{n^2}\Big)=O\Big(\frac{1}{\ln^2 N}\frac{1}{N/\ln N}\Big)=O\Big(\frac{1}{N\ln N}\Big).$$

Thus,

$$\frac{1}{N}\sum_{n\le N}\ln\frac{\varphi(n)}{n}=A+O\Big(\frac{\ln\ln N}{N}\Big).$$

§4.3. Wirsing's theorem

To obtain the principal terms in summation formulas for multiplicative functions we can make use of the following elementary theorem of Wirsing [154].

THEOREM. *Suppose $f(n)$ is a multiplicative function such that 1) $f(n)\ge 0$, $n=1,2,\dots$, and 2) $f(p^\nu)\le c_1c_2^\nu$ with $c_2<2$ for primes p and $\nu=2,3,\dots$. Assume that as $x\to\infty$*

$$\sum_{p\le x}f(p)=(\tau+o(1))\frac{x}{\ln x},$$

where $\tau\ge 0$ is a constant. Then as $x\to\infty$

$$\sum_{n\le x}f(n)=\Big(\frac{e^{-\gamma\tau}}{\Gamma(\tau)}+o(1)\Big)\frac{x}{\ln x}\prod_{p\le x}\Big(1+\frac{f(p)}{p}+\frac{f(p^2)}{p^2}+\cdots\Big),$$

where γ is Euler's constant and $\Gamma(\tau)$ denotes the gamma-function.

REMARK. We will show that the condition

$$\sum_{p\le x}f(p)=\tau\frac{x}{\ln x}+o\Big(\frac{x}{\ln x}\Big)$$

is equivalent to

$$\sum_{p\le x} f(p)\ln p = \tau x + o(x).$$

Indeed, suppose the former holds. We have

$$\begin{aligned}
\sum_{p\le x} f(p)\ln p &= \ln x \sum_{p\le x} f(p) - \sum_{p\le x} f(p)\ln\frac{x}{p} \\
&= \ln x \sum_{p\le x} f(p) - \sum_{p\le x} f(p)\int_p^x \frac{du}{u} \\
&= \tau x + o(x) - \int_2^x \frac{\sum_{p\le u} f(p)}{u}\,du \\
&= \tau x + o(x) + O\left(\int_2^x \frac{du}{\ln u}\right) = \tau x + o(x).
\end{aligned}$$

Suppose, conversely, that $\sum_{p\le x} f(p)\ln p = \tau x + o(x)$, $f(p)\ge 0$. Obviously,

$$\begin{aligned}
\sum_{p\le x} f(p) &= \frac{1}{\ln x}\left(\sum_{p\le x} f(p)\ln p + \sum_{p\le x} f(p)\ln\frac{x}{p}\right) \\
&= O\left(\frac{x}{\ln x}\right) + \frac{1}{\ln x}\sum_{p\le x} f(p)\ln\frac{x}{p} \\
&= O\left(\frac{x}{\ln x}\right) + \frac{1}{\ln x}\sum_{p\le x} f(p)\int_p^x \frac{du}{u} \\
&= O\left(\frac{x}{\ln x}\right) + \frac{1}{\ln x}\int_2^x \frac{\sum_{p\le u} f(p)}{u}\,du \\
&= O\left(\frac{x}{\ln x}\right) + O\left(\frac{1}{\ln x}\int_2^x \frac{\sum_{p\le u} f(p)\ln p}{u}\,du\right) \\
&= O\left(\frac{x}{\ln x}\right) + O\left(\frac{1}{\ln x}\int_2^x O(1)\,du\right) = O\left(\frac{x}{\ln x}\right).
\end{aligned}$$

We now obtain

$$\begin{aligned}
\sum_{p\le x} f(p) &= \frac{1}{\ln x}\left(\tau x + o(x) + \frac{1}{\ln x}\int_2^x \frac{\sum_{p\le u} f(p)}{u}\,du\right) \\
&= \frac{\tau x}{\ln x} + o\left(\frac{x}{\ln x}\right) + O\left(\frac{1}{\ln x}\int_2^x O(1)\,du\right) = \frac{\tau x}{\ln x} + o\left(\frac{x}{\ln x}\right).
\end{aligned}$$

Before proving the theorem, let us establish some notation. We will denote by $\mathfrak{P}$ the set of primes, and we let

$$M(x) = \sum_{n \le x} f(n), \qquad l(x) = \sum_{n \le x} f(n) \ln n,$$

$$m(x) = \sum_{n \le x} \frac{f(n)}{n}, \qquad \Pi(x) = \prod_{p \le x} \left(1 + \frac{f(p)}{p} + \frac{f(p^2)}{p^2} + \cdots \right),$$

$$T(x) = \sum_{p \le x} f(p), \qquad \vartheta(x) = \sum_{p^2 \le x} f(p) \ln p.$$

The symbol θ will denote numbers such that $|\theta| \le 1$.

REMARK. Since $f(p^\nu) \le c_1 c_2^\nu$, the product

$$\prod_p e^{-f(p)/p} \left(1 + \frac{f(p)}{p} + \frac{f(p^2)}{p^2} + \cdots \right) = P$$

converges. Thus, as $x \to \infty$

$$\Pi(x) \sim P \exp\left(\sum_{p \le x} \frac{f(p)}{p}\right).$$

But

$$\sum_{p \le x} \frac{f(p)}{p} = \int_1^x \frac{dT(t)}{t} = \frac{T(x)}{x} + \int_1^x \frac{T(t)}{t^2}\, dt.$$

Since $T(x) = \sum_{p \le x} f(p) = O(x/\ln x) = o(x)$, it follows that

$$\exp\left(\sum_{p \le x} \frac{f(p)}{p}\right) = \exp\left(\int_1^x \frac{T(t)}{t^2}\, dt + o(1)\right) = (1 + o(1)) \exp\left(\int_1^x \frac{T(t)}{t^2}\, dt\right).$$

Thus the assertion of the theorem can be expressed in the form

$$\sum_{n \le x} f(n) = \left(\frac{e^{-\gamma\tau}}{\Gamma(\tau)} + o(1)\right) \frac{Px}{\ln x} \exp\left(\int_1^x \frac{T(t)}{t^2}\, dt\right).$$

PROOF OF THE THEOREM. Consider the expression

$$\sum_{p \le x} f(p) \frac{\ln p}{p} = \int_1^x \frac{\ln t}{t}\, dT(t) = \frac{\ln x}{x} T(x) + \int_1^x T(t) \frac{\ln t - 1}{t^2}\, dt$$

$$= \int_1^x (\tau + o(1)) \frac{dt}{t} + O(1) = (\tau + o(1) \ln x.$$

We have also seen that

$$\sum_{p \le x} f(p) \ln p = (\tau + o(1))x.$$

We next establish a number of lemmas.

LEMMA 1. *The estimates*

$$\sum_{\substack{p^\nu \le x \\ \nu \ge 2}} f(p^\nu) \ln p^\nu = o(x), \tag{1}$$

$$\sum_{\substack{p^{\nu+\mu} \le x \\ \nu \ge 1,\ \mu \ge 1}} f(p^\nu) f(p^\mu) \ln p^\mu = o(x) \tag{2}$$

hold, the series

$$\sum_p \frac{f^2(p)}{p}, \tag{3}$$

$$\sum_{p,\ \nu \ge 2} \frac{f(p^\nu)}{p^\nu} \ln p \tag{4}$$

converge, and

$$\sum_{\substack{p \le x \\ \nu \ge 1,\ \mu \ge 1}} \frac{f(p^\nu) f(p^\mu)}{p^{\nu+\mu}} \ln p^\mu = o(\ln x). \tag{5}$$

PROOF. We have

$$\sum_{\substack{p^\nu \le x \\ \nu \ge 2}} f(p^\nu) \ln p^\nu \le \ln x \sum_{\substack{p^\nu \le x \\ \nu \ge 2}} f(p^\nu) = O\Bigl(\ln x \sum_{\nu \le \ln x/\ln 2} c_2^\nu\Bigr),$$

where $0 \le c_2 < 2$ (the latter inequality is strict). If $c_2 \le 1$, then (1) follows at once. If $1 < c_2 < 2$, then

$$\sum_{\substack{p^\nu \le x \\ \nu \ge 2}} f(p^\nu) \ln p^\nu = O(\ln x\, c_2^{\ln x/\ln 2}) = O(x^{\ln c_2/\ln 2} \ln x) = o(x),$$

since $\ln c_2/\ln 2 < 1$ (strict inequality).

Let us prove (2):

$$\sum_{\substack{p^{\nu+\mu} \le x \\ \nu \ge 1,\ \mu \ge 1}} f(p^\nu) f(p^\mu) \ln p^\mu \le \ln x \sum_{\substack{p^{\nu+\mu} \le x \\ \nu \ge 1,\ \mu \ge 1}} f(p^\nu) f(p^\mu)$$
$$= \ln x \Bigl(\sum_{\nu=\mu=1} + \sum_{\nu=1,\ \mu \ge 2} + \sum_{\nu \ge 2,\ \mu=1} + \sum_{\nu \ge 2,\ \mu \ge 2}\Bigr).$$

A bound for the first sum can be obtained easily:

$$\sum_{\nu=\mu=1} = \sum_{p^2 \le x} f^2(p) \le \Bigl(\sum_{p \le \sqrt{x}} f(p)\Bigr)^2 = O\Bigl(\frac{x}{\ln^2 x}\Bigr).$$

The sums $\sum_{\nu=1,\ \mu\geq 2}$ and $\sum_{\nu\geq 2,\ \mu=1}$ can be estimated in the same way. Thus we need only work with the sum $\sum_{\nu=1,\ \mu\geq 2}$:

$$\begin{aligned}\sum_{p^{1+\mu}\leq x} f(P)f(p^\mu) &\leq \sum_{p^3\leq x} f(p) \sum_{\mu\leq \ln x/\ln p} c_1c_2^\mu \\ &= O\Bigl(\sum_{p^3\leq x} f(p)c_2^{\ln x/\ln p}\Bigr) = O\Bigl(\sum_{p^3\leq x} f(p)x^{\ln c_2/\ln p}\Bigr) \\ &= O\Bigl(\sum_{p\leq c_2^2} f(p)x^{\ln c_2/\ln p}\Bigr) + O\Bigl(\sum_{c_2^2<p\leq \sqrt[3]{x}} f(p)x^{\ln c_2/\ln p}\Bigr) \\ &= O(x^{\ln C_2/\ln 2}) + O\Bigl(\sum_{p\leq \sqrt[3]{x}} f(p)x^{1/2}\Bigr) \\ &= O\Bigl(x^{\ln c_2/\ln 2}\Bigr) + O\Bigl(x^{1/2}\frac{\sqrt[3]{x}}{\ln x}\Bigr) = O(x^\alpha),\end{aligned}$$

where $\alpha < 1$. Finally, we consider the sum $\sum_{\nu\geq 2,\ \mu\geq 2}$:

$$\begin{aligned}\sum_{\nu\geq 2,\ \mu\geq 2} &\leq \sum_{p^4\leq x} \sum_{\nu+\mu\leq \ln x/\ln p} c_1^2c_2^{\nu+\mu} = O\Bigl(\ln^2 x\sum_{p^4\leq x} c_2^{\ln x/\ln p}\Bigr) \\ &= O\Bigl(\ln^2 x\sum_{p\leq c_2^2} c_2^{\ln x/\ln 2}\Bigr) + O\Bigl(\ln^2 x\sum_{c_2^2<p<\sqrt[4]{x}} x^{1/2}\Bigr) \\ &= O(x^{\ln c_2/\ln 2}\ln^2 x) + O\Bigl(x^{1/2}\ln^2 x\frac{\sqrt[4]{x}}{\ln\sqrt[4]{x}}\Bigr) = O(x^\alpha),\end{aligned}$$

where $0 < \alpha < 1$. Thus (2) is proved.

Since $\sum_{p\leq x} f(p) \sim (\tau + o(1))x/\ln x$, it follows that $f(p) = O(p/\ln p)$. Therefore,

$$\sum_{p\leq x}\frac{f^2(p)}{p^2} = O\Bigl(\sum_{p\leq x}\frac{f(p)}{p\ln p}\Bigr) = O\Bigl(\int_{3/2}^x \frac{dT(t)}{t\ln t}\Bigr).$$

Integration by parts yields

$$\int_{3/2}^x \frac{dT(t)}{t\ln t} = \int_{3/2}^x \frac{T(t)}{t^2\ln^2 t}(1+\ln t)\,dt = O(1).$$

Thus the partial sums of $\sum_p f^2(p)/p^2$ are bounded, hence the series converges.

We will now prove convergence of

$$\sum_{p,\ \nu\geq 2}\frac{f(p^\nu)}{p^\nu}\ln p^\nu.$$

We have

$$\sum_{\substack{\nu\geq 2\\ p^\nu\leq x}} \frac{f(p^\nu)}{p^\nu}\ln p^\nu \leq \sum_{p^2\leq x}\frac{f(p^2)}{p^2}\ln p^2 + \sum_{p^3\leq x}\frac{f(p^3)}{p^3}\ln p^3 + \cdots$$

$$= O\Big(\sum_{p\leq\sqrt{x}}\frac{c_2^2\ln p^2}{p^2} + \sum_{p\leq\sqrt[3]{x}}\frac{c_2^3\ln p^3}{p^3} + \cdots\Big)$$

$$= O\Big(\sum_{p\leq\sqrt{x}}\ln p\sum_{\mu=2}^{\infty}\mu\frac{1}{(p/c_2)^\mu}\Big) = O\Big(\sum_{p\leq\sqrt{x}}\frac{\ln p}{p^2}\Big) = O(1).$$

Thus series (4) converges.

Finally, we show that

$$\sum_{\substack{p\leq x\\ \nu\geq 1,\ \mu\geq 1}}\frac{f(p^\nu)f(p^\mu)}{p^{\nu+\mu}}\ln p^\mu = o(\ln x).$$

We have seen that $f(p) = O(p/\ln p) = o(p)$. For the case $\nu = 1, \mu = 1$ we have

$$\sum_{p\leq x}\frac{f^2(p)}{p^2}\ln p = \sum_{p\leq x} o\Big(\frac{f(p)}{p}\ln p\Big) = o(\ln x).$$

If $\nu = 1$ and $\mu \geq 2$ (the case $\mu = 1,\ \nu\geq 2$ is similar), then

$$\sum_{\substack{p\leq x\\ \nu=1,\ \mu\geq 2}}\frac{f(p^\nu)f(p^\mu)}{p^{\nu+\mu}}\ln p^\mu \leq \sum_{p\leq x}\frac{f(p)}{p}\ln p\sum_{\mu\geq 2}\mu c_1\Big(\frac{c_2}{p}\Big)^\mu$$

$$= \sum_{p\leq x} O\Big(\frac{f(p)}{p^3}\ln p\Big) = O\Big(\sum_p\frac{\ln p}{p^2}\Big) \quad (\text{since } f(p) = O(p)),$$

$$\sum_{\substack{p\leq x\\ \nu=1,\ \mu\geq 2}}\frac{f(p^\nu)f(p^\mu)}{p^{\nu+\mu}}\ln p^\mu = O(1).$$

Finally,

$$\sum_{\substack{p\leq x\\ \nu\geq 2,\ \mu\geq 2}}\frac{f(p^\nu)f(p^\mu)}{p^{\nu+\mu}}\ln p^\mu \leq \sum_{p\leq x}\ln p\sum_{\nu\geq 2,\ \mu\geq 2}\mu c_1^2\Big(\frac{c_2}{p}\Big)^{\nu+\mu}$$

$$= O\Big(\sum_{p\leq x}\ln p\Big(\sum_{\nu\geq 2}\Big(\frac{c_2}{p}\Big)^\nu\Big)\Big(\sum_{\mu\geq 2}\mu\Big(\frac{c_2}{p}\Big)^\mu\Big)\Big)$$

$$= O\Big(\sum_{p\leq x}\frac{\ln p}{p^4}\Big) = O(1).$$

Lemma 1 is proved.

LEMMA 2. *As* $x \to \infty$

$$\sum_{n\le x} f(n) = (\tau + o(1))\frac{x}{\ln x}\sum_{n\le x}\frac{f(n)}{n} \tag{6}$$

(*i.e.*, $M(x) = (\tau + o(1))(x/\ln x)m(x)$). *If* $\tau = 0$, *then*

$$M(x) = o\left(\frac{x}{\ln x}m(x)\right).$$

PROOF. We have

$$\begin{aligned}
l(x) &= \sum_{n\le x} f(n)\ln n = \sum_{n\le x} f(n) \sum_{\substack{p,\ \nu \\ p^\nu \mid n,\ p^{\nu+1}\nmid n}} \ln p^\nu \\
&= \sum_{p^\nu\le x} \ln p^\nu \sum_{\substack{n'\le x/p^\nu \\ (n',p)=1}} f(n'p^\nu) = \sum_{p^\nu \le x} \ln p^\nu f(p^\nu) \sum_{\substack{n\le x/p^\nu \\ (n,p)=1}} f(n) \\
&= \sum_{p^\nu\le x} f(p^\nu)\ln p^\nu\Bigg(\sum_{n\le x/p^\nu} f(n) - \sum_{\substack{n\le x/p^\nu \\ p\mid n}} f(n)\Bigg) \\
&= \sum_{p^\nu\le x} f(p^\nu)\ln p^\nu\Bigg(\sum_{n\le x/p^\nu} f(n) - \sum_{\substack{n\le x/p^\nu \\ p^\mu\mid n,\ p^{\mu+1}\nmid n}} f\left(p^\mu\frac{n}{p^\mu}\right)\Bigg) \\
&= \sum_{p^\nu\le x} f(p^\nu)\ln p^\nu\Bigg(\sum_{n\le x/p^\nu} f(n) - \sum_{\substack{n'\le x/(p^{\nu+\mu}) \\ (n',p)=1}} f(p^\mu)f(n')\Bigg) \\
&= \sum_{n\le x} f(n)\Bigg(\sum_{p^\nu\le x/n} f(p^\nu)\ln p^\nu - \sum_{\substack{p\nmid n \\ p^{\nu+\mu}\le x/n}} f(p^\nu)f(p^\mu)\ln p^\nu\Bigg)
\end{aligned}$$

(we have excluded those p that divide n). Thus,

$$\begin{aligned}
l(x) = \sum_{n\le x} f(n)\Bigg(&\sum_{p\le x/n} f(p)\ln p \\
&+ \sum_{\substack{p^\nu\le x/n \\ \nu\ge 2}} f(p^\nu)\ln p^\nu - \sum_{\substack{p\nmid n \\ p^{\nu+\mu}\le x/n}} f(p^\nu)f(p^\mu)\ln p^\nu\Bigg).
\end{aligned} \tag{7}$$

Take a function $h(x)$ that increases to infinity along with x; other conditions on the growth of $h(x)$ will be specified later. We continue the calculation:

$$\begin{aligned} l(x) = \sum_{n\le x/h(x)} f(n)\Bigg(&\sum_{p\le x/n} f(p)\ln p \\ &+ \sum_{\substack{p^\nu\le x/n\\ \nu\ge 2}} f(p^\nu)\ln p^\nu - \sum_{\substack{p\nmid n\\ p^{\nu+\mu}\le x/n}} f(p^\nu)f(p^\mu)\ln p^\nu\Bigg) \\ &+ \sum_{x/h(x)<n\le x} f(n)\Bigg(\sum_{p\le x/n} f(p)\ln p + \sum_{\substack{p^\nu\le x/n\\ \nu\ge 2}} f(p^\nu)\ln p^\nu \\ &\qquad - \sum_{\substack{p\nmid n\\ p^{\nu+\mu}\le x/n}} f(p^\nu)f(p^\mu)\ln p^\nu\Bigg). \end{aligned}$$

In the first sum, $x/n \ge h(x) \to \infty$; hence

$$\sum_{p\le x/n} f(p)\ln p = \frac{\tau}{n}x + o\left(\frac{x}{n}\right), \qquad \sum_{\substack{p^\nu\le x/n\\ \nu\ge 2}} f(p^\nu)\ln p^\nu = o\left(\frac{x}{n}\right),$$

$$\sum_{\substack{p\nmid n\\ p^{\nu+\mu}\le x/n}} f(p^\nu)f(p^\mu)\ln p^\nu = o\left(\frac{x}{n}\right).$$

In the second sum we employ rougher estimates:

$$\sum_{p\le x/n} f(p)\ln p = O\left(\frac{x}{n}\right), \qquad \sum_{\substack{p^\nu\le x/n\\ \nu\ge 2}} f(p^\nu)\ln p^\nu = O\left(\frac{x}{n}\right),$$

$$\sum_{\substack{p\nmid n\\ p^{\nu+\mu}\le x/n}} f(p^\nu)f(p^\mu)\ln p^\nu = O\left(\frac{x}{n}\right).$$

Thus we obtain

$$\begin{aligned} l(x) = x\tau \sum_{n\le x/h(x)} \frac{f(n)}{n} + o\left(x \sum_{n\le x/h(x)} \frac{f(n)}{n}\right) \\ + O\left(x \sum_{x/h(x)\le n\le x} \frac{f(n)}{n}\right) \end{aligned}$$

$$= x\tau \sum_{n \le x/h(x)} \frac{f(n)}{n} + o\left(x \sum_{n \le x} \frac{f(n)}{n}\right) + O\left(x \sum_{x/h(x) \le n \le x} \frac{f(n)}{n}\right)$$
$$= x\tau \sum_{n \le x} \frac{f(n)}{n} + o\left(x \sum_{n \le x} \frac{f(n)}{n}\right) + O\left(x \sum_{x/h(x) \le n \le x} \frac{f(n)}{n}\right).$$

Therefore,

$$l(x) = xm(x)(\tau + o(1)) + O\left(x \frac{h(x)}{x} \sum_{x/h(x) \le n \le x} f(n)\right)$$
$$= xm(x)(\tau + o(1)) + O\left(h(x) \sum_{n \le x} f(n)\right).$$

If $\sum_{n \le x} f(n)$ converges, we take $h(x) = o(x)$. Then $h(x) = o(xm(x))$ and $l(x) = xm(x)(\tau + o(1))$.

If $\sum_{n \le x} f(n)$ diverges,then

$$\sum_{n \le x} f(n) = o\left(\sum_{n \le x} f(n) \ln n\right).$$

Indeed, for a fixed M

$$\frac{\sum_{n \le x} f(n)}{\sum_{n \le x} f(n) \ln n} \le \frac{\sum_{n \le M} f(n) + \sum_{M < n \le x} f(n)}{\ln M \sum_{M < n \le x} f(n)}.$$

Since the series diverges, for a sufficiently large x we have

$$\sum_{n \le M} f(n) < \sum_{M < n \le x} f(n).$$

Therefore,

$$\frac{\sum_{n \le x} f(n)}{\sum_{n \le x} f(n) \ln n} \le \frac{2}{\ln M}.$$

But M can be arbitrarily large. Choose $h(x)$ so that $h(x) \to \infty$, but

$$h(x) \sum_{n \le x} f(n) = o(l(x)).$$

Then

$$l(x) = xm(x)(\tau + o(1)) + o(l(x)),$$
$$l(x)(1 + o(1)) = xm(x)(\tau + o(1)),$$
$$l(x) = xm(x)(\tau + o(1)).$$

Thus we always have $l(x) = (\tau + o(1))xm(x)$. Now

$$\begin{aligned} M(x) &= \int_2^x \frac{dl(t)}{\ln t} = \frac{l(x)}{\ln x} + \int_2^x \frac{l(t)}{t \ln^2 t}\, dt \\ &= (\tau + o(1))\frac{x}{\ln x} m(x) + O\left(\int_2^x \frac{m(t)}{\ln^2 t}\, dt\right) \\ &= (\tau + o(1))\frac{x}{\ln x} m(x) + m(x) O\left(\int_2^x \frac{dt}{\ln^2 t}\right) \\ &= (\tau + o(1))\frac{x}{\ln x} m(x). \end{aligned}$$

LEMMA 3. *Assume that in the original relation*

$$\sum_{p \le x} f(p) = (\tau + o(1))\frac{x}{\ln x}$$

τ *is such that* $0 < \tau \le \sqrt{2}/2$. *Then there exists* x_0 *such that for* $x \ge x_0$

$$(1 - 2\tau^2)\Pi(x) \le m(x) \le \Pi(x). \tag{8}$$

If $\tau = 0$, *then as* $x \to \infty$

$$m(x) \sim \Pi(x). \tag{9}$$

PROOF. The inequality

$$m(x) \le \Pi(x)$$

follows from the definition of $m(x)$ and $\Pi(x)$ and the multiplicativity of the function $f(n)$. Establishing the lower bound for $m(x)$ is the tricky part of the proof.

Suppose $X \ge 1$ is an arbitrary but sufficiently large number. Let us put

$$\tilde{f}(p^\nu) = \begin{cases} f(p^\nu) & \text{if } p \le X, \\ 0 & \text{if } p > X \end{cases}$$

and extend the domain of definition of this function to all natural numbers n be requiring that $\tilde{f}(n)$ be multiplicative. For a given $\varepsilon > 0$ we can find $x_1 = x_1(\varepsilon)$ such that for $x \ge x_1(\varepsilon)$

$$\sum_{p \le x} f(p)\frac{\ln p}{p} = (\tau + \varepsilon\theta) \ln x,$$

where $|\theta| \le 1$. If $X \ge x_1(\varepsilon)$, then for $x_1(\varepsilon) \le x \le X$ we have

$$\sum_{p\le x} \tilde{f}(p)\frac{\ln p}{p} = (\tau + \varepsilon\theta)\ln x,$$

and for $x \ge X$

$$\sum_{p\le x} \tilde{f}(p)\frac{\ln p}{p} = \sum_{p\le X} f(p)\frac{\ln p}{p} = (\tau + \varepsilon\theta)\ln X.$$

Since the function $\tilde{f}(n)/n$ is multiplicative, we can apply (7) (of course with $f(n)$ replaced by $\tilde{f}(n)/n$):

$$\begin{aligned}\sum_{n\le x}\frac{\tilde{f}(n)}{n}\ln n &= \sum_{n\le x}\frac{\tilde{f}(n)}{n}\sum_{p\le x/n}\frac{\tilde{f}(p)}{p}\ln p + \sum_{n\le x}\frac{\tilde{f}(n)}{n}\sum_{\substack{p^\nu\le x/n\\ \nu\ge 2}}\frac{\tilde{f}(p^\nu)}{p^\nu}\ln p^\nu\\ &\quad - \sum_{n\le x}\frac{\tilde{f}(n)}{n}\sum_{\substack{p\nmid n,\ \nu\ge 1,\ \mu\ge 1\\ p^{\nu+\mu}\le x/n}}\frac{\tilde{f}(p^\nu)\tilde{f}(p^\mu)}{p^{\nu+\mu}}\ln p^\nu\\ &= S + S_1 - S_2.\end{aligned}$$

We have

$$S_1 = \sum_{n\le x}\frac{\tilde{f}(n)}{n}\sum_{\substack{p^\nu\le x/n\\ \nu\ge 2}}\frac{\tilde{f}(p^\nu)}{p^\nu}\ln p^\nu \le \sum_{n\le x}\frac{\tilde{f}(n)}{n}\sum_{\substack{\nu\ge 2\\ p\le x}}\frac{f(p^\nu)}{p^\nu}\ln p^\nu,$$

$$\begin{aligned}S_2 &= \sum_{n\le x}\frac{\tilde{f}(n)}{n}\sum_{\substack{p\nmid n, p^{\nu+\mu}\le x/n\\ \nu\ge 1,\ \mu\ge 1}}\frac{\tilde{f}(p^\nu)}{p^\nu}\frac{\tilde{f}(p^\mu)}{p^\mu}\ln p^\nu\\ &\le \sum_{n\le x}\frac{\tilde{f}(n)}{n}\sum_{\substack{p\le X\\ \nu\ge 1,\ \mu\ge 1}}\frac{f(p^\nu)}{p^\nu}\frac{f(p^\mu)}{p^\mu}\ln p^\nu.\end{aligned}$$

Using (4) (in a weakened form) and (5), we see that for $X \ge x_2(\varepsilon)$

$$|S_1| \le \frac{\varepsilon}{2}\ln X\sum_{n\le x}\frac{\tilde{f}(n)}{n}, \qquad |S_2| \le \frac{\varepsilon}{2}\ln X\sum_{n\le x}\frac{\tilde{f}(n)}{n},$$

and hence

$$|S_1 - S_2| \le |S_1| + |S_2| \le \varepsilon\tilde{m}(x)\ln X,$$

where $\tilde{m}(x) = \sum_{n\le x} \tilde{f}(n)/n$. Also,

$$\begin{aligned}
S &= \sum_{n\le x} \frac{\tilde{f}(n)}{n} \sum_{p\le x/n} \tilde{f}(p)\frac{\ln p}{p} \\
&= \sum_{n\le x/X} \frac{\tilde{f}(n)}{n} \sum_{p\le x/n} \tilde{f}(p)\frac{\ln p}{p} + \sum_{x/X<n\le x/x_1} \frac{\tilde{f}(n)}{n} \sum_{p\le x/n} \tilde{f}(p)\frac{\ln p}{p} \\
&\quad + \sum_{x/x_1<n\le x} \frac{\tilde{f}(n)}{n} \sum_{p\le x/n} \tilde{f}(p)\frac{\ln p}{p} \\
&= \sum_{n\le x/X} \frac{\tilde{f}(n)}{n}(\tau+\varepsilon\theta)\ln X + \sum_{x/X\le n\le x/x_1} \frac{\tilde{f}(n)}{n}(\tau+\varepsilon\theta)\ln\frac{x}{n} \\
&\quad + \sum_{x/x_1<n\le x} \frac{\tilde{f}(n)}{n} \sum_{p\le x/n} \tilde{f}(p)\frac{\ln p}{p} \\
&= \sum_{n\le x/X} \frac{\tilde{f}(n)}{n}(\tau+\varepsilon\theta)\ln X + \sum_{x/X<n\le x} \frac{\tilde{f}(n)}{n}(\tau+\varepsilon\theta)\ln\frac{x}{n} \\
&\quad + \sum_{x/x_1<n\le x} \frac{\tilde{f}(n)}{n}\left(\sum_{p\le x/n} \tilde{f}(p)\frac{\ln p}{p} - (\tau+\varepsilon\theta)\ln\frac{x}{n}\right).
\end{aligned}$$

If $x/x_1 < n \le x$, then $x/n \le x_1$, and we have

$$\left|\sum_{p\le x/n} \tilde{f}(p)\frac{\ln p}{p} - (\tau+\varepsilon\theta)\ln\frac{x}{n}\right| \le \sum_{p\le x_1} \tilde{f}(p)\frac{\ln p}{p} + (\tau+\varepsilon)\ln x_1 < K(\varepsilon).$$

Thus,

$$\begin{aligned}
S = (\tau+\theta_1\varepsilon)\Bigg(\tilde{m}\left(\frac{x}{X}\right)\ln X + \tilde{m}(x)\ln x \\
- \tilde{m}\left(\frac{x}{X}\right)\ln x - \sum_{x/X<n\le x} \frac{\tilde{f}(n)}{n}\ln n\Bigg) + \theta_2 K(\varepsilon)\tilde{m}(x),
\end{aligned}$$

where $|\theta_1| \le 1$ and $|\theta_2| \le 1$. Choose X so that $\ln X \ge K(\varepsilon)/\varepsilon$. Under these conditions,

$$\begin{aligned}
\sum_{n\le x} \tilde{f}(n)\frac{\ln n}{n} = (\tau+\varepsilon\theta)\Bigg(\tilde{m}(x)\ln x - \tilde{m}\left(\frac{x}{X}\right)\ln\frac{x}{X} \\
- \sum_{x/X<n\le x} \tilde{f}(n)\frac{\ln n}{n}\Bigg) + 2\varepsilon\theta\tilde{m}(x)\ln X.
\end{aligned} \tag{10}$$

For $x = X$ we obtain

$$(1+\tau+\varepsilon\theta)\sum_{n\le X}\tilde{f}(n)\frac{\ln n}{n} = (\tau+3\varepsilon\theta)\tilde{m}(X)\ln X.$$

Thus,

$$\sum_{n\le X}\tilde{f}(n)\frac{\ln n}{n} > \frac{\tau-3\varepsilon}{1+\tau+\varepsilon}\tilde{m}(X)\ln X = \frac{\tau-3\varepsilon}{1+\tau+\varepsilon}m(X)\ln X$$

(since $\tilde{m}(t) = m(t)$ when $t \le X$).

If $x \ge X$, we use integration by parts:

$$\sum_{n\le x}\tilde{f}(n)\frac{\ln n}{n} = (\tau+\theta\varepsilon)\int_{x/X}^{x}\tilde{m}(t)\,d\ln t + 2\varepsilon\theta\tilde{m}(x)\ln X.$$

But

$$\tilde{m}(t) = \sum_{n\le t}\frac{\tilde{f}(n)}{n} \le \prod_{p\le X}\left(1+\frac{f(p)}{p}+\frac{f(p^2)}{p^2}+\cdots\right),$$

and hence

$$\sum_{n\le x}\tilde{f}(n)\frac{\ln n}{n} \le (\tau+\varepsilon)\Pi(X)\int_{x/X}^{x}d\ln t + 2\varepsilon\Pi(X)\ln X$$
$$= (\tau+3\varepsilon)\Pi(X)\ln X.$$

This upper bound does not depend on x; therefore $\sum\tilde{f}(n)(\ln n)/n$ converges and

$$\sum\frac{\tilde{f}(n)}{n}\ln n \le (\tau+3\varepsilon)\Pi(X)\ln X.$$

So for large X we have

$$\ln X\sum_{n>X}\frac{\tilde{f}(n)}{n} \le \sum_{n>X}\frac{\tilde{f}(n)}{n}\ln n$$
$$\le (\tau+3\varepsilon)\Pi(X)\ln X - \frac{\tau-3\varepsilon}{1+\tau+\varepsilon}m(x)\ln X.$$

It follows that

$$\sum_{n>X}\frac{\tilde{f}(n)}{n} \le (\tau+3\varepsilon)\Pi(X) - \frac{\tau-3\varepsilon}{1+\tau+\varepsilon}m(X).$$

On the other hand,

$$\sum_{n>X}\frac{\tilde{f}(n)}{n} = \Pi(X) - m(X).$$

Therefore,

$$(1-\tau-3\varepsilon)\Pi(X) \le m(X)\left(1-\frac{\tau-3\varepsilon)}{1+\tau+\varepsilon}\right).$$

Thus for sufficiently large $X \geq X_0$

$$m(X) \geq \Pi(X)\frac{(1+\tau+\varepsilon)(1-\tau-3\varepsilon)}{1+4\varepsilon}.$$

If $\tau = 0$, then, since ε can be arbitrarily small,

$$\varliminf_{X\to\infty} (m(X)/\Pi(X)) \geq 1.$$

This in conjunction with $m(X) \leq \Pi(X)$ proves the lemma for the case $\tau = 0$. If $\tau > 0$, then for a sufficiently small ε

$$\frac{(1+\tau+\varepsilon)(1-\tau-3\varepsilon)}{1+4\varepsilon} \geq 1 - 2\tau^2.$$

Lemma 3 is proved.

LEMMA 4. *Suppose $\delta > 0$ is fixed and $x^\delta \leq y \leq x$. Then as $x \to \infty$*

$$m(y) = (1+o(1))m(x)\left(\frac{\ln y}{\ln x}\right)^\tau. \tag{11}$$

The symbol $o(1)$ is uniform in y (when $x^\delta \leq y \leq x$).

PROOF. We put $x = X$ in (10) and integrate the resulting expression by parts:

$$\tilde{m}(X)\ln X - \int_1^X \tilde{m}(t)\,d\ln t = (\tau+\varepsilon\theta)\int_1^X \tilde{m}(t)\,d\ln t + 2\varepsilon\theta\tilde{m}(X)\ln X.$$

But since $\tilde{m}(t) = m(t)$ when $t \leq X$, it follows that

$$(1+2\varepsilon\theta)m(X)\ln X = (1+\tau+\varepsilon\theta)\int_1^X m(t)\,d\ln t$$

when $X > x_0(\varepsilon)$. Therefore as $x \to \infty$

$$m(x)\ln x \sim (1+\tau)\int_1^X m(t)\,d\ln t. \tag{12}$$

The function $\ln\int_1^x m(t)\,d\ln t$, $x > 1$, has the derivative

$$\frac{1}{\int_1^x m(t)\,d\ln t}\frac{m(x)}{x}$$

at nonintegral x. This derivative is continuous at nonintegral x. According to (12),

$$\frac{1}{\int_1^x m(t)\,d\ln t}\frac{m(x)}{x} = \frac{1+\tau+o(1)}{x\ln x}.$$

By integration,

$$\ln\frac{\int_1^x m(t)\,d\ln t}{\int_1^y m(t)\,d\ln t} = \int_y^x \frac{1+\tau+o(1)}{t\ln t}\,dt = (1+\tau+o(1))\ln\frac{\ln x}{\ln y}.$$

The symbol $o(1)$ in these equalities pertains to $y \to \infty$, but, inasmuch as $x^\delta \le y \le x$, $o(1)$ pertains also to $x \to \infty$. Furthermore,

$$\frac{\int_1^x m(t)\,d\ln t}{\int_1^y m(t)\,d\ln t} = \exp\left(\ln \frac{\ln x}{\ln y}(1+\tau)\right) \exp\left(o(1)\ln \frac{\ln x}{\ln y}\right).$$

But

$$1 \le \frac{\ln x}{\ln y} \le \frac{1}{\delta}, \qquad \exp\left(o(1)\ln\frac{\ln x}{\ln y}\right) = 1 + o(1),$$

$$\frac{\int_1^x m(t)\,d\ln t}{\int_1^y m(t)\,d\ln t} = \left(\frac{\ln x}{\ln y}\right)^{1+\tau}(1+o(1)).$$

Using (12), we obtain

$$\frac{m(x)}{m(y)} = (1+o(1))\frac{\ln y}{\ln x}\frac{\int_1^x m(t)\,d\ln t}{\int_1^y m(t)\,d\ln t} = (1+o(1))\left(\frac{\ln x}{\ln y}\right)^\tau.$$

Lemma 4 is proved.

LEMMA 5. *Suppose the set $\mathfrak{P}$ of primes is partitioned into r classes $\mathfrak{P}_1, \dots, \mathfrak{P}_r$ so that*

a) *the classes $\mathfrak{P}_i$, $i = 1, \dots, r$, are mutually disjoint ($\mathfrak{P}_i \cap \mathfrak{P}_j = \varnothing$), and*

b) *to each class $\mathfrak{P}_i$, $i = 1, \dots, r$, there is associated a function $f_i(n)$ of a natural argument defined as follows:*

$$f_i(n) = \begin{cases} f(n) & \text{if all prime divisors of } n \text{ belong to } \mathfrak{P}_i, \\ 0 & \text{otherwise.} \end{cases}$$

Assume that for each $i = 1, \dots, r$ the limit

$$\tau_i = \lim_{x\to\infty} \frac{1}{\ln x}\sum_{p \le x} f_i(p)\frac{\ln p}{p}$$

exists and $\tau_i \ne 0$. Then as $x \to \infty$

$$m(x)\Gamma(1+\tau) \sim \prod_{i=1}^r m_i(x)\Gamma(1+\tau_i)$$

(where $m_i(x) = \sum_{n \le x} f_i(n)/n$).

PROOF. The proof is easily effected by induction on r. It sufficies to prove the lemma for $r = 2$.

Let $n = n_1 n_2$, where all prime divisors of n_1 belong to $\mathfrak{P}_1$ and all prime divisors of n_2 belong to $\mathfrak{P}_2$ (it is possible that n_1 or n_2 is equal to 1). We have

$$f(n) = f(n_1)f(n_2) = f_1(n_1)f_2(n_2),$$

and hence

$$\begin{aligned} m(x) &= \sum_{n \le x} \frac{f(n)}{n} = \sum_{n_1 n_2 \le x} \frac{f_1(n_1) f_2(n_2)}{n_1 n_2} \\ &= \sum_{n_1 \le x} \frac{f_1(n_1)}{n_1} \sum_{n_2 \le x/n_1} \frac{f_2(n_2)}{n_2} = \sum_{n_1 \le x} \frac{f_1(n_1)}{n_1} m_2\left(\frac{x}{n_1}\right). \end{aligned}$$

Given $\varepsilon > 0$, we can write

$$\begin{aligned} m(x) &= \sum_{n_1 \le x^\varepsilon} \frac{f_1(n_1)}{n_1} m_2\left(\frac{x}{n_1}\right) + \sum_{x^\varepsilon < n_1 \le x^{1-\varepsilon}} \frac{f_1(n_1)}{n_1} m_2\left(\frac{x}{n_1}\right) \\ &\quad + \sum_{x^{1-\varepsilon} < n_1 \le x} \frac{f_1(n_1)}{n_1} m_2\left(\frac{x}{n_1}\right) \\ &= S_1 + S_2 + S_3. \end{aligned}$$

To estimate S_1 and S_3 we use (11):

$$\begin{aligned} S_1 = \sum_{n \le x^\varepsilon} \frac{f_1(n)}{n} m_2\left(\frac{x}{n}\right) &\le m_2(x) \sum_{n \le x^\varepsilon} \frac{f_1(n)}{n} \\ &= m_2(x) m_1(x^\varepsilon) = m_1(x) m_2(x) O(e^{\tau_1}). \end{aligned}$$

Since $n > x^{1-\varepsilon}$ implies $x/n < x^\varepsilon$, it follows that

$$\begin{aligned} S_3 = \sum_{x^{1-\varepsilon} < n \le x} \frac{f_1(n)}{n} m_2\left(\frac{x}{n}\right) &\le m_2(x^\varepsilon) \sum_{x^{1-\varepsilon} < n \le x} \frac{f_1(n)}{n} \\ &\le m_2(x^\varepsilon) m_1(x) \le m_1(x) m_2(x) O(\varepsilon^{\tau_2}). \end{aligned}$$

To investigate S_2 we must use (11) in a more subtle way. Since $n \le x^{1-\varepsilon}$ implies $x^\varepsilon \le x/n \le x$, it follows that

$$\begin{aligned} S_2 &= \sum_{x^\varepsilon < n \le x^{1-\varepsilon}} \frac{f_1(n)}{n} m_2\left(\frac{x}{n}\right) \\ &= (1 + o(1)) m_2(x) \sum_{x^\varepsilon < n \le x^{1-\varepsilon}} \frac{f_1(n)}{n} \left(\frac{\ln(x/n)}{\ln x}\right)^{\tau_2} \\ &= (1 + o(1)) m_2(x) \int_{x^\varepsilon}^{x^{1-\varepsilon}} \left(\frac{\ln(x/t)}{\ln x}\right)^{\tau_2} dm_1(t). \end{aligned}$$

Integrating by parts, we obtain (with $u = (\ln(x/t))/\ln x$)

$$\begin{aligned} S_2 &= (1 + o(1)) m_2(x) \\ &\quad \times \left(\varepsilon^{\tau_2} m_1(x^{1-\varepsilon}) - (1-\varepsilon)^{\tau_2} m_1(x^\varepsilon) - \int_{x^\varepsilon}^{x^{1-\varepsilon}} m_1(t)\, du^{\tau_2}\right). \end{aligned}$$

Applying (11) once again (to the function $m_1(t)$), we obtain

$$S_2 = (1+o(1))m_2(x)\Bigg((O(\varepsilon^{\tau_2}) + O(\varepsilon^{\tau_1}))m_1(x) - (1+o(1))m_1(x)\int_{x^\varepsilon}^{x^{1-\varepsilon}} \left(\frac{\ln t}{\ln x}\right)^{\tau_1} du^{\tau_2}\Bigg)$$

$$= (1+o(1))m_1(x)m_2(x) \times \left((1+o(1))\int_\varepsilon^{1-\varepsilon} (1-u)^{\tau_1}\, du^{\tau_2} + O(\varepsilon^{\tau_2}) + O(\varepsilon^{\tau_1})\right).$$

Thus,

$$\begin{aligned}\frac{m(x)}{m_1(x)m_2(x)} &= O(\varepsilon^{\tau_1} + \varepsilon^{\tau_2}) + (1+o(1))\int_\varepsilon^{1-\varepsilon} (1-u)^{\tau_1}\, du^{\tau_2} \\ &= O(\varepsilon^{\tau_1} + \varepsilon^{\tau_2}) + (1+o(1)) \\ &\quad \times \left(\int_0^1 (1-u)^{\tau_1}\, du^{\tau_2} + O\left(\varepsilon^{\tau_1}\int_{1-\varepsilon}^1 du^{\tau_2} + \int_0^{\varepsilon} du^{\tau_2}\right)\right) \\ &= O(\varepsilon^{\tau_1} + \varepsilon^{\tau_2}) + (1+o(1))\int_0^1 (1-u)^{\tau_1}\, du^{\tau_2} \\ &= \frac{\Gamma(\tau_1+1)\Gamma(\tau_2+1)}{\Gamma(\tau_1+\tau_2+1)} + o(1) + O(\varepsilon^{\tau_1} + \varepsilon^{\tau_2}).\end{aligned}$$

Since ε is arbitrary and $\tau_1 + \tau_2 = \tau$, it follows that

$$\frac{m(x)}{m_1(x)m_2(x)} \to \frac{\Gamma(1+\tau_1)\Gamma(1+\tau_2)}{\Gamma(1+\tau)}$$

as $x \to \infty$. Lemma 5 is proved.

LEMMA 6. *Given a sequence of nonnegative numbers a_n, suppose it is known that $\sum_{n\le x} a_n \sim \tau \ln x$ as $x\to\infty$, where $\tau > 0$. Let there be given a number β such that $0<\beta<1$. Then there exists a subsequence $a_{n'}$ such that*

$$\sum_{n\le x} a_{n'} \sim \beta\tau \ln x.$$

PROOF. As the first member of the subsequence we take a_2. We next consider two possibilities: if $a_2 \le \beta\tau \ln 2$, then as the second member of the subsequence we take a_3: if $a_2 > \beta\tau\ln 2$, we skip a_3. Let us assume we know which members of the original sequence belong to the subsequence up to the index $N-1$:

$$a_1\chi(1),\ a_2\chi(2),\dots,a_{N-1}\chi(N-1),$$

where

$$\chi(i) = \begin{cases} 1, & \text{if } a_i \text{ remains,} \\ 0, & \text{if } a_i \text{ was skipped.}\end{cases}$$

We put

$$\chi(N) = \begin{cases} 1, & \text{if } \sum_{n=1}^{N-1} a_n\chi(n) \le \beta\tau \ln(N-1), \\ 0, & \text{if } \sum_{n=1}^{N-1} a_n\chi(n) > \beta\tau \ln(N-1). \end{cases}$$

We will prove that

$$\sum_{n=1}^{N} a_n\chi(n) \sim \beta\tau \ln N$$

as $N \to \infty$. We will say that N is a *working index* if $\chi(N) = 1$ and that N is *idle* if $\chi(N) = 0$. We will call N a *skip index* if one of the following two conditions is satisfied:

$$\text{A}_1) \quad \sum_{n=1}^{N-2} a_n\chi(n) \le \beta\tau \ln(N-2), \quad \text{but} \quad \sum_{n=1}^{N-1} a_n\chi(n) > \beta\tau \ln(N-1);$$

$$\text{A}_2) \quad \sum_{n=1}^{N-2} a_n\chi(n) > \beta\tau \ln(N-2), \quad \text{but} \quad \sum_{n=1}^{N-1} a_n\chi(n) \le \beta\tau \ln(N-1).$$

Since for a fixed N and for $M \to \infty$ we have $\sum_{N\le n\le M} a_n \sim \tau \ln M > \beta\tau \ln M$ (strict inequality), there exist infinitely many skip indices.

Let $\varepsilon_N = \sup_{n\ge N} a_n/\ln n$. Then $\varepsilon_N \to 0$ as $N \to \infty$. Indeed, for any $\varepsilon > 0$ there exists N_0 such that for $n \ge N_0$

$$\left| \frac{1}{\ln(n+1)} \sum_{j\le n+1} a_j - \frac{1}{\ln n} \sum_{j\le n} a_n \right| \le \frac{\varepsilon}{2}.$$

Therefore,

$$\left| \left(\frac{1}{\ln n} - \frac{1}{\ln(n+1)} \right) \sum_{j\le n} a_n - \frac{a_{n+1}}{\ln(n+1)} \right| \le \frac{\varepsilon}{2}.$$

For $n \ge N_1$

$$\frac{1}{\ln n} \sum_{j\le n} a_n \le 2\tau.$$

For $n \ge \max(N_0, N_1)$

$$\frac{a_{n+1}}{\ln(n+1)} \le \frac{\varepsilon}{2} + \frac{2\tau}{\ln(n+1)}(\ln(n+1) - \ln n).$$

For $n \ge N_2$ we have $a_n/\ln n \le \varepsilon$.

Consider $\sum_{n\le N} a_n\chi(n)$.

a) Suppose N is an idle index. Then

$$\sum_{n\le N} a_n\chi(n) = \sum_{n\le N-1} a_n\chi(n) > \beta\tau \ln(N-1).$$

Let N' be the nearest skip index less than N. Then

$$\beta\tau \ln(N'-1) + a_{N'} > \sum_{n\le N} a_n\chi(n) > \beta\tau \ln(N-1),$$

$$\beta\tau + \frac{a_{N'}}{\ln N} > \frac{1}{\ln N}\sum_{n\le N} a_n\chi(n) > \beta\tau\frac{\ln(N-1)}{\ln N},$$

$$\frac{a_{N'}}{\ln N} \le \frac{a_{N'}}{\ln N'} \to 0,$$

since $N' \to \infty$ as $N \to \infty$.

b) Suppose N is a working index and N' is the nearest skip index less than N. Then

$$\begin{aligned}&\beta\tau \ln(N-1) + a_N \\ &\quad \ge \sum_{n\le N} a_n\chi(n) \ge \beta\tau \ln(N'-1) + a_{N'} + a_{N'+1} + \cdots + a_N.\end{aligned}$$

Since $a_N/\ln N \to 0$, it follows that

$$\overline{\lim_{N\to\infty}} \frac{1}{\ln N}\sum_{n\le N} a_n\chi(n) \le \beta\tau.$$

Let

$$\varepsilon_{N'} = \sup_{N>N'}\left|\frac{1}{\ln N}\sum_{n\le N} a_n - \frac{1}{\ln N'}\sum_{n\le N'} a_n\right|.$$

By the Cauchy criterion, $\varepsilon_{N'} \to 0$ as $N' \to \infty$. We have

$$\left(\frac{1}{\ln N'} - \frac{1}{\ln N}\right)\sum_{n\le N'} a_n - \frac{1}{\ln N}\sum_{N'<n\le N} a_n = \theta\varepsilon_{N'}, \qquad |\theta| \le 1,$$

$$\sum_{N'\le n\le N} a_n = a_{N'} + \theta \ln N\varepsilon_{N'} + \frac{\ln N - \ln N'}{\ln N'}\sum_{n\le N} a_n.$$

Also,

$$\begin{aligned}\frac{1}{\ln N}\sum_{n\le N} a_n\chi(n) \ge \beta\tau\frac{\ln(N'-1)}{\ln N} + \theta\varepsilon_{N'} + \frac{a_{N'}}{\ln N}& \\ + \frac{\ln N - \ln N'}{\ln N}\frac{1}{\ln N'}\sum_{n\le N'} a_n&.\end{aligned}$$

For sufficiently large N

$$\frac{1}{\ln N'}\sum_{n\le N'} a_n > \beta\tau.$$

Consequently,

$$\frac{1}{\ln N}\sum_{n\le N} a_n\chi(n) > \beta\tau\left(\frac{\ln(N-1)}{\ln N}+\frac{\ln(N'-1)-\ln N'}{\ln N}\right)+\frac{a_N}{\ln N}+\theta\varepsilon_{N'}$$

and

$$\underline{\lim}_{N\to\infty}\frac{1}{\ln N}\sum_{n\le N} a_n\chi(n)\ge\beta\tau.$$

Lemma 6 is proved.

LEMMA 7. *As* $x\to\infty$

$$m(x)\sim\frac{e^{-\gamma\tau}}{\Gamma(1+\tau)}\Pi(x).$$

PROOF. When $\tau=0$ this assertion is contained in Lemma 3.

Suppose $\tau>0$. Then

$$\sum_{p\le x} f(p)\frac{\ln p}{p}\sim\tau\,\ln x.$$

Put

$$a_n=\begin{cases} f(p)\ln p/p, & \text{if } n=p \text{ is prime,}\\ 0 & \text{if } n \text{ is composite.}\end{cases}$$

In this notation, $\sum_{n\le x} a_n\sim\tau\,\ln x$.

Given a natural number r. By Lemma 6, we can partition the set of primes $\mathfrak{P}$ into r mutually disjoint classes $\mathfrak{P}_1,\dots,\mathfrak{P}_r$ such that for each $i=1,\dots,r$

$$\sum_{p\le x, p\in\mathfrak{P}_i} f(p)\frac{\ln p}{p}\sim\frac{\tau}{r}\,\ln x.$$

We apply Lemma 5. As $x\to\infty$,

$$m(x)\Gamma(1+\tau)\sim\left(\Gamma\left(1+\frac{\tau}{r}\right)\right)^r\prod_{i=1}^r m_i(x).$$

We apply Lemma 3 to each $m_i(x)$:

$$m(x)\Gamma(1+\tau)=(1+o(1))\Gamma\left(1+\frac{\tau}{r}\right)^r\prod_{i=1}^r\left(1+2\theta_i\frac{\tau^2}{r^2}\right)\Pi_i(x),$$

where

$$\Pi_i(x)=\prod_{p\le x, p\in\mathfrak{P}_i}\left(1+\frac{f(p)}{p}+\frac{f(p^2)}{p^2}+\cdots\right),\qquad |\theta_i|\le 1.$$

Obviously $\prod_{i=1}^r\Pi_i(x)=\Pi(x)$. Also, as $r\to\infty$

$$\prod_{i=1}^r\left(1+2\theta_i\frac{\tau^2}{r^2}\right)=1+O\left(\frac{1}{r}\right).$$

Since

$$\lim_{x\to\infty}\frac{1}{x}\ln\Gamma(1+\tau x)=\frac{\tau\Gamma'(1)}{\Gamma(1)}=-\gamma\tau,$$

it follows that

$$\lim_{r\to\infty}\Gamma\left(1+\frac{\tau}{r}\right)^r=e^{-\gamma\tau}.$$

Passing to the limit as $r\to\infty$, we obtain as $x\to\infty$

$$m(x)\sim\frac{e^{-\gamma\tau}}{\Gamma(1+\tau)}\Pi(x).$$

Lemma 7 is proved.

Combining Lemmas 2 and 7, we see that for $\tau>0$

$$M(x)\sim\frac{\tau x}{\ln x}m(x)\sim\frac{\tau e^{-\gamma\tau}}{\Gamma(1+\tau)}\frac{x}{\ln x}\Pi(x)=\frac{e^{-\gamma\tau}}{\Gamma(\tau)}\frac{x}{\ln x}\Pi(x),$$

and for $\tau=0$

$$M(x)=o\left(\frac{x}{\ln x}m(x)\right)=o\left(\frac{x}{\ln x}\Pi(x)\right).$$

This proves Wirsing's theorem.

Levin and Faĭnleĭb [95] gave the following generalization of Wirsing's theorem.

THEOREM. *Suppose $f(n)$ is a complex-valued multiplicative function. Assume that $x\to\infty$*

$$\sum_{p\le x}f(p)=(\tau+o(1))\frac{x}{\ln x}.$$

Assume also that

$$\sum_{p\le x}|f(p)|=O\left(\frac{x}{\ln x}\right)$$

and, finally, that for any $r\ge 2$

$$f(p^r)=O((2p)^{c_0 r}),\qquad 0<c_0<1/2.$$

Then as $x\to\infty$

$$\sum_{n\le x}f(n)=\frac{e^{-\gamma\tau}}{\Gamma(\tau)}\frac{x}{\ln x}\prod_{p\le x}\left(1+\frac{f(p)}{p}+\frac{f(p^2)}{p^2}+\cdots\right)$$
$$+o\left(\frac{x}{\ln x}\prod_{p\le x}\left(1+\frac{|f(p)|}{p}+\frac{|f(p^2)|}{p^2}+\cdots\right)\right).$$

§4.4. The Turán-Kubilius inequality

We consider here the problem of the distribution of values of the function $\nu(n)$, the number of different prime divisors of n.

Note that the function $\nu(n)$ is strongly additive, i.e., if n and m are relatively prime, then $\nu(nm) = \nu(n) + \nu(m)$, and for any prime p and natural number r, $\nu(p^r) = \nu(p)$ (in this case, $\nu(p^r) = \nu(p) = 1$).

We estimate the order of growth of the mean value of this function:

$$\frac{1}{N}\sum_{n\le N}\nu(n) = \frac{1}{N}\sum_{n\le N}\sum_{p\mid n} 1 = \frac{1}{N}\sum_{p\le N}\left[\frac{N}{p}\right]$$
$$= \sum_{p\le N}\frac{1}{p} + O\left(\frac{1}{\ln N}\right) = \ln\ln N + O(1).$$

There naturally arises the question of how greatly the function $\nu(n)$ oscillates in the mean about $\ln\ln n$. Let

$$P_N\left(|\nu(m) - \ln\ln N| \le \psi(N)\sqrt{\ln\ln N}\right)$$

denote the number of integers $m = 1, \dots, N$ for which the inequality

$$|\nu(m) - \ln\ln N| \le \psi(N)\sqrt{\ln\ln N}$$

holds for a given positive function $\psi(N)$. We also introduce a notation for the frequency:

$$V_N\left(|\nu(m) - \ln\ln N| \le \psi(N)\sqrt{\ln\ln N}\right)$$
$$= \frac{P_N\left(|\nu(m) - \ln\ln N| \le \psi(N)\sqrt{\ln\ln N}\right)}{N}.$$

We are interested in the behavior of $V_N(\cdots)$ as $N \to \infty$.

The first result in this direction was due to Hardy and Ramanujan [68], who proved in 1917 that for any positive function $\psi(N)$ that increases without bound as $N \to \infty$

$$\lim_{N\to\infty} V_N\left(|\nu(m) - \ln\ln N| \le \psi(N)\sqrt{\ln\ln N}\right) = 1.$$

Another proof of this result was given by Turán [140]. An interesting feature of this proof is that there can be seen in it a sharp analogy to Tchebycheff's proof of the law of large numbers. Turán's method was generalized by Kubilius [87]. We will give a proof of the following inequality, known as the *Turán-Kubilius inequality.*

THEOREM. *For a given complex additive function $f(n)$ let*

$$A(N) = \sum_{p\le N}\frac{f(p)}{p} \quad \text{and} \quad D^2(N) = \sum_{p^\alpha \le N}\frac{|f(p^\alpha)|^2}{p^\alpha}.$$

Then

$$\frac{1}{N}\sum_{n=1}^{N}|f(n)-A(N)|^2 < C_1D^2(N),$$

where C_1 is a positive constant that does not depend on $f(n)$.

PROOF. We write $p^\alpha\|m$ if $p^\alpha|m$ but $p^{\alpha+1}\nmid m$. It is clear that

$$\frac{1}{N}\sum_{m=1}^{N}f(m)=\frac{1}{N}\sum_{m=1}^{N}\sum_{p^\alpha\|m}f(p^\alpha)=\frac{1}{N}\sum_{p^\alpha\le N}f(p^\alpha)P_N(p^\alpha\|m),$$

where $P_N(p^\alpha\|m)$ denotes the number of integers not exceeding N for which $p^\alpha\|m$. But

$$P_N(p^\alpha\|m)=\left[\frac{N}{p^\alpha}\right]-\left[\frac{N}{p^{\alpha+1}}\right].$$

Consequently,

$$\begin{aligned}\frac{1}{N}\sum_{m=1}^{N}f(m)&=\frac{1}{N}\sum_{p^\alpha\le N}f(p^\alpha)\left(\left[\frac{N}{p^\alpha}\right]-\left[\frac{N}{p^{\alpha+1}}\right]\right)\\&=\sum_{p^\alpha\le N}\left(1-\frac{1}{p}\right)\frac{f(p^\alpha)}{p^\alpha}+\frac{2\theta}{N}\sum_{p^\alpha\le N}|f(p^\alpha)|,\qquad |\theta|\le 1.\end{aligned}$$

Also, by the Cauchy-Schwarz-Bunyakovskiĭ inequality,

$$\begin{aligned}\sum_{p^\alpha\le N}|f(p^\alpha)|&\le\sqrt{\sum_{p^\alpha\le N}p^\alpha\cdot\sum_{p^\alpha\le N}\frac{|f(p^\alpha)|^2}{p^\alpha}}\\&=D(N)\Bigl(\sum_{p^\alpha\le N}p^\alpha\Bigr)^{1/2}\le\sqrt{N}\,D(N)\sqrt{\sum_{p^\alpha\le N}1}.\end{aligned}$$

But

$$\begin{aligned}\sum_{p^\alpha\le N}1&\le\sum_{p\le N}1+\sum_{p\le\sqrt{N}}\ \sum_{2\le\alpha\le\ln N/\ln p}1\\&=O\left(\frac{N}{\ln N}\right)+O\Bigl(\sum_{p\le\sqrt{N}}\ln N\Bigr)=O\left(\frac{N}{\ln N}\right).\end{aligned}$$

Thus,

$$\sum_{p^\alpha\le N}|f(p^\alpha)|=O\left(\frac{ND(N)}{\sqrt{\ln N}}\right)$$

with an absolute constant implied by the symbol O. We obtain

$$\frac{1}{N}\sum_{m=1}^{N}f(m)=\sum_{p^\alpha\le N}\left(1-\frac{1}{p}\right)\frac{f(p^\alpha)}{p^\alpha}+O\left(\frac{D(N)}{\sqrt{\ln N}}\right).$$

Also,

$$\frac{1}{N}\sum_{m=1}^{N} f(m) = A(N) + \sum_{\substack{p^\alpha \le N\\ \alpha>1}} \frac{f(p^\alpha)}{p^\alpha} - \sum_{p^\alpha\le N} \frac{f(p^\alpha)}{p^{\alpha+1}} + O\left(\frac{D(N)}{\sqrt{\ln N}}\right),$$

and hence

$$\left|\frac{1}{N}\sum_{m=1}^{N} f(m) - A(N)\right| \le \left|\sum_{\substack{p^\alpha \le N\\ \alpha>1}} \frac{f(p^\alpha)}{p^\alpha}\right| + \left|\sum_{p^\alpha\le N} \frac{f(p^\alpha)}{p^{\alpha+1}}\right| + O\left(\frac{D(N)}{\sqrt{\ln N}}\right).$$

By the Cauchy-Schwarz-Bunyakovskiĭ inequality,

$$\left|\sum_{\substack{p^\alpha \le N\\ \alpha>1}} \frac{f(p^\alpha)}{p^\alpha}\right| \le \sqrt{\sum_{\substack{p^\alpha \le N\\ \alpha>1}} \frac{1}{p^\alpha} \sum_{p^\alpha\le N} \frac{|f(p^\alpha)|^2}{p^\alpha}} = O(D(N))$$

with an absolute constant implied by the symbol O. Analogously,

$$\left|\sum_{p^\alpha\le N} \frac{f(p^\alpha)}{p^{\alpha+1}}\right| \le \sqrt{\sum_{p^\alpha\le N} \frac{1}{p^{\alpha+2}} \sum_{p^\alpha\le N} \frac{|f(p^\alpha)|^2}{p^\alpha}} = O(D(N)).$$

Thus,

$$\left|\frac{1}{N}\sum_{m=1}^{N} f(m) - A(N)\right| = O(D(N)).$$

Therefore,

$$\begin{aligned}\frac{1}{N}\sum_{n=1}^{N} &|f(n) - A(N)|^2 \\ &\le \frac{2}{N}\sum_{n=1}^{N}\left|f(n) - \frac{1}{N}\sum_{m=1}^{N} f(m)\right|^2 + 2\left|\frac{1}{N}\sum_{m=1}^{N} f(m) - A(N)\right|^2 \\ &= \frac{2}{N}\sum_{n=1}^{N}\left|f(n) - \frac{1}{N}\sum_{m=1}^{N} f(m)\right|^2 + O(D^2(N)).\end{aligned}$$

Obviously

$$\frac{1}{N}\sum_{n=1}^{N}\left|f(n) - \frac{1}{N}\sum_{m=1}^{N} f(m)\right|^2 = \frac{1}{N}\sum_{m=1}^{N} |f(m)|^2 - \frac{1}{N^2}\left|\sum_{m=1}^{N} f(m)\right|^2.$$

We have

$$\begin{aligned}\sum_{m=1}^{N} |f(m)|^2 &= \sum_{m=1}^{N} \sum_{p^\alpha \| m} \sum_{q^\beta \| m} f(p^\alpha)\overline{f(q^\beta)} \\ &= \sum_{p^\alpha \le N} |f(p^\alpha)|^2 P_N(p^\alpha \| m) \\ &\qquad + \sum_{\substack{p^\alpha q^\beta \le N \\ p \ne q}} f(p^\alpha)\overline{f(q^\beta)} P_N(p^\alpha \| m, q^\beta \| m).\end{aligned}$$

If $p \neq q$,

$$\begin{aligned}P_N(p^\alpha \| m, q^\beta \| m) &= P_N(p^\alpha q^\beta | m) - P_N(p^{\alpha+1} q^\beta | m) \\ &\qquad - P_N(p^\alpha q^{\beta+1} | m) + P_N(p^{\alpha+1} q^{\beta+1} | m) \\ &= \left[\frac{N}{p^\alpha q^\beta}\right] - \left[\frac{N}{p^{\alpha+1} q^\beta}\right] - \left[\frac{N}{p^\alpha q^{\beta+1}}\right] + \left[\frac{N}{p^{\alpha+1} q^{\beta+1}}\right] \\ &= \frac{N}{p^\alpha q^\beta}\left(1 - \frac{1}{p}\right)\left(1 - \frac{1}{q}\right) + 4\theta,\end{aligned}$$

where $|\theta| \le 1$. Since $P_N(p^\alpha \| m) \le N/p^\alpha$, it follows that

$$\begin{aligned}\sum_{m=1}^{N} |f(m)|^2 = O(ND^2(N)) + N \sum_{\substack{p^\alpha q^\beta \le N \\ p \ne q}} \frac{f(p^\alpha)\overline{f(q^\beta)}}{p^\alpha q^\beta}\left(1 - \frac{1}{p}\right)\left(1 - \frac{1}{q}\right) \\ + O\Bigg(\sum_{p^\alpha q^\beta \le N} |f(p^\alpha)|\,|f(q^\beta)|\Bigg).\end{aligned}$$

Let us estimate the last sum. By the Cauchy-Schwarz-Bunyakovskiĭ inequality,

$$\begin{aligned}\sum_{p^\alpha q^\beta \le N} |f(p^\alpha) f(q^\beta)| &\le \Bigg(\sum_{p^\alpha q^\beta \le N} p^\alpha q^\beta\Bigg)^{1/2} \Bigg(\sum_{p^\alpha q^\beta \le N} \frac{|f(p^\alpha)|^2 |f(q^\beta)|^2}{p^\alpha q^\beta}\Bigg)^{1/2} \\ &\le \sqrt{N} D^2(N) \Bigg(\sum_{p^\alpha q^\beta \le N} 1\Bigg)^{r} 1/2 = O(ND^2(N)).\end{aligned}$$

Thus,

$$\sum_{m=1}^{N} |f(m)|^2 = O(ND^2(N)) + N \sum_{p^\alpha q^\beta \le N} \frac{f(p^\alpha)\overline{f(q^\beta)}}{p^\alpha q^\beta}\left(1-\frac{1}{p}\right)\left(1-\frac{1}{q}\right) - N \sum_{\substack{p^\alpha q^\beta \le N \\ p=q}} \frac{f(p^\alpha)\overline{f(q^\beta)}}{p^\alpha q^\beta}\left(1-\frac{1}{p}\right)\left(1-\frac{1}{q}\right).$$

By analogy with the preceding calculations,

$$\begin{aligned}\Bigg| \sum_{\substack{p^\alpha q^\beta \le N \\ p=q}} \frac{f(p^\alpha)\overline{f(q^\beta)}}{p^\alpha q^\beta}\left(1-\frac{1}{p}\right)\left(1-\frac{1}{q}\right)\Bigg| &\le \Bigg(\sum_{\substack{p^\alpha q^\beta \le N \\ p=q}} \frac{|f(p^\alpha)|^2}{p^\alpha q^\beta}\Bigg)^{1/2} \Bigg(\sum_{\substack{p^\alpha q^\beta \le N \\ p=q}} \frac{|f(q^\beta)|^2}{p^\alpha q^\beta}\Bigg)^{1/2} \\ &= O\Bigg(\sum_{p^\alpha \le N} \frac{|f(p^\alpha)|^2}{p^{\alpha+1}}\Bigg) = O(D^2(N)).\end{aligned}$$

These estimates imply

$$\sum_{m=1}^{N} |f(m)|^2 = N \sum_{p^\alpha q^\beta \le N} \frac{f(p^\alpha)\overline{f(q^\beta)}}{p^\alpha q^\beta}\left(1-\frac{1}{p}\right)\left(1-\frac{1}{q}\right) + O(ND^2(N)).$$

We obtain the inequality

$$\frac{1}{N}\sum_{n=1}^{N} |f(n) - A(N)|^2 \le 2\Bigg(\sum_{p^\alpha q^\beta \le N} \frac{f(p^\alpha) f(q^\beta)}{p^\alpha q^\beta}\left(1-\frac{1}{p}\right)\left(1-\frac{1}{q}\right) - \frac{1}{N^2}\Bigg|\sum_{m=1}^{N} f(m)\Bigg|^2\Bigg) + O(D^2(N))$$

with an absolute constant implied by the symbol O. But we have already seen that

$$\frac{1}{N}\sum_{m=1}^{N} f(m) = \sum_{p^\alpha \le N}\left(1-\frac{1}{p}\right)\frac{f(p^\alpha)}{p^\alpha} + O\left(\frac{D(N)}{\sqrt{\ln N}}\right).$$

Thus,

$$\begin{aligned}\frac{1}{N^2}\left|\sum_{m=1}^{N} f(m)\right|^2 &= \left|\sum_{p^\alpha\le N}\left(1-\frac{1}{p}\right)\frac{f(p^\alpha)}{p^\alpha}\right|^2 + O\left(\frac{D^2(N)}{\ln N}\right)\\ &\quad + O\left(\frac{D(N)}{\sqrt{\ln N}}\left|\sum_{p^\alpha\le N}\left(1-\frac{1}{p}\right)\frac{f(p^\alpha)}{p^\alpha}\right|^2\right)\\ &= \left|\sum_{p^\alpha\le N}\left(1-\frac{1}{p}\right)\frac{f(p^\alpha)}{p^\alpha}\right|^2\\ &\quad + O\left(\frac{D^2(N)}{\ln N}\right) + O\left(\frac{D^2(N)}{\sqrt{\ln N}}\sqrt{\ln\ln N}\right),\end{aligned}$$

since, again by the Cauchy-Schwarz-Bunyakovskiĭ inequality,

$$\begin{aligned}&\left(\sum_{p^\alpha\le N}\left(1-\frac{1}{p}\right)\frac{|f(p^\alpha)|}{p^\alpha}\right)^2\\ &\qquad\le \left(\sum_{p^\alpha\le N}\frac{1}{p^\alpha}\right)\left(\sum_{p^\alpha\le N}\frac{|f(p^\alpha)|^2}{p^\alpha}\right) = O(D^2(N)\ln\ln N).\end{aligned}$$

Therefore,

$$\begin{aligned}&\frac{1}{N}\sum_{n=1}^{N}|f(n)-A(N)|^2\\ &\qquad\le 2\sum\sum_{\substack{p^\alpha q^\beta>N\\ p^\alpha\le N,\ q^\beta\le N}}\frac{f(p^\alpha)\overline{f(q^\beta)}}{p^\alpha q^\beta}\left(1-\frac{1}{p}\right)\left(1-\frac{1}{q}\right)+O(D^2(N)).\end{aligned}$$

We apply the Cauchy-Schwarz-Bunyakovskiĭ inequality once more, obtaining

$$\begin{aligned}&\sum_{\substack{p^\alpha q^\beta>N\\ p^\alpha\le N,\ q^\beta\le N}}\frac{f(p^\alpha)\overline{f(q^\beta)}}{p^\alpha q^\beta}\left(1-\frac{1}{p}\right)\left(1-\frac{1}{q}\right)\\ &\qquad\le\sqrt{\sum_{\substack{p^\alpha q^\beta>N\\ p^\alpha\le N,\ q^\beta\le N}}\frac{1}{p^\alpha}\frac{1}{q^\beta}}\sqrt{\sum_{\substack{p^\alpha\le N\\ q^\beta\le N}}\frac{|f(p^\alpha)|^2|f(q^\beta)|^2}{p^\alpha q^\beta}}\\ &\qquad= D^2(N)\sqrt{\sum_{\substack{p^\alpha q^\beta>N\\ p^\alpha\le N,\ q^\beta\le N}}\frac{1}{p^\alpha}\frac{1}{q^\beta}}.\end{aligned}$$

Obviously

$$\sum_{\substack{p^\alpha q^\beta > N \\ p^\alpha \le N,\ q^\beta \le N}} \frac{1}{p^\alpha}\frac{1}{q^\beta} = 2 \sum_{p^\alpha \le \sqrt{N}} \frac{1}{p^\alpha} \sum_{N/p^\alpha < q^\beta \le N} \frac{1}{q^\beta} + O\Bigg(\sum_{\sqrt{N} \le p^\alpha \le N} \frac{1}{p^\alpha}\Bigg)$$

$$= 2 \sum_{p^\alpha \le \sqrt{N}} \frac{1}{p^\alpha} \left(\ln \frac{\ln N}{\ln N - \ln p^\alpha} + O\left(\frac{1}{\ln N}\right)\right) + O(1)$$

$$= O\Bigg(\frac{1}{\ln N} \sum_{p^\alpha \le \sqrt{N}} \frac{\ln p^\alpha}{p^\alpha}\Bigg) + O(1) = O(1).$$

Thus,

$$\frac{1}{N} \sum_{n=1}^{N} |f(n) - A(N)|^2 = O(D^2(N))$$

with an absolute constant implied by the symbol O. This proves the Turán-Kubilius inequality.

COROLLARY. *Suppose $f(n)$ is a strongly additive function. Let*

$$B^2(N) = \sum_{p \le N} \frac{|f(p)|^2}{p}.$$

Then

$$\frac{1}{N} \sum_{n=1}^{N} |f(n) - A(N)|^2 \le C_E 1 B^2(N).$$

Indeed, for strongly additive functions we have

$$D^2(N) = \sum_{p^\alpha \le N} \frac{|f(p^\alpha)|^2}{p^\alpha} = \sum_{p^\alpha \le N} \frac{|f(p)|^2}{p^\alpha}$$

$$\le \sum_{p \le N} \frac{|f(p)|^2}{p} \left(1 + \frac{1}{p} + \frac{1}{p^2} + \cdots\right)$$

$$\le \left(1 + \frac{1}{2} + \frac{1}{2^2} + \cdots\right) B^2(N).$$

We will show that the Turán-Kubilius inequality implies the theorem of Hardy and Ramanujan on $\nu(n)$. Since for $\nu(n)$ we have

$$A(N) = \sum_{p \le N} \frac{1}{p} = \ln\ln N + O(1)$$

and

$$B^2(N) = \sum_{p \le N} \frac{1}{p} = \ln\ln N + O(1) = O(\ln\ln N),$$

it follows from the Turán-Kubilius inequality for strongly additive functions that

$$\frac{1}{N}\sum_{n=1}^{N}|\nu(n)-\ln\ln N|^2 \le \frac{2}{N}\sum_{n=1}^{N}|\nu(n)-A(N)|^2+O(1)$$
$$= O(B^2(N))+O(1)=O(\ln\ln N).$$

Now suppose there is given a positive function $\psi(n)$ that tends to infinity as $N\to\infty$. We have

$$P_N(|\psi(n)-\ln\ln N|\ge\psi(N)\sqrt{\ln\ln N})$$
$$= \sum_{\substack{n=1\\|\psi(n)-\ln\ln N|\ge\psi(N)\sqrt{\ln\ln N}}}^{N} 1$$
$$\ge \frac{1}{\psi^2(N)\ln\ln N}\sum_{n=1}^{N}|\psi(N)-\ln\ln N|^2=\left(\frac{N}{\psi^2(N)}\right).$$

Thus as $N\to\infty$

$$V_N(|\psi(n)-\ln\ln N|\ge\psi(N)\sqrt{\ln\ln N})\to 0,$$

as required.

§4.5. Delange's theorems

We will consider here two theorems of Delange—a direct and an inverse theorem. We first state and prove Delange's direct theorem.

THEOREM 1. *Suppose $f(n)$ is a multiplicative function such that $|f(n)|\le 1$ for all natural numbers n. Assume the limit*

$$\lim_{N\to\infty}\frac{1}{N}\sum_{n\le N}f(n)=M(f)$$

exists and $M(f)\ne 0$. Then the following assertions are true:

1) *The series $\sum_p(1-f(p))/p$ converges.*
2) *There exists a natural number r such that $f(2^r)\ne -1$.*

PROOF. For any multiplicative function $f(n)$ we define the numbers $c_r^{(p)}(f)$ ($r=1,2,\dots$ and p ranges over the primes) as follows:

a) for $r=1$

$$c_1^{(p)}(f)=f(p); \tag{1}$$

b) for $r>1$

$$c_r^{(p)}(f)=rf(p^r)-\sum_{j=1}^{r-1}c_j^{(p)}(f)f(p^r-j). \tag{1'}$$

Then for all $n \geq 1$

$$f(n)\ln n = \sum_{p^j \mid n} c_j^{(p)}(f) f\left(\frac{n}{p^j}\right) \ln p. \tag{2}$$

Indeed, suppose $n = p_0^r m$ and $(m, p_0) = 1$. If $r = 1$, then in the right-hand side of the equality we have the term

$$c_1^{(p_0)}(f) f(m) \ln p_0 = f(p_0) f(m) \ln p_0 = f(n) \ln p_0.$$

If $r > 1$, then the right-hand side of (2) contains the sum

$$\sum_{j=1}^{r} c_j^{(p_0)}(f) f(p_0^{r-j} m) \ln p_0 = f(m) \ln p_0 \left(\sum_{j=1}^{r-1} c_j^{(p_0)}(f) f(p_0^{r-j}) + c_r^{(p_0)}(f) \right)$$
$$= f(m) \ln p_0 r f(p_0^r) = f(m p_0^r) \ln p_0^r = f(n) \ln p_0^r.$$

Suppose $f(n)$ is a multiplicative function such that $|f(n)| \leq 1$. We prove by induction that

$$|c_r^{(p)}(f)| \leq 2^r - 1. \tag{3}$$

This is obvious for $r = 1$. Also,

$$|c_r^{(p)}(f)| \leq r + \sum_{j=1}^{r-1} (2^j - 1) = 2^r - 1.$$

For each prime p the power series

$$F_p(z) = 1 + \sum_{j=1}^{\infty} f(p^j) z^j$$

converges for $|z| < 1$. The series

$$c_p(z) = \sum_{j=1}^{\infty} \frac{c_j^{(p)}(f)}{j} z^j$$

converges (in view of (3)) for $|z| < 1/2$.

LEMMA 1. *For* $|z| < 1/2$

$$F_p(z) = e^{c_p(z)}. \tag{4}$$

PROOF. It follows from (1) and (1′) that

$$c_p'(z) F_p(z) = F_p'(z) \tag{5}$$

(equalities (1) and (1′) can be obtained by equating the coefficients of identical powers of z in (5)). Integrating, we obtain

$$F_p(z) = c e^{c_p(z)},$$

where c is a constant. But $F_p(0) = 1$ and $c_p(0) = 0$ imply $c = 1$.

It follows from

$$1 + \sum_{j=1}^{\infty} f(p^j) z^j = \exp\left(\sum_{j=1}^{\infty} \frac{c_j^{(p)}(f)}{j} z^j \right) \tag{4}$$

and

$$1 - z = e^{\ln(1-z)} = \exp\left(-\sum_{j=1}^{\infty} \frac{z^j}{j} \right)$$

that for $|z| < 1/2$

$$(1-z)\left(1 + \sum_{j=1}^{\infty} f(p^j) z^j\right) = \exp\left(\sum_{j=1}^{\infty} \frac{c_j^{(p)}(f) - 1}{j} z^j \right). \tag{4'}$$

LEMMA 2. *Suppose $f(n)$ is a multiplicative function that satisfies the conditions of Theorem 1. Let s approach 1 along the real axis through values greater than 1 (we write $s \to 1+0$). Then the expression*

$$\sum_p \frac{1 - f(p)}{p^s}$$

approaches a finite limit.

PROOF. Since $f(n)$ is multiplicative, for $\operatorname{Re} s > 1$ we have

$$\sum_{n=1}^{\infty} \frac{f(n)}{n^s} = \prod_p \left(1 + \sum_{j=1}^{\infty} \frac{f(p^j)}{p^{js}} \right).$$

Therefore,

$$\frac{1}{\zeta(s)} \sum_{n=1}^{\infty} \frac{f(n)}{n^s} = \prod_p \left(1 - \frac{1}{p^s}\right)\left(1 + \sum_{j=1}^{\infty} \frac{f(p^j)}{p^{js}} \right),$$

and, in view of (4′),

$$\frac{1}{\zeta(s)} \sum_{n=1}^{\infty} \frac{f(n)}{n^s} = \left(1 - \frac{1}{2^s}\right)\left(1 + \sum_{j=1}^{\infty} \frac{f(2^j)}{2^{js}} \right)$$
$$\times \exp\left(-\sum_{p>2} \frac{1 - f(p)}{p^s} + \sum_{p>2,\ j>1} \frac{c_j^{(p)}(f) - 1}{j p^{js}} \right).$$

For a real $s > 1$

$$\operatorname{Re} \sum_{p>2,\ j>1} \frac{c_j^{(p)}(f) - 1}{j p^{js}} \le \left| \sum_{p>2,\ j>1} \frac{c_j^{(p)}(f) - 1}{j p^{js}} \right| \le \sum_{p>2,\ j>1} \frac{2^j}{j p^j}$$
$$\le \sum_{p>2,\ j>1} \frac{2^{j-1}}{p^j} = \sum_{p>2} \frac{2}{p^2} \frac{1}{1 - 2/p} = \sum_{p>2} \frac{2}{p(p-2)}.$$

Consequently, for a real $s > 1$

$$\left|\exp\left(-\sum_{p>2}\frac{1-f(p)}{p^s}+\sum_{p>2,\ j>1}\frac{c_j^{(p)}(f)-1}{jp^{js}}\right)\right| \le \exp\left[a-\sum_{p>2}\frac{1-\operatorname{Re} f(p)}{p^s}\right],$$

where $a = \sum_{p>2} 2/p(p-2)$.

By hypothesis, $1 - \operatorname{Re} f(p) \ge 0$; hence

$$\left|\exp\left[-\sum_{p>2}\frac{1-f(p)}{p^s}+\sum_{p>2,\ j>1}\frac{c_j^{(p)}(f)-1}{jp^{js}}\right]\right| \le e^a. \tag{6}$$

We now use the hypothesis of the theorem which says that as $x \to \infty$

$$\sum_{n\le x} f(n) \sim M(f)x \tag{7}$$

and $M(f) \ne 0$. Using Abel's transformation, we see easily from (7) that as $s \to 1+0$

$$\sum_{n=1}^{\infty}\frac{f(n)}{n^s} \sim \frac{M(f)}{s-1}. \tag{8}$$

Thus as $s \to 1+0$ the quantity

$$\left(1-\frac{1}{2^s}\right)\left(1+\sum_{j=1}^{\infty}\frac{f(2^j)}{2^{js}}\right)\exp\left[-\sum_{p>2}\frac{1-f(p)}{p^s}+\sum_{p>2,\ j>1}\frac{c_j^{(p)}(f)-1}{jp^{js}}\right]$$

tends to the limit $M(f) \ne 0$. As $s \to 1+0$

$$\left(1-\frac{1}{2^s}\right)\left(1+\sum_{j=1}^{\infty}\frac{f(2^j)}{2^{js}}\right) \to \frac{1}{2}\left(1+\sum_{j=1}^{\infty}\frac{f(2^j)}{2^j}\right).$$

In view of (6), the expression

$$\exp\left[-\sum_{p>2}\frac{1-f(p)}{p^s}+\sum_{p>2,\ j>1}\frac{c_j^{(p)}(f)-1}{jp^{js}}\right]$$

is uniformly bounded for $\operatorname{Re} s > 1$. It follows that

$$1+\sum_{j=1}^{\infty}\frac{f(2^j)}{2^j} \ne 0. \tag{9}$$

If it were true that $f(2^j) = -1$ for all integers $j \ge 1$, then

$$1+\sum_{j=1}^{\infty}\frac{f(2^j)}{2^j} = 1-\sum_{j=1}^{\infty}\frac{1}{2^j} = 0,$$

which contradicts (9). We have proved, incidentally, the second assertion of the theorem.

Thus the function

$$\exp\left[-\sum_{p>2}\frac{1-f(p)}{p^s}+\sum_{p>2,\ j>1}\frac{c_j^{(p)}(f)-1}{jp^{js}}\right]$$

tends to a nonzero limit as $s\to 1+0$. In view of (3), the expression

$$\sum_{p>2,\ j>1}\frac{c_j^{(p)}(f)-1}{jp^{js}}$$

approaches

$$\sum_{p>2,\ j>1}\frac{c_j^{(p)}(f)-1}{jp^{j}}.$$

Therefore the expression

$$\exp\left[-\sum_{p>2}\frac{1-f(p)}{p^s}\right]$$

approaches a nonzero limit as $s\to 1+0$, i.e., the series

$$\sum_{p>2}\frac{1-f(p)}{p^s}$$

approaches a finite limit as $s\to 1+0$. Lemma 2 is proved.

Lemma 2 can be expressed as follows: As $s\to 0+0$ (this means that the real variable s tends to zero from the right) the expression

$$\sum_{p}\frac{1-f(p)}{p^{1+s}}$$

approaches a finite limit. We denote this limit by l. We must now prove that the series

$$\sum_{p}\frac{1-\operatorname{Re}f(p)}{p}\quad\text{and}\quad -\sum\frac{\operatorname{Im}f(p)}{p}$$

converge (the first series to $\operatorname{Re}l$, the second to $\operatorname{Im}l$). By definition of the Stieltjes integral,

$$\sum_{p}\frac{1-\operatorname{Re}f(p)}{p^{s+1}}=\int_0^\infty e^{-st}\,d\alpha_1(t),$$

where

$$\alpha_1(t)=\sum_{\ln p\le t}\frac{1-\operatorname{Re}f(p)}{p}=\sum_{p\le e^t}\frac{1-\operatorname{Re}f(p)}{p},$$

and

$$-\sum_p \frac{\operatorname{Im} f(p)}{p^{s+1}} = \int_0^\infty e^{-st}\, d\alpha_2(t),$$

where

$$\alpha_2(t) = \sum_{\ln p \le t} -\frac{\operatorname{Im} f(p)}{p} = \sum_{p \le e^t} -\frac{\operatorname{Im} f(p)}{p}.$$

The deduction of Delange's theorem from Lemma 2 is effected by means of the generalized Littlewood theorem, which was stated in §1.1. To complete the proof we must show that the functions $\alpha_1(t)$ and $\alpha_2(t)$ are slowly oscillating. We use the classical formula

$$\sum_{p \le x} \frac{1}{p} = \ln\ln x + B + O\left(\frac{1}{\ln x}\right).$$

Suppose $t' = \lambda t$, $\lambda > 1$, and $t \to \infty$. We have

$$|\alpha_1(t') - \alpha_1(t)| = \left| \sum_{e^t < p \le e^{\lambda t}} \frac{1 - \operatorname{Re} f(p)}{p} \right| \le 2 \sum_{e^t < p \le e^{\lambda t}} \frac{1}{p} = 2\ln\lambda + o(1)$$

and

$$|\alpha_2(t') - a_2(t)| = \left| \sum_{e^t < p \le e^{\lambda t}} \frac{\operatorname{Im} f(p)}{p} \right| \le \sum_{e^t < p \le e^{\lambda t}} \frac{1}{p} = \ln\lambda + o(1).$$

Therefore, as $\lambda \to 1$ and $t \to \infty$,

$$\alpha_1(t') - \alpha_1(t) \to 0 \quad \text{and} \quad \alpha_2(t') - \alpha_2(t) \to 0.$$

Theorem 1 is proved.

THEOREM 2. *Suppose $g(n)$ is a multiplicative function such that*
1) $|g(n)| \le 1$, $n = 1, 2, \ldots$, *and*
2) *the series* $\sum_p (1 - g(p))/p$ *converges.*
Then the limit

$$\lim_{N\to\infty} \frac{1}{N} \sum_{n=1}^N g(n) = M(g)$$

exists, and

$$M(g) = \prod_p \left(1 - \frac{1}{p}\right)\left(1 + \sum_{j=1}^\infty \frac{g(p^j)}{p^j}\right).$$

REMARK. Under a condition stronger than condition 2) of the theorem, namely under the condition that $\sum_p (g(p) - 1)/p$ be absolutely convergent, the assertion is trivial. Indeed, consider the multiplicative function $\Phi(n)$ defined at prime powers by

$$\Phi(p^k) = \begin{cases} 1, & k = 0, \\ g(p^k) - g(p^{k-1}), & k \geq 1. \end{cases}$$

We have

$$\sum_{d|n} \Phi(d) = g(n),$$

$$\sum_{n \leq N} g(n) = N \sum_{n=1}^{N} \frac{\Phi(n)}{n} + O\Big(\sum_{n \leq N} |\Phi(n)| \Big).$$

The absolute convergence of $\sum_p (g(p) - 1)/p$ implies the convergence of the product

$$\prod_p \left(1 + \frac{|\Phi(p)|}{p} + \frac{|\Phi(p^2)|}{p^2} + \cdots \right) = \prod_p \left(1 + \frac{|g(p) - 1|}{p} + O\left(\frac{1}{p^2} \right) \right).$$

Thus $\sum_1^\infty |\Phi(n)|/n$ converges. Using Abel's transformation, it is easy to see that $\sum_{n \leq N} |\Phi(n)| = o(N)$. Thus

$$\begin{aligned} \frac{1}{N} \sum_{n \leq N} g(n) &= \sum_{n=1}^{\infty} \frac{\Phi(n)}{n} + o(1) \\ &= \prod_p \left(1 + \frac{g(p) - 1}{p} + \frac{g(p^2) - g(p)}{p^2} + \cdots \right) + o(1) \\ &= \prod_p \left(1 - \frac{1}{p} \right) \left(1 + \frac{g(p)}{p} + \frac{g(p^2)}{p^2} + \cdots \right) + o(1). \end{aligned}$$

Delange gave two proofs of his theorem [28], [29]. We will present here the proof of Rényi [124].

We begin by proving

LEMMA 3. *Suppose $g_1(n)$ and $g_2(n)$ are multiplicative functions such that $|g_1(n)| \leq 1$ and $|g_2(n)| \leq 1$. Assume that*

1) *$M(g_1)$ exists,*

2) *$\sum_p |g_1(p) - g_2(p)|/p < \infty$, and*

3) *if $|g_1(2)| = 1$ and $g_1(2^r) = (-1)^{r+1} g_1^r(2)$ for all $r > 1$, then $g_2(2^r) = g_1(2^r)$ for all $r = 1, 2, \ldots$.*

Then $M(g_2)$ exists and

$$M(g_2) = M(g_1) \prod_p \frac{1 + \sum_{j=1}^{\infty} g_2(p^j)/p^j}{1 + \sum_{j=1}^{\infty} g_1(p^j)/p^j}.$$

(*If the factor corresponding to* $p = 2$ *has the form* $0/0$, *it is assumed equal to* 1.)

The product occurring in the right-hand side is absolutely convergent.

PROOF. We introduce, for each p, two functions defined for $|z| < 1$:

$$F_p(z) = 1 + \sum_{j=1}^{\infty} g_2(p^j)z^j \quad \text{and} \quad G_p(z) = 1 + \sum_{j=1}^{\infty} g_1(p^j)z^j.$$

It is clear that $G_p(z) \neq 0$ for $|z| < 1/2$, since for $|z| < 1/2$ we have

$$\left|\sum_{j=1}^{\infty} g_1(p^j)z^j\right| \leq \sum_{j=1}^{\infty} |z|^j = \frac{|z|}{1-|z|} < 1.$$

The function $F_p(z)/G_p(z)$ is obviously equal to 1 at $z = 0$, is meromorphic to $|z| < 1$, and is analytic for $|z| < 1/2$.

For $|z| < 1/2$ we put

$$\frac{F_p(z)}{G_p(z)} = 1 + \sum_{j=1}^{\infty} h_j^{(p)} z^j$$

and define a multiplicative function $h(n)$ by means of the equalities $h(p^r) = h_r^{(p)}$. For all $n \geq 1$

$$g_2(n) = \sum_{d|n} h(d) g_1\left(\frac{n}{d}\right).$$

Indeed, for any prime p and any natural number $r \geq 1$

$$\begin{aligned}\sum_{d|p^r} h(d) g_1\left(\frac{n}{d}\right) &= \sum_{j=0}^{r} h(p^j) g_1(p^{r-j}) \\ &= g_1(p^r) + \sum_{j=1}^{r} h_j^{(p)} g_1(p^{r-j}) = g_2(p^r).\end{aligned}$$

Moreover the function $\sum_{d|n} h(d)g_1(n/d)$ is multiplicative (by the generalized Wintner theorem).

Note that $|h(p^r)| \leq 2^r$ for any prime p and for $r \geq 1$. This is a consequence of the fact that the coefficients of $F_p(z)/G_p(z)$ are majorized by those of the series

$$p\frac{1 + \sum_{j=1}^{\infty} z^j}{1 - \sum_{j=1}^{\infty} z^j} = \frac{1}{1-2z}.$$

Furthermore, $h(p) = h_1(p) = g_2(p) - g_1(p)$. It follows that for any $p > 2$ the series $\sum_1^{\infty} h(p^j)/p^j$ is absolutely convergent and is majorized by

$$\frac{|g_1(p) - g_2(p)|}{p} + \sum_{j=2}^{\infty} \frac{2^j}{p^j} = \frac{|g_1(p) - g_2(p)|}{p} + \frac{4}{p(p-2)}.$$

We will prove that the series

$$\sum_{j=1}^{\infty} \frac{|h(2^j)|}{2^j} = \sum_{j=1}^{\infty} \frac{|h_j^{(2)}|}{2^j}$$

converges. It suffices to show that the function $F_2(z)/G_2(z)$ is analytic in a disc with center at the origin and with radius greater than 1/2. Suppose $G_2(z) = 0$ for some z such that $|z| = 1/2$. Putting $z = e^{i\theta}/2$ and $g_1(2^j) = \rho_j e^{i\theta_j}, 0 \leq \rho_j < 1$, we have

$$\sum_{j=1}^{\infty} \frac{\rho_j}{2^j} e^{i(j\theta+\theta_j)} = -1,$$

or, taking real parts,

$$\sum_{j=1}^{\infty} \frac{\rho_j}{2^j} \cos(j\theta + \theta_j) = -1.$$

For the left-hand side to be equal to the right it is necessary and sufficient that $\rho_j = 1$, $j = 1, 2, \ldots$, and $\theta_j + \theta_j(2k_j + 1)\pi$, $j = 1, 2, \ldots$, with certain integers k_j. The last relation implies that

$$j\theta_1 - \theta_j \equiv \pi(j+1) \pmod{2\pi}.$$

Thus if $G_2(z) = 0$ with $|z| = 1/2$, we must have $|g_1(2)| = 1$, and for all $j \geq 1$

$$g_1(2^j) = (-1)^{j+1} g_1(2)^j.$$

But then, by hypothesis, $F_2(z)/G_2(z) = 1$. Therefore the double series

$$\sum_p \sum_j \frac{h(p^j)}{p^j}$$

is absolutely convergent. Since for any $x > 0$

$$\sum_{n \leq x} \frac{|h(n)|}{n} \leq \prod_{p \leq x} \left(1 + \sum_{p^j \leq x} \frac{|h(p^j)|}{p^j}\right)$$

and the infinite product

$$\prod_p \left(1 + \sum_{j=1}^{\infty} \frac{|h(p^j)|}{p^j}\right)$$

converges, the series $\sum_1^{\infty} |h(n)|/n$ also converges. By the generalized Wintner theorem, the mean value

$$M(g_2) = \lim_{N \to \infty} \frac{1}{N} \sum_{n \leq N} g_2(n)$$

exists, and

$$M(g_2) = M(g_1) \sum_{n=1}^{\infty} \frac{h(n)}{n}.$$

Also, since $h(n)$ is multiplicative and the series in the left-hand side of

$$\sum_{n=1}^{\infty} \frac{h(n)}{n} = \prod\left(1 + \sum_{j=1}^{\infty} \frac{h(p^j)}{p^j}\right) = \prod\left(1 + \sum_{j=1}^{\infty} \frac{h_j^{(p)}}{p^j}\right)$$

is absolutely convergent, the product in the right-hand side is absolutely convergent,

If $p > 2$, then, since $1/p < 1/2$,

$$1 + \sum_{j=1}^{\infty} \frac{h_j^{(p)}}{p^j} = \frac{F_p(1/p)}{G_p(1/p)} = \frac{1 + \sum_{j=1}^{\infty} g_2(p^j)/p^j}{1 + \sum_{j=1}^{\infty} g_1(p^j)/p^j}.$$

If $p = 2$ and $G_2(1/2) \neq 0$, we have analogously

$$1 + \sum_{j=1}^{\infty} \frac{h_j^{(2)}}{2^j} = \frac{F_2(1/2)}{G_2(1/2)} = \frac{1 + \sum_{j=1}^{\infty} g_2(2^j)/2^j}{1 + \sum_{j=1}^{\infty} g_1(2^j)/2^j}.$$

If $G_2(1/2) = 0$, then $F_2(z) = G_2(z)$ and, in particular, $F_2(1/2) = 0$. Then, by agreement,

$$\frac{0}{0} = \frac{1 + \sum_{j=1}^{\infty} g_2(2^j)/2^j}{1 + \sum_{j=1}^{\infty} g_1(2^j)/2^j} = 1.$$

The lemma is proved.

A multiplicative function $g(n)$ will be called *strongly multiplicative* if, for any prime p and natural number r, $g(p^r) = g(p)$.

We will prove Delange's theorem for the special case of strongly multiplicative functions.

LEMMA 4. *Suppose $g(n)$ is a strongly multiplicative function such that $|g(n)| \leq 1$, $n = 1, 2, \ldots$, and the series $\sum_p (g(p) - 1)/p$ converges. Then the limit*

$$\lim_{N\to\infty} \frac{1}{N} \sum_{n=1}^{N} g(n) = M(g)$$

exists, and

$$M(g) = \prod_p \left(1 + \frac{g(p) - 1}{p}\right).$$

PROOF. By hypothesis, $\sum_p (1 - g(p))/p$ converges. Therefore the series

$$\sum_p \frac{1 - \operatorname{Re} g(p)}{p}$$

converges also. But the condition $|g(n)| \leq 1$ implies that all terms of the latter series are nonnegative, i.e. this series is absolutely convergent. Moreover, the series

$$\sum_{\operatorname{Re} g(p) \leq 1/2} \frac{1}{p}$$

converges, since

$$\sum_{\operatorname{Re} g(p) \leq 1/2} \frac{1}{p} \leq \frac{1}{2} \sum_{\operatorname{Re} g(p) \leq 1/2} \frac{1 - \operatorname{Re} g(p)}{p} \leq \frac{1}{2} \sum_{p} \frac{1 - \operatorname{Re} g(p)}{p}.$$

Since

$$\sum_{\operatorname{Re} g(p) \leq 1/2} \frac{1}{p}$$

converges, by Lemma 3 it suffices to prove our assertion for the case where $\operatorname{Re} g(p) > 1/2$ for all primes p.

We will prove the lemma by the method of "curtailment". Instead of $g(n)$ we consider the family of axiliary functions $g_N(n)$, $N = 1, 2, \ldots$, defined as follows:

1) $g_N(n)$ is strongly multiplicative;

2)
$$g_N(p) = \begin{cases} g(p), & p \leq \ln N, \\ 1, & p > \ln N. \end{cases}$$

Thus,

$$g_N(n) = \prod_{\substack{p|n \\ p \leq \ln N}} g(p).$$

Consider $(1/N) \sum_{n \leq N} g_N(n)$. To calculate it we introduce the function

$$h_N(n) = \sum_{d|n} \mu\left(\frac{n}{d}\right) g_N(d).$$

The function $h_N(n)$ is multiplicative (but not strongly multiplicative, since $h_N(p) = g_N(p) - 1$ and $h_N(p^r) = 0$ for $r \geq 2$). According to the Tchebycheff-Möbius inversion formula,

$$g_N(n) = \sum_{d|n} h_N(d).$$

The series $\sum_{n=1}^{\infty} h_N(n)/n$ contains a finite number of terms and is therefore convergent. We represent this series as a product:

$$\sum_{n=1}^{\infty} \frac{h_N(n)}{n} = \prod_{p \leq \ln N} \left(1 + \frac{g(p) - 1}{p}\right).$$

Now

$$\frac{1}{N}\sum_{n\le N} g_N(n) = \frac{1}{N}\sum_{n=1}^{N}\sum_{d|n} h_N(d) = \frac{1}{N}\sum_{d=1}^{N} h_N(d)\left[\frac{N}{d}\right]$$
$$= \sum_{d=1}^{\infty}\frac{h_N(d)}{d} + R_N,$$

where

$$R_N \le \frac{1}{N}\sum_{d=1}^{\infty}|h_N(d)| \le \frac{1}{N}\prod_{p\le \ln N}(1+|g(p)-1|) \le \frac{2^{\pi(\ln N)}}{N},$$

and hence $\lim_{N\to\infty} R_N = 0$. Since

$$\lim_{N\to\infty}\prod_{p\le \ln N}\left(1+\frac{g(p)-1}{p}\right) = \prod_{p}\left(1+\frac{g(p)-1}{p}\right),$$

and the product appearing in the right-hand side converges by virtue of the restriction $|g(p)-1| < 2$ and the convergence of $\sum(g(p)-1)/p$, it follows that

$$\lim_{N\to\infty}\frac{1}{N}\sum_{n=1}^{N} g_N(n) = \prod_{p}\left(1+\frac{g(p)-1}{p}\right).$$

To prove the lemma it suffices to show that

$$\lim_{N\to\infty}\frac{1}{N}\sum_{n=1}^{N}(g(n)-g_N(n)) = 0.$$

This is the most difficult part of the proof. Let

$$D_N = \frac{1}{N}\sum_{n=1}^{N}(g(n)-g_N(n)).$$

It is clear that

$$D_N = \frac{1}{N}\sum_{n=1}^{N} g_N(n)\Big(\prod_{\substack{p|n\\ p>\ln N}} g(p) - 1\Big).$$

Put

$$f_N(n) = \sum_{\substack{p|n\\ p>\ln N}} \ln g(p)$$

(if n has no divisors greater than $\ln N$, we put $f_N(n) = 0$). We have

$$|D_N| \le \frac{1}{N}\sum_{n=1}^{N}|e^{f_N(n)} - 1|.$$

For subsequent estimates we require the following inequality: for $\operatorname{Re} z \leq 0$

$$|e^z - 1| \leq |z|;$$

the proof is obvious:

$$|e^z - 1| = \left|\int_0^z e^\xi \, d\xi\right| \leq \int_0^{|z|} dx = |z|.$$

Since $\operatorname{Re} f_N(n) \leq 0$, we obtain

$$|D_N|^2 \leq \frac{1}{N}\sum_{n=1}^{N} |f_N(n)|^2.$$

Since $|a+b|^2 \leq 2(|a|^2 + |b|^2)$, it follows that

$$\frac{1}{N}\sum_{n=1}^{N} |f_N(n)|^2 \leq \frac{2}{N}\sum_{n=1}^{N}\left|f_N(n) - \sum_{p\leq N}\frac{f_N(p)}{p}\right|^2 + 2\left|\sum_{p\leq N}\frac{f_N(p)}{p}\right|^2.$$

Now remember our restriction $\operatorname{Re} g(p) > 1/2$. Put $|g(p)| = r(p)$ and $\arg g(p) = \theta(p)$, $-\pi < \theta(p) \leq \pi$. Since $\operatorname{Re} g(p) > 1/2$, we have $|r(p)| > 1/2$.

We will prove several preliminary propositions.

REMARK 1. The series

$$\sum_p \frac{\ln(1/r(p))}{p}$$

converges.

Indeed, the inequality $\ln(1/x) \leq 2(1-x)$, which holds for $x \geq 1/2$, implies that

$$\sum_p \frac{\ln(1/r(p))}{p} \leq 2\sum_p \frac{1-r(p)}{p} \leq 2\sum_p \frac{1-\operatorname{Re} g(p)}{p}.$$

REMARK 2. The series

$$\sum_p \frac{\ln^2 r(p)}{p}$$

converges.

Indeed, $\ln^2 x = O(1-x)$ for $x \geq 1/2$.

REMARK 3. The series $\sum_p \theta^2(p)/p$ converges.

PROOF. Since $\operatorname{Re} g(p) \geq 1/2$, it follows that

$$\theta^2(p) = O(\sin^2\theta(p)).$$

We must prove convergence of $\sum_p \sin^2\theta(p)/p$. We have

$$\frac{\sin^2\theta(p)}{1-\operatorname{Re} g(p)} \leq \frac{\sin^2\theta(p)}{1-\cos\theta(p)} = \frac{4\sin^2(\theta(p)/2)\cos^2(\theta(p)/2)}{2\sin^2(\theta(p)/2)}$$
$$= 2\cos^2(\theta(p)/2) < c_1,$$

and hence $\sin^2\theta(p) = O(1-\operatorname{Re} g(p))$.

REMARK 4. The series $\sum_p \theta(p)/p$ converges (but not necessarily absolutely).

Indeed, since $|\theta - \sin\theta| \le c\theta^2$ and $\sum_p \theta^2(p)/p$ converges, it suffices to prove convergence of $\sum_p (\sin\theta(p))/p$. We have

$$\sum_{p\le N} \frac{\sin\theta(p)}{p} = \sum_{p\le N} \frac{r(p)\sin\theta(p)}{p} + \sum_{p\le N} \frac{(1-r(p))\sin\theta(p)}{p}.$$

Since, by hypothesis, $\sum_p (1-g(p))/p$ converges, it follows by looking at the imaginary part of this series that $\sum_p (r(p)\sin\theta(p))/p$ also converges. Furthermore,

$$\sum_{p\le N} \frac{(1-r(p))\sin\theta(p)}{p} \le \sum_{p\le N} \frac{1-r(p)}{p} \le \sum_{p\le N} \frac{1-\operatorname{Re} g(p)}{p},$$

i.e., the second series is absolutely convergent.

We have seen that

$$\frac{1}{N}\sum_{n=1}^{N} |f_N(n)|^2 \le \frac{2}{N}\sum_{n=1}^{N}\left| f_N(n) - \sum_{p\le N}\frac{f_N(p)}{p}\right|^2 + 2\left|\sum_{p\le N}\frac{f_N(p)}{p}\right|^2.$$

We first estimate $\sum_{p\le N} f_N(p)/p$. We have

$$\left|\sum_{p\le N}\frac{f_N(p)}{p}\right|^2 \le 2\left|\sum_{\ln N\le p\le N}\frac{\ln r(p)}{p}\right|^2 + 2\left|\sum_{\ln N\le p\le N}\frac{\theta(p)}{p}\right|^2.$$

Since $\sum_p (\ln(1/r(p))/p$ and $\sum_p \theta(p)/p$ converge, it follows that

$$\lim_{N\to\infty}\left|\sum_{p\le N}\frac{f_N(p)}{p}\right|^2 = 0;$$

$f_N(n)$ is a strongly additive function. By the Turán-Kubilius inequality,

$$\begin{aligned}\frac{1}{N}\sum_{n=1}^{N}\left| f_N(n) - \sum_{p\le N}\frac{f_N(p)}{p}\right|^2 &\le c_1\sum_{p\le N}\frac{|f_N(p)|^2}{p}\\ &= c_1\sum_{\ln N<p\le N}\frac{\ln^2 r(p) + \theta^2(p)}{p}.\end{aligned}$$

Since the series $\sum(\ln^2 r(p))/p$ and $\sum \theta^2(p)/p$ converge, it follows that

$$\lim_{N\to\infty}\frac{1}{N}\sum_{n=1}^{N}\left| f_N(n) - \sum_{p\le N}\frac{f_N(p)}{p}\right|^2 = 0,$$

and hence $\lim_{N\to\infty} D_N^2 = 0$, as required. Lemma 4 is proved.

Suppose $g(n)$ is a given multiplicative function such that

1) $|g(n)| \le 1$, $n = 1, 2, \ldots$, and

2) the series $\sum_p (1 - g(p))/p$ converges.

We construct a strongly multiplicative function $\overline{g}(n)$ defined for primes $p > 2$ by the rule $\overline{g}(p) = g(p)$. If $g(2) \ne -1$, we put $\overline{g}(2) = g(2)$; if $g(2) = -1$, we put $\overline{g}(2) = 1$.

By Lemma 3, which is applicable in view of the absolute convergence of the series

$$\sum_p \frac{g(p) - \overline{g}(p)}{p},$$

we have

$$\begin{aligned} M(g) &= M(\overline{g}) \frac{\prod_p (1 + \sum_{j=1}^{\infty} g(p^j)/p^j)}{\prod_p (1 + \sum_{j=1}^{\infty} \overline{g}(p)/p^j)} \\ &= M(\overline{g}) \frac{\prod_p (1 - 1/p)(1 + \sum_{j=1}^{\infty} g(p^j)/p^j)}{\prod_p (1 + (g(p) - 1)/p)}. \end{aligned}$$

By Lemma 4,

$$M(\overline{g}) = \prod_p \left(1 + \frac{g(p) - 1}{p}\right).$$

Thus,

$$M(g) = \prod_p \left(1 - \frac{1}{p}\right) \left(1 + \sum_{j=1}^{\infty} \frac{g(p^j)}{p^j}\right),$$

as required.

A multidimensional variant of Delange's theorem was given by Kátai (see [81], Theorem 1).

For applications to the question of the distribution of values of additive functions we need the following modification of Delange's theorem.

THEOREM 2′. *Suppose t is a parameter belonging to some finite interval of the real line, and $g_t(n)$ is a family of multiplicative functions such that*

1) $|g_t(n)| \le 1$ *for each* t,

2) $g_t(n)$ *is a continuous function of* t *for each* $n = 1, 2, \ldots$, *and*

3) *the series* $\sum_p (1 - g_t(p))/p$ *converges uniformly in* t.

Then as $N \to \infty$ *the mean value*

$$\frac{1}{N} \sum_{n \le N} g_t(n)$$

converges uniformly in t to a limit (and is therefore a continuous function of t).

The proof is based on the following principle: If an absolutely convergent series whose terms depend on a parameter t is majorized by a series convergent with respect to t, then it is also uniformly convergent in t (since the remainder of the series tends to zero uniformly in t). We first make the following observation. Suppose $g_t(n)$ and $h_t(n)$ are two families of functions of a natural argument. Let

$$f_t(n) = \sum_{d|n} h_t(n) g_t\left(\frac{n}{d}\right).$$

Suppose $M(g_t(n))$ exists for each t, and

$$\frac{1}{N}\sum_{n\le N} g_t(n) - M(g_t(n)) \to 0$$

uniformly in t and $\sum_{n=1}^{\infty} h_t(n)/n$ converges absolutely and uniformly in t. Then the mean $M(f_t)$ exists for each t, and

$$M(f_t) - \frac{1}{N}\sum_{n\le N} f_t(n) \to 0$$

uniformly in t.

We can now establish the following variant of Lemma 3.

LEMMA 3′. *Suppose two families of multiplicative functions, $g_t^{(1)}(n)$ and $g_t^{(2)}(n)$, are given. Assume that*

1) $M(g_t^{(1)}(n))$ exists for each t,

2) the series $\sum_p |g_t^{(1)}(p) - g_t^{(2)}(p)/p$ converges uniformly in t, and

3) if, for some t, $|g_t^{(1)}(2)| = 1$ and $g_t^{(1)}(2^r) = (-1)^{r+1}(g_t^{(1)}(2))^r$ for all $r > 1$, then $g_t^{(2)}(2^r) = g_t^{(1)}(2^r)$ for all r.

Then $M(g_t^{(2)}(n))$ exists for each t, and

$$\frac{1}{N}\sum_{n\le N} g_t^{(2)}(n) - M(g_t^{(2)}(n)) \to 0$$

uniformly in t.

There is also an analogue of Lemma 4.

LEMMA 4′. *Suppose $g_t(n)$ is a family of strongly multiplicative functions, $|g_t(n)| \le 1$, $n = 1, 2, \dots$, and the series $\sum_p (g_t(p) - 1)/p$ converges uniformly in t. Then*

$$\frac{1}{N}\sum_{n\le N} g_t(n) - M(g_t(n)) \to 0$$

uniformly in t.

PROOF. By the hypothesis of the lemma, the series

$$\sum_p \frac{1 - \operatorname{Re} g_t(p)}{p} \quad \text{and} \quad \sum_{\operatorname{Re} g_t(p) \le 1/2} \frac{1}{p}$$

converge uniformly in t. In view of Lemma $3'$, it suffices to prove the lemma for the case where $\operatorname{Re} g_t(p) > 1/2$ for all primes p (and for all values of the parameter t). By the hypothesis of the lemma and the condition $\operatorname{Re} g_t(p) > 1/2$, the series

$$\sum_p \frac{\ln(1/r_t(p))}{p}, \qquad \sum_p \frac{\ln^2 r_t(p)}{p}, \qquad \sum_p \frac{\theta_t^2(p)}{p},$$
$$\sum_p \frac{\theta_t(p) - \sin\theta_t(p)}{p}, \qquad \sum_p \frac{(1 - r_t(p))\sin\theta_t(p)}{p}$$

converge uniformly in t. Therefore $\sum_p \theta_t(p)/p$ converges uniformly in t.

We introduce the strongly multiplicative functions

$$g_{t,N}(p) = \begin{cases} g_t(p), & p \le \ln N, \\ 1, & p > \ln N. \end{cases}$$

It is easy to see that the transition to the limit

$$\lim_{N\to\infty} \frac{1}{N} \sum_{n \le N} g_{t,N}(n) = \prod_p \left(1 + \frac{g(p) - 1}{p}\right)$$

is uniform in t.

Put

$$f_{t,N}(n) = \sum_{\substack{p|n \\ p > \ln N}} \ln g(p).$$

Since

$$\left|\sum_{p \le N} \frac{f_{t,N}(p)}{p}\right|^2 \le 2\left|\sum_{\ln N \le p \le N} \frac{\ln(1/r_t(p))}{p}\right|^2 + 2\left|\sum_{\ln N \le p \le N} \frac{\theta_t(p)}{p}\right|^2$$

and the series

$$\sum_p \frac{\ln(1/r_t(p))}{p} \quad \text{and} \quad \sum \frac{\theta_t(p)}{p}$$

converge uniformly in t, it follows that as $N \to \infty$

$$\sum_{p \le N} \frac{f_{t,N}(p)}{p} \to 0$$

uniformly in t. Since $\sum(\ln^2 r_t(p))/p$ and $\sum \theta_t^2(p)/p$ converge uniformly in t and, by the Turán-Kubilius inequality,

$$\frac{1}{N}\sum_{n=1}^{N}\left|f_{t,N}(n)-\sum_{p\le N}\frac{f_{t,N}(p)}{p}\right|^2 \le c \sum_{\ln N<p\le N}\frac{\ln^2 r_t(p)+\theta_t^2(p)}{p},$$

we conclude that

$$\frac{1}{N}\sum_{n=1}^{N}\left|f_{t,N}(n)-\sum_{p\le N}\frac{f_{t,N}(p)}{p}\right|^2 \to 0$$

uniformly in t. It follows from these two assertions that

$$\lim_{N\to\infty}\frac{1}{N}\sum_{n\le N}(g_{t,N}(n)-g_t(n))\to 0$$

uniformly in t. This proves Lemma 4′.

The proof of the theorem can now be completed by applying Lemma 3′ and proceeding in the usual way.

§4.6. Remarks on summation of multiplicative functions

In this section we will present a survey of some theorems on multiplicative functions $f(n)$ in which the behavior of these functions for natural values of the argument can be "reconstructed" from conditions on their behavior on the primes.

Delange [27] proved the following result.

THEOREM. *Suppose $f(n)$ is a multiplicative function such that $|f(n)|\le 1$ for all n and*

$$\sum_{p\le x} f(p)\sim\rho\frac{x}{\ln x}, \tag{1}$$

where $\rho\ne 1$. Then $\sum_{n\le x}f(n)=o(x)$.

Corrádi and Kátai ([21], Theorem G) proved the next result.

THEOREM. *Suppose $f(n)$ is a multiplicative function such that $|f(n)|\le 1$ for all n. Assume that*

$$\sum_{p\le x}\frac{|f(p)-\rho|}{p}=o(\ln\ln x), \tag{2}$$

where $|\rho|\le 1$ and $\rho\ne 1$. Then $\sum_{n\le x}f(n)=o(x)$.

In the case $|\rho|=1$, $\rho\ne 1$, Delange's result is a consquence of the Corrádi-Kátai theorem. Indeed, this will be proved once we show that (1) implies

$$\sum_{p\le x}|f(p)-\rho|=o\left(\frac{x}{\ln x}\right),$$

since we can obtain (2) from this relation by partial summation. Put $\overline{\rho} f(p) = r_p e^{i\varphi_p}$, $0 \le r_p \le 1$. Then (1) implies

$$\sum_{p\le x}(1 - r_p \cos\varphi_p) \le \left|\sum_{p\le x}(1 - \overline{\rho} f(p))\right| = o\left(\frac{x}{\ln x}\right).$$

Also, from

$$\begin{aligned}\sum_{p\le x}(1 - r_p\cos\varphi_p) = &\sum_{\substack{p\le x\\ |\varphi_p|<\pi/2}}(1 - r_p\cos\varphi_p)\\ &+ \sum_{\substack{p\le x\\ |\varphi_p|\ge\pi/2}}(1 + r_p|\cos\varphi_p|)\\ = &\sum_{\substack{p\le x\\ |\varphi_p|<\pi/2}}(1 - r_p\cos\varphi_p) + \sum_{\substack{p\le x\\ |\varphi_p|\ge\pi/2}} 1 + \sum_{\substack{p\le x\\ |\varphi_p|\ge\pi/2}} r_p|\cos\varphi_p|,\end{aligned}$$

inasmuch as all three sums consist of nonnegative terms, we obtain

$$\sum_{\substack{p\le x\\ |\varphi_p|\ge\pi/2}} 1 = o\left(\frac{x}{\ln x}\right).$$

We have

$$\begin{aligned}\sum_{p\le x}|f(p) - \rho| &= \sum_{p\le x}|1 - \overline{\rho} f(p)|\\ &\le \sum_{p\le x}(1 - r_p\cos\varphi_p + r_p|\sin\varphi_p|)\\ &\le \sum_{p\le x}|\sin\varphi_p| + o\left(\frac{x}{\ln x}\right)\\ &\le O\left(\left(\frac{x}{\ln x}\right)^{1/2}\left(\sum_{p\le x}\sin^2\varphi_p\right)^{1/2}\right) + o\left(\frac{x}{\ln x}\right)\\ &= O\left(\left(\frac{x}{\ln x}\right)^{1/2}\left(\sum_{p\le x}(1 - \cos\varphi_p)\right)^{1/2}\right) + o\left(\frac{x}{\ln x}\right)\\ &= O\left(\left(\frac{x}{\ln x}\right)^{1/2}\left(\sum_{p\le x}(1 - r_p\cos\varphi_p) + \sum_{\substack{p\le x\\ |\varphi_p|\ge\pi/2}} 2\right)^{1/2}\right) + o\left(\frac{x}{\ln x}\right)\\ &= o\left(\frac{x}{\ln x}\right),\end{aligned}$$

as required.

As we know, there is in Wirsing's theorem the restrictive condition

$$\sum_{p\le x} f(p) = (\tau + o(1))\frac{x}{\ln x}.$$

There have been several investigations whose aim is to determine the asymptotic behavior of the sum $\sum_{n\le x} f(n)$ under a weaker condition on $\sum_{p\le x} f(p)$.

In [157] Wirsing proved the following statement.

THEOREM. *Suppose $f(n)$ is a multiplicative function such that*

$$\sum_{p\le x} \frac{\ln p}{p} f(p) \sim \tau \ln x$$

with $\tau > 0$ and $|f(p)| \le C$, where C is a constant. Suppose also that

$$\sum_{p,\ \nu,\ \nu\ge 2} \frac{1}{p^\nu}|f(p^\nu)| < \infty$$

and, if $\tau \le 1$, that

$$\sum_{\substack{p,\ \nu,\ \nu\ge 2\\ p^\nu\le x}} f(p^\nu) = O\left(\frac{x}{\ln x}\right).$$

Finally, suppose

$$\sum_p \frac{1}{p}(|f(p)| - \operatorname{Re} f(p)) < \infty.$$

Then as $x \to \infty$

$$\sum_{n\le x} f(n) \sim \frac{e^{-\gamma\tau}}{\Gamma(\tau)}\frac{x}{\ln x}\prod_{p\le x}\left(1 + \frac{f(p)}{p} + \cdots + \frac{f(p^n)}{p^n} + \cdots\right).$$

We will show that Delange's theorem is a consequence of this theorem of Wirsing (the proof is due to B. V. Levin). Suppose a multiplicative function $f(n)$ satisfies the conditions of Delange's theorem. Then it satisfies the conditions of the theorem of Wirsing stated above:

a) $f(n)$ is multiplicative in both Delange's theorem and Wirsing's theorem.

b) $|f(n)| \le 1$, and hence $|f(n)| \le C$.

c) Since $\sum_p (1 - f(p))/p$ converges, i.e.,

$$\sum_{p\le x} \frac{1 - f(p)}{p} = l + \varepsilon(x),$$

where $\varepsilon(x) \to 0$ as $x \to \infty$, it follows that

$$\sum_{p\le x} \frac{f(p)}{p} = \sum_{p\le x} \frac{1}{p} - l + \varepsilon(x),$$

and hence

$$\sum_{p\le x}\frac{\ln p}{p}f(p)=\int_2^x \ln y\, d\Big(\sum_{p\le y}\frac{f(p)}{p}\Big)$$
$$=\int_2^x \ln y\, d\Big(\sum_{p\le y}\frac{1}{p}-l+\varepsilon(y)\Big)$$
$$=\int_2^x \ln y\, d(\ln\ln y+C+\varepsilon_1(y)),$$

where $\varepsilon_1(y)\to 0$ as $y\to\infty$. Consequently,

$$\sum_{p\le x}\frac{\ln p}{p}f(p)=\int_2^x \ln y\, d(\ln\ln y+C+\varepsilon_1(y))$$
$$=\int_2^x\frac{dy}{y}+\int_2^x \ln y\, d(\varepsilon_1(y))$$
$$=\ln x-\ln 2+\varepsilon_1(y)\ln y|_2^x-\int_2^x\frac{\varepsilon_1(y)}{y}\,dy$$
$$=\ln x+o(\ln x).$$

Thus $f(n)$ satisfies the first condition of Wirsing's theorem with $\tau=1$.

Consider the expression $\sum_{p,\nu:\nu\ge 2,p^\nu\le x}f(p^\nu)$. We have

$$\Big|\sum_{\substack{p,\ \nu\\ \nu\ge 2,\ p^\nu\le x}}f(p^\nu)\Big|\le\sum_{\substack{p,\ \nu\\ \nu\ge 2,\ p^\nu\le x}}1=\sum_{p\le\sqrt{x}}\ \sum_{2\le\nu\le\ln x/\ln p}1$$
$$\le\ln x\sum_{p\le\sqrt{x}}\frac{1}{\ln p}=O\Big(\frac{x}{\ln x}\Big).$$

Let us verify the last condition of Wirsing's theorem. It is obvious that under the conditions of Delange's theorem

$$\sum_p\frac{1}{p}(|f(p)|-\operatorname{Re}f(p))\le\sum_p\frac{1}{p}(1-\operatorname{Re}f(p))\le\Big|\sum_p\frac{1}{p}(1-f(p))\Big|<\infty.$$

Thus if $f(n)$ satisfies the conditions of Delange's theorem, then it satisfies those of Wirsing's theorem. Consequently, by Wirsing's theorem,

$$\sum_{n\le x}f(n)\sim e^{-\gamma}\frac{x}{\ln x}\prod_{p\le x}\Big(1+\frac{f(p)}{p}+\cdots+\frac{f(p^n)}{p^n}+\cdots\Big).$$

Since by Mertens' formula

$$\prod_{p\le x}\Big(1-\frac{1}{p}\Big)\sim\frac{e^{-\gamma}}{\ln x},$$

it follows that

$$\sum_{n\le x} f(n) \sim x \prod_{p\le x}\left(1-\frac{1}{p}\right)\left(1+\frac{f(p)}{p}+\cdots+\frac{f(p^n)}{p^n}+\cdots\right).$$

By the hypothesis of Delange's theorem, $\sum_p (f(p)-1)/p$ converges. Therefore,

$$\lim_{x\to\infty}\prod_{p\le x}\left(1-\frac{1}{p}\right)\left(1+\frac{f(p)}{p}+\frac{f(p^2)}{p^2}+\cdots\right)$$
$$=\prod_{p}\left(1-\frac{1}{p}\right)\left(1+\frac{f(p)}{p}+\frac{f(p^2)}{p^2}+\cdots\right).$$

Thus,

$$\sum_{n\le x} f(n) \sim \left(\prod_{p}\left(1-\frac{1}{p}\right)\left(1+\frac{f(p)}{p}+\frac{f(p^2)}{p^2}+\cdots\right)\right)x.$$

Results of a general nature pertaining to the summation of multiplicative functions were obtained by Halász [63]. We mention some of them.

THEOREM. *Suppose $f(n)$ is any multiplicative function such that $|f(n)| \le 1$. Then there exist a complex number C_0, a real number a_0, and a function $L_0(u)$ with $|L_0(u)| = 1$ and $L_0(u_1)/L_0(u) \to 1$ uniformly as $u\to\infty$, $u \le u_1 \le 2u$, such that*

$$\sum_{n\le x} f(n) = C_0 L_0(\ln x)x^{1+ia_0} + o(x).$$

This theorem implicitly contains the following corollary.

THEOREM. *Suppose $f(n)$ is a multiplicative function, $|f(n)|\le 1$ for all n, and there exists $r\ge 1$ such that $f(2^r)\ne -1$. Then the estimate*

$$\sum_{n\le x} f(n) = o(x)$$

holds if and only if the series

$$\sum_p \frac{1-\operatorname{Re}(f(p)e^{-ib})}{p}$$

diverges for each real b.

Let us examine the relation between this assertion and Halász's theorem. From the restriction $|f(n)|\le 1$ we obtain the conditions of Theorem 2 of [63]. The condition that there exists r such that $f(2^r)\ne -1$ is equivalent to $H(1+it)\ne 0$. Suppose that for each real b the series

$$\sum_p \frac{1-\operatorname{Re}(f(p)e^{-ib})}{p}$$

diverges; then for each t the series (in the notation of [63])

$$-\sum_{n=1}^{\infty} \frac{\lambda(n)}{n^\sigma}(1 - \operatorname{Re}(f^*(n)e^{-it}))$$

diverges, and then formula (23) of [63] shows that $F^*(s) = o(1/(\sigma - 1))$, which implies $\sum_{n\le x} f(n) = o(n)$.

Assume there exists a real number b for which

$$\sum_p \frac{1 - \operatorname{Re}(f(p)e^{ib})}{p} < \infty;$$

then there exists a real number a_0 such that

$$\gamma_0 = \sum_{n=1}^{\infty} \frac{\lambda(n)}{n}(1 - \operatorname{Re}(f^*(n)n^{-ia_0})) < \infty,$$

and therefore, as shown by Halász, there exist a constant $C_0 = e^{-\gamma_0} \neq 0$ and a function $L_0(u)$ with $|L_0(u)| = 1$ and $L_0(u_1)/L_0(u) \to 1$ uniformly as $u \to \infty$, $u \le u_1 \le 2u$, such that

$$F^*(s) = \frac{C_0 L_0(1/(\sigma - 1))}{s - (1 + ia_0)} + o\left(\frac{1}{\sigma - 1}\right),$$

and hence

$$\sum_{n\le x} f(n) = \overline{C} L_0(\ln x) x^{1+ia_0} + o(x)$$

with $\overline{C} \neq 0$.

As an example, consider the Möbius function $f(n) = \mu(n)$. If $b = 0$, then

$$\sum_p \frac{1 - \operatorname{Re}(f(p)e^{-ib})}{p} = \sum_p \frac{2}{p} = \infty.$$

If $b \neq 0$ is a real number, we have

$$\sum_{p\le n} \frac{1 - \operatorname{Re}(f(p)e^{-ib})}{p} = \sum_{p\le n} \frac{1 + \cos b \ln p}{p} = \sum_{p\le n} \frac{1}{p} + \operatorname{Re} \sum_{p\le n} \frac{1}{p^{1+ib}}.$$

In the theory of the zeta-function (see, for example, Titchmarsh [139], p. 58) it is proved that $\sum_p 1/p^{1+ib}$ converges when $b \neq 0$. Therefore, since $\sum_p 1/p$ diverges, the series

$$\sum_p \frac{1 + \cos b \ln p}{p}$$

also diverges. Thus,

$$\sum_{n\le x} \mu(n) = o(x).$$

Halász [63] proved the following

THEOREM. *Suppose a multiplicative function $f(n)$ satisfies the following conditions:*

1) *$f(p)$ is bounded in modulus, or even*

$$\sum_p \frac{|f(p)|^{1+\delta}}{p^\sigma} \ln p = O\left(\frac{1}{\sigma - 1}\right) .$$

with some $\delta > 0$.

2) $\sum_{p,k:k\geq 2} f(p^k)/p^k < \infty$.

3) *None of the expressions*

$$1 + \frac{f(p)}{p^s} + \frac{f(p^2)}{p^{2s}} + \cdots$$

vanishes on the line $\operatorname{Re} s = 1$.

4) $$\sum_{n=1}^{\infty} \frac{f(n)}{n} = \frac{C}{s-1} + o\left(\frac{|s|}{\sigma - 1}\right) \quad \text{uniformly for } \operatorname{Re} s = \sigma.$$

Then

$$\sum_{n\leq x} f(n) = Cx + o(x).$$

§4.7. The Erdős-Wintner theorem

In 1939 Erdős and Wintner [41] proved the following result.

THEOREM. *For any real additive function $f(n)$, convergence of the three series*

$$\sum_{|f(p)|\geq 1} \frac{1}{p}, \qquad \sum_{|f(p)|<1} \frac{f(p)}{p}, \qquad \sum_{|f(p)|<1} \frac{f^2(p)}{p} \tag{1}$$

is necessary and sufficient for weak convergence of the sequence of distribution functions

$$F_N(x) = \frac{1}{N} \sum_{\substack{n\leq N \\ f(n)\leq x}} 1$$

as $N \to \infty$.

Sufficiency was proved in 1938 by Erdős [34].

The Erdős-Wintner theorem can be obtained as a consequence of Delange's theorems. We will prove sufficiency of the conditions (1).

Suppose $f(n)$ is an additive function. The characteristic function of the distribution function

$$F_N(x) = \frac{1}{N} \sum_{\substack{n\leq N \\ f(n)\leq x}} 1$$

is equal to

$$\varphi_N(t) = \int_{-\infty}^{\infty} e^{itx}\, dF_N(x) = \frac{1}{N} \sum_{n \le N} e^{itf(n)}.$$

By the inverse limit theorem (Gnedenko [61], §38), a necessary and sufficient condition for weak convergence of the sequence of distribution functions $N^{-1} \sum_{n \le N; f(n) \le x} 1$ to a limit distribution function is uniform convergence of their characteristic functions $\varphi_N(t)$, $t \in [-T, T]$, for any fixed T. Let $g_t(n) = e^{itf(n)}$ and $\varphi_N(t) = \sum_{n \le N} g_t(n)$. When $f(n)$ is an additive function, $g_t(n)$ is a multiplicative function. Let us verify that the conditions of the modified Delange theorem are satisfied.

1) Obviously $|g_t(n)| \le 1$, $n = 1, 2, \ldots$, for any t.

2) For any $n = 1, 2, \ldots$ the function $g_t(n) = e^{itf(n)}$ is continuous.

3) We will prove that under the conditions (1) the sequence

$$\sum_{p<N} \frac{1 - e^{itf(p)}}{p}$$

converges uniformly in $t \in [-T, T]$. The series $\sum_p (1 - e^{itf(p)})/p$ can be represented in the form

$$\sum_p \frac{1 - e^{itf(p)}}{p} = \sum_{|f(p)|<1} \frac{1 - e^{itf(p)}}{p} + \sum_{|f(p)| \ge 1} \frac{1 - e^{itf(p)}}{p}.$$

Since $|e^{iy} - 1 - iy| \le y^2/2$ for any real y, it follows that

$$\left| \sum_{p<N} \frac{1 - e^{itf(p)}}{p} + it \sum_{\substack{|f(p)|<1 \\ p<N}} \frac{f(p)}{p} \right| < \frac{T^2}{2} \sum_{\substack{|f(p)|<1 \\ p<N}} \frac{f^2(p)}{p} + C \sum_{\substack{|f(p)| \ge 1 \\ p<N}} \frac{1}{p}$$

for $|t| \le T$. This inequality, together with (1), implies convergence of

$$\sum_p \frac{1 - e^{itf(p)}}{p} + it \sum_{|f(p)|<1} \frac{f(p)}{p}$$

uniformly in t for $|t| \le T$. Let

$$B(t) = \sum_p \frac{1 - e^{itf(p)}}{p} \quad \text{and} \quad B = \sum_{|f(p)|<1} \frac{f(p)}{p}.$$

We have

$$\left| \sum_{p<N} \frac{1 - e^{itf(p)}}{p} - B(t) \right|$$
$$\le \left| \sum_{p<N} \frac{1 - e^{itf(p)}}{p} + it \sum_{\substack{|f(p)|<1 \\ p<N}} \frac{f(p)}{p} - B(t) - itB \right| + T \left| \sum_{\substack{|f(p)|<1 \\ p<N}} \frac{f(p)}{p} - B \right|.$$

Therefore as $N \to \infty$ the sequence $\sum_{p<N}(1-e^{itf(p)})/p$ tends to $B(t)$ uniformly in t for $|t| \leq T$.

By the modified Delange theorem, the sequence

$$F_N(x) = \frac{1}{N} \sum_{n \leq N,\, f(n) \leq x} 1$$

is weakly convergent to a limit distribution function:

$$\frac{1}{N} \sum_{n \leq N,\, f(n) \leq x} 1 \Rightarrow F(x).$$

By Delange's theorem, the characteristic function of the distribution function $F(x)$ is

$$\prod_p \left(1 - \frac{1}{p}\right)\left(1 + \sum_{\alpha=1}^{\infty} \frac{e^{itf(p^\alpha)}}{p^\alpha}\right).$$

The theorem is proved.

A different proof of the Erdős-Wintner theorem was given by Kubilius [88].

Consider the function $\varphi(n)/n$. Its logarithm $f(n) = \ln(\varphi(n)/n)$ is a strongly additive function. We have $|f(p)| < 1$ for $p > 2$. Also, the series

$$\sum_p \frac{\ln(1-1/p)}{p} \quad \text{and} \quad \sum_p \frac{\ln^2(1-1/p)}{p}$$

converge. Therefore, by the Erdős-Wintner theorem, the sequence of distribution functions

$$F_N(u) = \sum_{\ln \varphi(n)/n < u,\, n \leq N} 1$$

is weakly convergent to a limit function $F(u)$.

Let $F_N(\ln u) = V_N(u)$ and $F(\ln u) = V(u)$. It is clear that

$$V_N(\lambda) = \frac{1}{N} \sum_{n \leq N,\, \varphi(n)/n \leq \lambda} 1$$

and that $V_N(\lambda) \Rightarrow V(\lambda)$. Obviously $V(0) = 0$ and $V(1) = 1$.

Since $V_N(\lambda) \Rightarrow V(\lambda)$, it follows that for any continuous function $R(x)$ on $[0,1]$ the limit

$$\lim_{N\to\infty} \frac{1}{N} \sum_{n \leq N} R\left(\frac{\varphi(n)}{n}\right) = A(R)$$

exists and $A(R) = \int_0^1 R(x)\, dV(x)$.

A multidimensional generalization of the Erdős-Wintner theorem was given by Kátai ([81], Theorem 3).

We now turn to the question of the nature of the limit distribution law for a given additive function $f(n)$.

The sum of the jumps of a distribution function will be called its *total jump*. The total jump of a continuous distribution function is equal to zero.

If a distribution function is discontinuous, we will call it *purely discontinuous* or *mixed* according to whether its total jump is equal to 1 or less than 1.

We will prove the following special case of a theorem of Lévy [98].(1)

THEOREM. *Suppose F is an infinite convergent convolution of purely discontinuous distribution functions F_k; that is, $F = F_1 * F_2 * \cdots$. Let d_k be the maximal jump of F_k. If $\sum_{k=1}^{\infty}(1-d_k)$ diverges, then the limit distribution function is continuous; if it converges, then the limit distribution function is purely discontinuous.*

PROOF. We consider the space of all sequences $x_1, x_2, \ldots$ of real numbers. Let us look at the subset of these sequences defined by the inequalities $a_1 \le x_1 < b_1, \ldots, a_k \le x_k < b_k$, where k is any natural number and $a_1, b_1, \ldots, a_k, b_k$ are arbitrary real numbers. For each such subset we construct the number

$$(F_1(b_1) - F_1(a_1))(F_2(b_2) - F_2(a_2)) \cdots (F_k(b_k) - F_k(a_k)).$$

These subsets generate a σ-algebra of sets in the usual way (see Halmos [64]), and the function constructed on the subsets can be extended to a countably additive set function on this algebra, i.e., to a measure. This measure will be called a probability. The measure of a set of sequences defined by certain conditions will be denoted by $\mathsf{P}(\cdots)$, with the given conditions in parentheses. It is clear that

$$\mathsf{P}(x_1 + \cdots + x_n < x) = F_1(x) * \cdots * F_n(x)$$

and

$$\mathsf{P}(x_1 + x_2 + \cdots < x) = F_1(x) * F_2(x) * \cdots .$$

We will prove the first part of the theorem. Let $S_n = x_1 + \cdots + x_n$ and $S = x_1 + x_2 + \cdots$. Suppose $F(x)$ is discontinuous, i.e., there exists a value S' of S that has probability $\alpha > 0$. Choose l so that

$$\mathsf{P}(|S - S'| \le 2l) < \alpha + \varepsilon.$$

We next choose a natural number n' so that $n > n'$ implies

$$\mathsf{P}(|S - S_n| > l) < \varepsilon.$$

---

(1)In the proof of this theorem given by Lévy [98] there is a difficulty caused by a misprint on p. 128. The condition $\sum \mathsf{P}(x_\nu \neq \bar{x}_\nu) < \infty$ is needed in the definition of equivalence of sequences.

We have

$$\mathsf{P}(|S-S_n|\le l,\ |S_n-S'|\le l)\le \mathsf{P}(|S-S'|\le 2l)<\alpha+\varepsilon,$$

and hence

$$\mathsf{P}(|S_n-S'|\le l)<\frac{\alpha+\varepsilon}{1-\varepsilon}.$$

We see from the formula

$$\begin{aligned}\alpha=\mathsf{P}(S=S')&=\sum_x \mathsf{P}(S_n=x)\mathsf{P}(S-S_n=S'-x)\\ &\le \sum_{S'-l<x<S'+l}\mathsf{P}(S_n=x)\mathsf{P}(S-S_n=S'-x)+\mathsf{P}(|S-S'|\ge l)\\ &\le \sum_{S'-l<x<S'+l}\mathsf{P}(S_n=x)\mathsf{P}(S-S_n=S'-x)+\varepsilon\end{aligned}$$

that among the values lying in the interval $(S'-l,S'+l)$ there is at least one x such that $\mathsf{P}(S_n=x)>\alpha-\varepsilon$—otherwise

$$\alpha\le(\alpha-\varepsilon)\mathsf{P}(|S-S_n|\le l)+\varepsilon\le(\alpha-\varepsilon)(1-\varepsilon)+\varepsilon<\alpha.$$

Let S_n' be such a value, $|S-S_n'|<l$. We have

$$\alpha-\varepsilon<\alpha_n=\mathsf{P}(S_n=S_n')<\frac{\alpha+\varepsilon}{1-\varepsilon}.$$

Now

$$\begin{aligned}\mathsf{P}(S=S',\ S_n\ne S_n')&=\mathsf{P}(S=S')-\mathsf{P}(S=S',\ S_n=S_n')\\ &=\alpha-\mathsf{P}(S-S_n=S'-S_n',\ S_n=S_n').\end{aligned}$$

Since the quantities $S-S_n=x_{n+1}+x_{n+2}+\cdots$ and $S_n=x_1+\cdots+x_n$ are independent,

$$\mathsf{P}(S-S_n=S'-S_n',\ S_n=S_n')=\alpha_n\mathsf{P}(S-S_n=S'-S_n').$$

Let $\beta_n=\mathsf{P}(S-S_n=S'-S_n')$. We have

$$\mathsf{P}(S=S',S_n\ne S_n')=\alpha-\alpha_n\beta_n.$$

Also,

$$\begin{aligned}\mathsf{P}(S=S',S_n\ne S_n')&=\mathsf{P}(S=S',\ |S_n-S'|\le l,\ S_n\ne S_n')\\ &\quad+\mathsf{P}(S=S',\ |S_n-S'|>l,\ S_n\ne S_n')\\ &\le \mathsf{P}(|S_n-S'|\le l,\ S_n\ne S_n')+\mathsf{P}(|S_n-S|>l)\\ &=\mathsf{P}(|S_n-S'|\le l)-\mathsf{P}(S_n=S_n')+\mathsf{P}(|S_n-S|>l)\\ &\le\frac{\alpha+\varepsilon}{1-\varepsilon}-(\alpha-\varepsilon)+\varepsilon.\end{aligned}$$

Thus,

$$\alpha - \alpha_n\beta_n < \varepsilon\left(2 + \frac{1+\alpha}{1-\varepsilon}\right).$$

Consequently,

$$\beta_n = \mathsf{P}(S - S_n = S' - S'_n) > \frac{\alpha - \varepsilon(2 + (1+\alpha)/(1+\varepsilon))}{\alpha_n}$$
$$> \frac{(1-\varepsilon)(\alpha - \varepsilon(2 + (1+\alpha)/(1-\varepsilon)))}{\alpha + \varepsilon} > 1 - C\varepsilon.$$

We conclude that the maximal jump of the distribution function of the random variable $S - S_n$ exceeds $1 - C\varepsilon$.

The maximal jump of the distribution function of a sum of independent random variables does not exceed the maximal jump of the distribution functions of the individual summands. Suppose $N > n$. Obviously $S - S_n = S_N - S_n + S - S_N$. Therefore the maximal jump of the distribution function of $S_N - S_n$ exceeds $1 - C\varepsilon$. This means that for any a

$$\mathsf{P}\left(\sum_{\nu=n+1}^{N} x_\nu \neq a\right) \leq C\varepsilon.$$

We now take a sequence of numbers ε_l such that $\sum \varepsilon_l$ converges, and we define a sequence of natural numbers n_l so that the maximal jump of the sum of the random variables $\sum_{n_l+1}^{n_{l+1}} x_\nu$ is at most $C\varepsilon_l$.

We now require the following

LEMMA. *Let there be given n independent random variables $x_1, \dots, x_n$. Denote by d_ν $(\nu = 1, \dots, n)$ the maximal jump of the distribution function of the random variable x_ν, and by f_ν $(\nu = 1, \dots, n)$ the maximal jump of the distribution function of the random variable $x_1 + \cdots + x_\nu$. Then*

$$1 - d_1 d_2 \cdots d_n \leq \frac{1 - f_n}{f_n^2}.$$

To prove this we choose constants $a_1, \dots, a_n$ so that

$$\mathsf{P}(x_1 + x_2 + \cdots + x_\nu = a_1 + a_2 + \cdots + a_\nu) = f_\nu$$

for $\nu = 1, \dots, n$.

Now suppose ν is any fixed number among $1, \dots, n-1$. Obviously

$$\begin{aligned}
&\mathsf{P}(x_1 - a_1 + x_2 - a_2 + \cdots + x_\nu - a_\nu \\
&\quad = x_1 - a_1 + x_2 - a_2 + \cdots + x_n - a_n) \\
&\quad \geq \mathsf{P}(x_1 + x_2 + \cdots + x_\nu = a_1 + a_2 + \cdots + a_\nu) \\
&\qquad \times \mathsf{P}(x_1 + x_2 + \cdots + x_n = a_1 + a_2 + \cdots + a_n) = f_\nu f_n.
\end{aligned}$$

Since as the sum of independent random variables increases the maximal jump of the distribution function does not increase, it follows that $f_\nu \geq f_n$, and hence

$$\mathsf{P}(x_1 - a_1 + x_2 - a_2 + \cdots + x_\nu - a_\nu \\ = x_1 - a_1 + x_2 - a_2 + \cdots + x_n - a_n) \geq f_n^2. \tag{2}$$

Let T denote the event (i.e., the set of sequences $x_1, \ldots, x_n$) for which at least one of the sums

$$x_1 - a_1 + x_2 - a_2 + \cdots + x_\nu - a_\nu, \qquad \nu = 1, 2, \ldots, n,$$

is different from zero, and let $\mathsf{P}(T)$ denote its probability. The complementary event $\overline{T}$ is obviously that $x_1 = a_1, \ldots, x_n = a_n$. Clearly $\mathsf{P}(\overline{T}) \leq d_1 \cdots d_n$. Since $\mathsf{P}(T) + \mathsf{P}(\overline{T}) = 1$, it follows that

$$\mathsf{P}(T) \geq 1 - d_1 \cdots d_n. \tag{3}$$

The joint occurrence of the events T and $x_1 + \cdots + x_n = a_1 + \cdots + a_n$ happens as follows: the event T occurs, i.e., there exists ν, $1 \leq \nu \leq n-1$, such that $x_1 + \cdots + x_\nu \neq a_1 + \cdots + a_\nu$, and the event $x_1 - a_1 + \cdots + x_\nu - a_\nu \neq x_1 - a_1 + \cdots + x_n - a_n$. In view of (2), we obtain

$$\mathsf{P}(T, x_1 + \cdots + x_n = a_1 + \cdots + a_n) \leq \mathsf{P}(T)(1 - f_n^2). \tag{4}$$

Since the event $x_1 + \cdots + x_n = a_1 + \cdots + a_n$ is a consequence of the event $\overline{T}$,

$$\mathsf{P}(\overline{T}, x_1 + \cdots + x_n = a_1 + \cdots + a_n) = \mathsf{P}(T) = 1 - \mathsf{P}(T). \tag{5}$$

Adding (4) and (5), we obtain

$$f_n = \mathsf{P}(x_1 + \cdots + x_n = a_1 + \cdots + a_n) \\ \leq \mathsf{P}(T)(1 - f_n^2) + 1 - \mathsf{P}(T) = 1 - \mathsf{P}(T) f_n^2,$$

and hence $\mathsf{P}(T) \leq (1 - f_n)/f_n^2$, or, by (3),

$$1 - d_1 d_2 \cdots d_n \leq \frac{1 - f_n}{f_n^2}.$$

The lemma is proved.

We apply this lemma to each of the sums $\sum_{n_l+1}^{n_{l+1}} x_\nu$. We obtain

$$1 - d_{n_l+1} \cdots d_{n_{l+1}} \leq C\varepsilon_1/(1 - \varepsilon_l)^2,$$

and therefore, starting from some $l = \tau$,

$$\prod_{l=\tau}^{\infty} \left(1 - \frac{C\varepsilon_l}{(1 - \varepsilon_l)^2}\right) \leq \prod_{j=\tau}^{\infty} d_j,$$

i.e., the product

$$\prod_{j=n_\tau+1}^{\infty} d_j = \prod_{j=n_\tau+1}^{\infty} (1 - (1 - d_j))$$

converges (by a convention regarding infinite products, the value of this infinite product is nonzero). By a criterion for convergence of infinite products, the series $\sum(1-d_j)$ is convergent.

Thus we have proved that if $\sum_1^\infty(1-d_k)$ diverges, then $F(x)$ is continuous.

We will now prove the second part of the theorem. Suppose the $F_k(x)$ are distribution functions of purely discontinuous random variables x_k such that $\sum_1^\infty(1-d_k)$ converges. Replacing x_k by x_k-a_k, where a_k is a suitably chosen constant, we can assure that the maximal jump of each function $F_k(x)$ occurs at $x=0$. For sufficiently large n

$$\begin{aligned}\mathsf{P}(x_n+x_{n+1}+\cdots\neq 0)&\leq \mathsf{P}(x_n\neq 0)+\mathsf{P}(x_{n+1}\neq 0)+\cdots\\&=\sum_{k=n}^{\infty}(1-d_k)<\varepsilon,\end{aligned}$$

i.e., $\mathsf{P}(S\neq S_n)<\varepsilon$. Consequently, for any β we have

$$\begin{aligned}\mathsf{P}(S_n=\beta)(1-\varepsilon)&\leq \mathsf{P}(S=\beta)\\&=\mathsf{P}(S_n\neq\beta,\ S-S_n\neq 0)+\mathsf{P}(S_n=\beta,\ S-S_n=0)\\&\leq\varepsilon+\mathsf{P}(S_n=\beta).\end{aligned}\tag{6}$$

This shows that all points β of positive probability for the sum S are points of positive probability for the sum S_n (for sufficiently large n), and all points β of positive probability for S_n are points of positive probability for S. We denote by Σ the total jump of the distribution function of S, and by $\Sigma^{(n)}$ the total jump of the distribution function of S_n. Summing the inequalities (6) over the points of positive probability (of the sums S_n and S), we obtain $\Sigma\geq\Sigma^{(n)}(1-\varepsilon)$. But $\Sigma^{(n)}=1$, since all of the distribution functions $F_k(x)$ are purely discontinuous; thus, $\Sigma\geq 1-\varepsilon$. The value ε can be taken to be arbitrarily small. Thus $\Sigma=1$, which proves the second part of the theorem.

A different proof of Lévy's criterion was given by van Kampen [79].

In [41] Erdős and Wintner proved the following result.

THEOREM. *A distribution function that is the asymptotic distribution function for an additive function $f(n)$ is continuous if and only if the series $\sum_{f(p)\neq 0}1/p$ diverges.*

Indeed, we have already seen that the limit distribution function $F(x)$ has a characteristic function of the form

$$\prod_p\left(1-\frac{1}{p}+\left(1-\frac{1}{p}\right)\sum_{\alpha=1}^{\infty}\frac{e^{itf(p^\alpha)}}{p^\alpha}\right).$$

Thus if for a prime p we define the distribution function $F^{(p)}(x)$ of the random variable $\rho(p)$ that assumes the values

$$\begin{cases} 0 & \text{with probability } 1-1/p, \\ f(p^\alpha)\ (\alpha=1,2,\dots) & \text{with probability } (1/p^\alpha)(1-1/p), \end{cases}$$

then $F(x)$ can be represented as the infinite convolution

$$F^{(2)}(x) * F^{(3)}(x) * F^{(5)}(x) * \cdots .$$

The maximal jump of the distribution function $F^{(p)}(x)$ is equal to

$$1-\frac{1}{p} \le 1-\frac{1}{p} + \sum_{\substack{\alpha=2 \\ f(p^\alpha)=0}}^{\infty} \frac{1-1/p}{p^\alpha} \le 1-\frac{1}{p}+\frac{1}{p^2}, \quad \text{if } f(p)\neq 0,$$

$$\begin{aligned} 1-\frac{1}{p^2} \le 1-\frac{1}{p} + \sum_{\substack{\alpha=1 \\ f(p^\alpha)=0}}^{\infty} \frac{1-1/p}{p^\alpha} &\le 1-\frac{1}{p}+\frac{1-1/p}{p}+\frac{1}{p^2} \\ &= 1-\frac{1}{p^2}+\frac{1}{p^2}, \quad \text{if } f(p)=0. \end{aligned}$$

Therefore,

$$\sum_p (1-d_p) = \sum_{f(p)\neq 0} \frac{1}{p} + O\left(\sum_p \frac{1}{p^2}\right),$$

and the theorem is reduced to Lévy's theorem.

EXAMPLE. Consider the additive function $f(n)=\ln(\varphi(n)/n)$. We have

$$\sum_{\ln(1-1/p)\neq 0} \frac{1}{p} = \sum_p \frac{1}{p} = \infty.$$

Thus the asymptotic distribution function for $\ln(\varphi(n)/n)$ is continuous, and therefore so is the asymptotic distribution function for $\varphi(n)/n$.

§4.8. On the distribution of values of Euler's function

We have already mentioned that for each λ such that $0\le\lambda\le 1$ the limit

$$\lim_{N\to\infty} \frac{1}{N} \sum_{n\le N,\ \varphi(n)<n\lambda} 1 = V(\lambda)$$

exists. We have also shown that the function $V(\lambda)$ is continuous at each point.

Davenport [23] suggested a method for obtaining an explicit expression for the function

$$V(\lambda) = \lim_{N\to\infty} \frac{1}{N} \sum_{n\le N,\ \varphi(n)<n\lambda} 1.$$

We will now give an account of Davenport's method.

Consider a number λ such that $0 < \lambda < 1$; we will call a natural number n a *λ-number* if $\varphi(n)/n < \lambda$. It is obvious that for any $m \geq 1$

$$\varphi(nm)/nm \leq \varphi(n)/n,$$

i.e., all multiplicities of a λ-number are λ-numbers. In view of this fact, it is appropriate to introduce the concept of a primitive λ-number: a number n is called a *primitive λ-number* if no proper divisor is a λ-number. It is obvious that in a primitive λ-number the prime factors can only occur to the first power. Thus for a finite set of primes $p_1, \ldots, p_r$ there is only a finite set of primitive λ-numbers that can be formed only from $p_1, \ldots, p_r$. Finally, since $\prod_{p \leq N}(1 - 1/p) \to 0$, there exist infinitely many primitive λ-numbers for each λ, $0 < \lambda < 1$.

Suppose P is a large natural number. We define a function $\varepsilon_P(m)$ as follows:

$$\varepsilon_P(m) = \begin{cases} 1, & \text{if all prime divisors of } m \text{ are less than } P, \\ 0, & \text{if } m \text{ has a prime divisor greater than } P. \end{cases}$$

Since the function $\varepsilon_P(m)$ is completely multiplicative,

$$\sum_{m=1}^{\infty} \frac{\varepsilon_P(m)}{m} = \prod_{p \leq P} \frac{1}{1 - 1/p}.$$

For brevity we put $\Pi(P) = \prod_{p \leq P}(1 - 1/p)$.

LEMMA. *For $V(\lambda)$*

$$V(\lambda) = \lim_{P \to \infty} \sum_{\substack{m=1 \\ \varphi(m) \leq m\lambda}}^{\infty} \frac{\varepsilon_P(m)}{m}.$$

PROOF. By Delange's theorem, if $\operatorname{Re} s \geq 0$, then

$$\lim_{N \to \infty} \frac{1}{N} \sum_{n=1}^{N} \left(\frac{\varphi(n)}{n} \right)^s = \prod_p \left(1 - \frac{1}{p} + \frac{1}{p} \left(1 - \frac{1}{p} \right)^s \right).$$

We will use this equality for natural numbers s.

We also have

$$\begin{aligned}
&\lim_{P\to\infty} \Pi(P) \sum_{m=1}^{\infty} \frac{\varepsilon_P(m)}{m}\left(\frac{\varphi(m)}{m}\right)^s \\
&= \lim_{P\to\infty} \prod_{p\le P}\left(1-\frac{1}{p}\right)\left(1+\sum_{\alpha=1}^{\infty}\left(\frac{\varphi(p^\alpha)}{p^\alpha}\right)^s \frac{1}{p^\alpha}\right) \\
&= \lim_{P\to\infty} \prod_{p\le P}\left(1-\frac{1}{p}\right)\left(1+\left(1-\frac{1}{p}\right)^s \sum_{\alpha=1}^{\infty} \frac{1}{p^\alpha}\right) \\
&= \lim_{P\to\infty} \prod_{p\le P}\left(1-\frac{1}{p}\right)\left(1+\frac{1}{p}\left(1-\frac{1}{p}\right)^{s-1}\right) \\
&= \prod_{p}\left(1-\frac{1}{p}+\frac{1}{p}\left(1-\frac{1}{p}\right)^s\right).
\end{aligned}$$

Thus for any $s = 0, 1, 2, \ldots$

$$\lim_{P\to\infty} \Pi(P) \sum_{m=1}^{\infty} \frac{\varepsilon_P(m)}{m}\left(\frac{\varphi(m)}{m}\right)^s = \lim_{N\to\infty} \frac{1}{N} \sum_{n=1}^{N}\left(\frac{\varphi(n)}{n}\right)^s.$$

Consequently, for any polynomial $P(x)$

$$\lim_{P\to\infty} \Pi(P) \sum_{m=1}^{\infty} \frac{\varepsilon_P(m)}{m} P\left(\frac{\varphi(m)}{m}\right) = \lim_{N\to\infty} \frac{1}{N} \sum_{m\le N} P\left(\frac{\varphi(m)}{m}\right).$$

Using Weierstrass's theorem on approximating a function by polynomials, we see that for any continuous function $F(x)$ on $[0,1]$

$$\lim_{P\to\infty} \Pi(P) \sum_{m=1}^{\infty} \frac{\varepsilon_P(m)}{m} F\left(\frac{\varphi(m)}{m}\right) = \lim_{N\to\infty} \frac{1}{N} \sum_{n\le N} F\left(\frac{\varphi(n)}{n}\right).$$

Taking the usual approximation of the function

$$F(x) = \begin{cases} 1, & x \le \lambda, \\ 0, & x > \lambda, \end{cases}$$

by continuous functions, we obtain

$$\lim_{P\to\infty} \Pi(P) \sum_{\substack{m=1 \\ \varphi(m)\le m\lambda}}^{\infty} \frac{\varepsilon_P(m)}{m} = \lim_{N\to\infty} \frac{1}{N} \sum_{\substack{n\le N \\ \varphi(n)/n\le\lambda}} 1 = V(\lambda).$$

THEOREM. *Suppose $a_1, \ldots, a_\nu$ is the beginning of the sequence of primitive λ-numbers, and $[a_{s_1}, \ldots, a_{s_k}]$ is the least common multiple of $a_{s_1}, \ldots, a_{s_k}$. Let*

$$A_\nu = \frac{1}{a_\nu} - \sum_{\mu<\nu} \frac{1}{[a_\mu, a_\nu]} + \sum_{\lambda<\mu<\nu} \frac{1}{[a_\lambda, a_\mu, a_\nu]} - \cdots$$

(the density of the numbers divisible by a_ν, but not divisible by any of $a_1, \dots, a_{\nu-1}$). Then $V(\lambda) = \sum_1^\infty A_\nu$.

PROOF. The quantity $\sum_1^k A_\nu$ is the density of the numbers divisible by at least one of $a_1, \dots, a_k$. Since all of these numbers are λ-numbers, $\sum_1^k A_\nu \le V(\lambda)$. Therefore $\sum_1^\infty A_\nu$ converges, and $\sum_1^\infty A_\nu \le V(\lambda)$. We must prove the reverse inequality.

Suppose $b_1, \dots, b_h$ is the sequence of primitive λ-numbers divisible only by the primes not exceeding P. A number n formed from primes not exceeding P is a λ-number if and only if n is a multiple of one of $b_1, \dots, b_h$ and n/b_j is formed only from primes not exceeding P. Therefore,

$$\begin{aligned}
&\sum_{\substack{n=1\\ \varphi(n)\le n\lambda}}^{\infty} \frac{\varepsilon_P(n)}{n}\\
&\quad= -\sum_{k=1}^{\infty}\sum_{j=1}^{h}(-1)^j \sum_{b_{i_1}<b_{i_2}<\cdots<b_{i_j}} \frac{\varepsilon_P(k[b_{i_1}, b_{i_2}, \dots, b_{i_j}])}{k[b_{i_1}, b_{i_2}, \dots, b_{i_j}]}\\
&\quad= -\sum_{k=1}^{\infty}\frac{\varepsilon_P(k)}{k}\sum_{j=1}^{h}(-1)^j \sum_{b_{i_1}<b_{i_2}<\cdots<b_{i_j}} \frac{\varepsilon_P([b_{i_1}, b_{i_2}, \dots, b_{i_j}])}{[b_{i_1}, b_{i_2}, \dots, b_{i_j}]}\\
&\quad= \Pi(P)^{-1}\left(-\sum_{j=1}^{h}(-1)^j \sum_{b_{i_1}<b_{i_2}<\cdots<b_{i_j}} \frac{1}{[b_{i_1}, b_{i_2}, \dots, b_{i_j}]}\right)\\
&\quad= \Pi(P)^{-1}V(b_1, b_2, \dots, b_h),
\end{aligned}$$

where $V(b_1, b_2, \dots, b_h)$ is the density of the set of those numbers divisible by at least one of $b_1, b_2, \dots, b_h$. Suppose $b_1, b_2, \dots, b_h$ is a subset of $a_1, a_2, \dots, a_k$, where a_k is the largest of $b_1, b_2, \dots, b_h$. Obviously

$$V(b_1, b_2, \dots, b_h) \le V(a_1, a_2, \dots, a_k) = \sum_{\nu=1}^{k} A_\nu.$$

Since

$$V(b_1, b_2, \dots, b_h) = \Pi(P) \sum_{n=1,\, \varphi(n)\le n\lambda}^{\infty} \frac{\varepsilon_P(n)}{n},$$

it follows that $\lim_{P\to\infty} V(b_1, \dots, b_h) = V(\lambda)$. Thus, $\sum_1^\infty A_\nu = V(\lambda)$. The theorem is proved.

Using Davenport's method, Venkov [146] established the following properties of the function

$$V(\lambda) = \lim_{N\to\infty} \frac{1}{N} \sum_{n\le N,\ \varphi(n)<\lambda n} 1.$$

a) $V(\lambda)$ is an increasing function on $[0,1]$, i.e., if $1 > \lambda_1 > \lambda_2 > 0$, then $V(\lambda_1) > V(\lambda_2)$.

b) $V(\lambda)$ has a left derivative equal to infinity on an everywhere dense set of points in $[0,1]$ (in Venkov's paper this set was defined constructively).

Erdős [37] obtained the following result on the distribution of values of the function $\varphi(n)/n$.

THEOREM. *Let $F(x)$ denote the distribution function for $n/\varphi(n)$. Then for any $\varepsilon > 0$ and sufficiently large x*

$$\exp(-\exp[(1+\varepsilon)ax]) < 1 - F(x) < \exp(-\exp[(1-\varepsilon)ax]),$$

where $a = \exp(-\gamma)$.

We now turn to the question of the rate at which the function

$$\frac{1}{N} \sum_{\varphi(n)<n\lambda,\ n\le N} 1 = \Phi_N(\lambda)$$

approaches its limit $\Phi(\lambda)$.

The first results in this direction were obtained by Tyan [141]. We present here a proof of a stronger result of Faĭnleĭb [44].

THEOREM.

$$\Phi_N(x) = \Phi(x) + O(1/\ln\ln N)$$

uniformly in $x \in [0,1]$.

PROOF. Let t be a real number. We will study the asymptotic behavior of the sum

$$\frac{1}{N}\sum_{n\le N}\left(\frac{\varphi(n)}{n}\right)^{it}.$$

Consider the function

$$h(n) = \sum_{d|n} \mu\left(\frac{n}{d}\right)\left(\frac{\varphi(d)}{d}\right)^{it}.$$

We have

$$\frac{1}{N}\sum_{n\le N}\left(\frac{\varphi(n)}{n}\right)^{it} = \frac{1}{N}\sum_{n\le N}\sum_{d|n} h(d) = \frac{1}{N}\sum_{n\le N} h(n)\left[\frac{N}{n}\right]$$
$$= \sum_{n\le N}\frac{h(n)}{n} + O\left(\frac{1}{N}\sum_{n\le N}|h(n)|\right),$$

$h(n)$ is a multiplicative function, $h(p) = (1 - 1/p)^{it} - 1$, and $h(p^s) = 0$ for $s \geq 2$. Also,

$$|h(p)| = |e^{it \ln(1-1/p)} - 1| \leq |t| \ln \frac{1}{1 - 1/p} \leq \frac{2|t|}{p}.$$

The product

$$\prod_p \left(1 + \frac{|h(p)|}{p} + \frac{h(p^2)}{p^2} + \cdots \right) = \prod_p \left(1 + \frac{|h(p)|}{p}\right)$$

is absolutely convergent. Therefore $\sum_1^\infty |h(n)|/n$ is absolutely convergent. Also,

$$\frac{1}{N} \sum_{n \leq N} \left(\frac{\varphi(n)}{n}\right)^{it} = \sum_{n=1}^{\infty} \frac{h(n)}{n} + O\left(\sum_{n>N} \frac{|h(n)|}{n} + \frac{1}{N} \sum_{n \leq N} |h(n)|\right).$$

Let

$$f(t) = \sum_{n=1}^{\infty} \frac{h(n)}{n} = \prod_p \left(1 - \frac{1}{p} + \frac{1}{p}\left(1 - \frac{1}{p}\right)^{it}\right).$$

We will need two estimates of the quantity

$$\sum_{n>N} \frac{|h(n)|}{n} + \frac{1}{N} \sum_{n \leq N} |h(n)|,$$

one for $|t| \leq 1$ and one for $|t| > 1$.

Suppose $|t| \leq 1$. Then

$$|h(n)| \leq \mu^2(n) \frac{(2|t|)^{\nu(n)}}{n} \leq |t| \frac{\tau(n)}{n}.$$

Therefore,

$$\sum_{n>N} \frac{|h(n)|}{n} + \frac{1}{N} \sum_{n \leq N} |h(n)| \leq |t| \left(\sum_{n \geq N} \frac{\tau(n)}{n^2} + \frac{1}{N} \sum_{n \leq N} \frac{\tau(n)}{n}\right) \leq C|t|,$$

where the constant C does not depend on N.

Suppose $|t| > 1$. Let $\alpha = 1 - 1/\ln(|t|+3)$. Since $0 < \alpha < 1$, it follows that

$$\begin{aligned}\sum_{n>N} \frac{|h(n)|}{n} + \frac{1}{N}\sum_{n\le N} |h(n)| &\le \sum_{n>N} \frac{|h(n)|}{n}\left(\frac{n}{N}\right)^{1-\alpha} + \frac{1}{N}\sum_{n\le N} |h(n)|\left(\frac{N}{n}\right)^{\alpha} \\ &= N^{\alpha-1}\sum_{n=1}^{\infty} \frac{|h(n)|}{n^\alpha} = N^{\alpha-1}\prod_p\left(1+\frac{|h(p)|}{p^\alpha}\right) \\ &\le N^{\alpha-1}\exp\left(\sum_p \frac{|h(p)|}{p^\alpha}\right) \\ &= N^{\alpha-1}\exp\left(\sum_{p\le|t|+3} \frac{|h(p)|}{p^\alpha} + \sum_{p>|t|+3}\frac{|h(p)|}{p^\alpha}\right).\end{aligned}$$

For $p \le |t| + 3$

$$\begin{aligned}\sum_{p\le|t|+3} \frac{|h(p)|p^{1-\alpha}}{p} &= \sum_{p\le|t|+3} \frac{|h(p)|e^{\ln p/\ln(|t|+3)}}{p} \\ &\le 2e\sum_{p\le|t|+3}\frac{1}{p} < 6\ln\ln(|t|+3) + O(1)\end{aligned}$$

(we have used the estimate $|h(p)| \le 2$). Also,

$$\sum_{p>|t|+3} \frac{|h(p)|}{p^\alpha} \le 2\sum_{p>|t|+3} \frac{|t|}{p^{\alpha+1}} = O((|t|+3)^{1-\alpha}) = O(1).$$

Therefore,

$$\sum_{n>N} \frac{|h(n)|}{n} + \frac{1}{N}\sum_{n\le N}|h(n)| = O(N^{\alpha-1}\ln^6(|t|+3)).$$

Thus for $|t| \le 1$ we have, uniformly in N,

$$\frac{1}{N}\sum_{n\le N}\left(\frac{\varphi(n)}{n}\right)^{it} = f(t) + O(|t|),$$

and for $|t| > 1$

$$\frac{1}{N}\sum_{n\le N}\left(\frac{\varphi(n)}{n}\right)^{it} = f(t) + O\left(\exp\left(-\frac{\ln N}{\ln(|t|+3)}\right)\ln^6(|t|+3)\right).$$

Let $F(x) = \Phi(e^x)$. This $F(x)$ is also a distribution function, whose characteristic function is $f(t)$. Consider the concentration function

$$Q(h) = \sup_x (F(x+h) - F(x)).$$

The expression for $f(t)$ shows that $F(x)$ is the distribution function of the sum of independent random variables

$$\eta = \xi_2 + \xi_3 + \xi_5 + \cdots + \xi_p + \cdots$$

(p is a prime), where

$$\xi_p = \begin{cases} \ln(1 - 1/p) & \text{with probability } 1/p, \\ 0 & \text{with probability } 1 - 1/p. \end{cases}$$

For a given h we can find an index r such that

$$\frac{1}{p_1 p_2 \cdots p_r p_{r+1}} < h \le \frac{1}{p_1 p_2 \cdots p_r},$$

$$\sum_{i=1}^{r} \ln p_i \le \ln \frac{1}{h} < \sum_{i=1}^{r+1} \ln p_i.$$

Let $F^{(r)}(x)$ denote the distribution function of the sum of independent random variables $\xi_{p_1} + \cdots + \xi_{p_r}$. The distribution function $F^{(r)}(x)$ has maximal jump equal to $\prod_1^r (1 - 1/p_i)$ (this jump occurs at $x = 0$). We must now find a lower bound for the distance between two points at which jumps occur. Suppose $p_1, \ldots, p_l$ and $q_1, \ldots, q_s$ are two different sets of primes, where $p_i \le p_r$ ($i = 1, \ldots, l$) and $q_j \le p_r$ ($j = 1, \ldots, s$). Note that in this case

$$\left(1 - \frac{1}{p_1}\right) \cdots \left(1 - \frac{1}{p_l}\right) \ne \left(1 - \frac{1}{q_1}\right) \cdots \left(1 - \frac{1}{q_s}\right).$$

Indeed, if

$$\left(1 - \frac{1}{p_1}\right) \cdots \left(1 - \frac{1}{p_l}\right) = \left(1 - \frac{1}{q_1}\right) \cdots \left(1 - \frac{1}{q_s}\right),$$

then

$$(p_l - 1) \cdots (p_1 - 1) q_1 \cdots q_s = p_1 \cdots p_l (q_1 - 1) \cdots (q_s - 1).$$

Suppose p_l is the largest among $p_1, \ldots, p_l, q_1, \ldots, q_s$. In view of the uniqueness of decomposition into prime factors, $p_l | q_1 \cdots q_s$; hence $p_l = q_s$. Cancelling $1 - 1/p_l$ and repeating this argument, we obtain

$$\left(1 - \frac{1}{p_1}\right) \cdots \left(1 - \frac{1}{p_{l-s}}\right) = 1,$$

which is impossible if $l \ne s$. So if the sets of primes are different, then, by a theorem of Lagrange,

$$\begin{aligned} &\left| \ln\left(1 - \frac{1}{p_1}\right) \cdots \left(1 - \frac{1}{p_l}\right) - \ln\left(1 - \frac{1}{q_1}\right) \cdots \left(1 - \frac{1}{q_s}\right) \right| \\ &\quad \ge \left| \left(1 - \frac{1}{p_1}\right) \cdots \left(1 - \frac{1}{p_l}\right) - \left(1 - \frac{1}{q_1}\right) \cdots \left(1 - \frac{1}{q_s}\right) \right| \\ &\quad \ge \frac{1}{p_1 \cdots p_r}, \end{aligned}$$

since the absolute value of the difference of two distinct integers is at least 1. Thus the minimal distance between jump points is at least $1/p_1 \cdots p_r$, and, by a previous condition, h is less than this quantity. Therefore,

$$\sup_x (F^{(r)}(x+h) - F^{(r)}(x)) \leq \prod_{i=1}^{r} \left(1 - \frac{1}{p_l}\right) \leq \frac{C}{\ln\ln(1/h)}.$$

We write $\eta = \xi_2 + \cdots + \xi_p + \cdots = \eta_1 + \eta_2$, where η_1 and η_2 are independent random variables:

$$\eta_1 = \xi_2 + \cdots + \xi_{p_r}, \qquad \eta_2 = \xi_{p_{r+1}} + \cdots.$$

Let $\overline{F}^{(r)}(x)$ be the distribution function of η_2. We have

$$\mathsf{P}(u - x \leq \eta_1 \leq u - x + h) \leq \sup_x (F^{(r)}(x+h) - F^{(r)}(x)).$$

Therefore,

$$\begin{aligned} \mathsf{P}(u < \eta < u + h) &\leq \int_{-\infty}^{\infty} \sup_x (F^{(r)}(x+h) - F^{(r)}(x))\, d\overline{F}^{(r)}(x) \\ &= \sup_x (F^{(r)}(x+h) - F^{(r)}(x)). \end{aligned}$$

Thus (we have repeated the proof of Lévy's concentration lemma [98]),

$$\sup_x (F(x+h) - F(x)) \leq \sup_x (F^{(r)}(x+h) - F^{(r)}(x)) \leq \frac{C}{\ln\ln(1/h)}.$$

Let $F_N(x)$ be the distribution function equal by definition to $\Phi_N(e^x)$. The characteristic function of $F_N(x)$ is

$$\frac{1}{N} \sum_{n \leq N} \left(\frac{\varphi(n)}{n}\right)^{it}.$$

To estimate the deviation of $F_N(x)$ from $F(x)$ we apply the generalized Esseen theorem:

$$\begin{aligned} &\sup_x |F_N(x) - F(x)| \\ &\quad \leq C\left[\sup_x \left(F\left(x + \frac{1}{T}\right) - F(x)\right) + \int_0^T \frac{|(1/N)\sum_{n \leq N}(\varphi(n)/n)^{it} - f(t)|}{t}\, dt\right]. \end{aligned}$$

Put $y = \ln^7 T$. We have

$$\begin{aligned}
&\sup_x |F_N(x) - F(x)| \\
&\quad = O\left(\frac{1}{\ln\ln T}\right) + O\left(\int_0^{1/y} \frac{|(1/N)\sum_{n\le N}(\varphi(n)/n)^{it} - f(t)|}{t}\,dt\right) \\
&\qquad + O\left(\int_{1/y}^{T} \frac{|(1/N)\sum_{n\le N}(\varphi(n)/n)^{it} - f(t)|}{t}\,dt\right) \\
&\quad = O\left(\frac{1}{\ln\ln T}\right) + O\left(\frac{1}{y}\right) + O\left(\int_{1/y}^{T} \frac{e^{-\ln N/\ln(|t|+3)}\ln^6(|t|+3)}{t}\,dt\right) \\
&= O\left(\frac{1}{\ln\ln T}\right) + O\left(\frac{1}{y}\right) + O\left(e^{-\ln N/(4\ln T)}\left(\int_{1/y}^{1}\frac{dt}{t} + \int_1^T \frac{\ln^6 t}{t}\,dt\right)\right) \\
&= O\left(\frac{1}{\ln\ln T}\right) + O\left(\frac{1}{\ln^7 T}\right) + O\left(e^{-\ln N/(4\ln T)}(\ln\ln T + \ln^7 T)\right).
\end{aligned}$$

Putting $T = \exp\sqrt{\ln N}$, we obtain

$$\sup_x |F_N(x) - F(x)| = O(1/\ln\ln N).$$

Since the estimate is uniform in x, it follows that

$$\Phi_N(x) = \Phi(x) + O(1/\ln\ln N),$$

as required.

Faĭnleĭb [45] proved the following result.

THEOREM. *Suppose $\psi(m)$ is a real additive function such that*

1) $\sum_{p^r} \psi^2(p^r)/p^r < \infty$, *and*

2) $|\psi(n) - \psi(m)| \ge (nm)^{-a}$ *if $n \ne m$ are square-free numbers, where a is a positive constant.*

Then, uniformly in x,

$$\frac{1}{N}\sum_{\psi(n)\le x + \sum_{p,n\le N}\psi(p)/p} 1 = F(x) + O\left(\frac{\ln\ln(1/\rho_N)}{\ln(1/\rho_N)\ln\ln\ln(1/\rho_N)}\right),$$

where $F(x)$ is the distribution function defined by the characteristic function

$$f(t) = \prod_p \left(1 - \frac{1}{p}\right)\left(1 + \sum_{r=1}^{\infty}\frac{e^{it\psi(p^r)}}{p^r}\right)e^{-it\psi(p)/p},$$

and where

$$\rho_N = \sum_{p > \exp((\ln N \ln\ln\ln N)/\ln\ln N)} \frac{\psi^2(p)}{p}.$$

When applied to the function $\ln(\varphi(n)/n)$, this theorem yields the following result:

$$\Phi_N(x) = \Phi(x) + O\left(\frac{1}{\ln N}\left(\frac{\ln\ln N}{\ln\ln\ln N}\right)^2\right)$$

uniformly in $x \in [0,1]$.

Suppose $R(x)$ is a continuous function on $[0,1]$. Tyan [141], Il′yasov [73], and Faĭnleĭb [46] considered the question of the remainder term in the relation

$$\lim_{N\to\infty} \frac{1}{N}\sum_{n=1}^{N} R\left(\frac{\varphi(n)}{n}\right) = A(R).$$

Let us examine the special case $R(x) = x^s$, where s is a natural number. We will prove that

$$\sum_{n=1}^{N}\left(\frac{\varphi(n)}{n}\right)^s = N\prod_p\left(1 - \frac{1}{p} + \frac{1}{p}\left(1-\frac{1}{p}\right)^s\right) + O(3^s(1+\ln N)^s).$$

PROOF. It is easy to see that

$$\sum_{d\mid n}\Phi(d) = \left(\frac{\varphi(n)}{n}\right)^s,$$

where

$$\Phi(d) = \begin{cases} 0, & \text{if } d \text{ is divisible by the square of a natural number greater than 1,} \\ \prod_{p\mid d}\left(\left(1-\frac{1}{p}\right)^s - 1\right), & \text{if } d \text{ is square-free.} \end{cases}$$

We have

$$\sum_{n\le N}\left(\frac{\varphi(n)}{n}\right)^s = \sum_{n\le N}\sum_{d\mid n}\Phi(d) = \sum_{d\le N}\Phi(d)\left[\frac{N}{d}\right]$$
$$= N\sum_{n\le N}\frac{\Phi(n)}{n} + O\left(\sum_{n\le N}|\Phi(n)|\right).$$

It is clear that for a square-free n

$$|\Phi(n)| = \left|\prod_{p\mid n}\left(\left(1-\frac{1}{p}\right)^s - 1\right)\right| \le \prod_{p\mid n}\frac{s}{p} = \frac{s^{\nu(n)}}{n},$$

where $\nu(n)$ is the number of different prime divisors of n. We will prove that

$$\sum_{n\le N}\frac{s^{\nu(n)}}{n} \le (1+\ln N)^s.$$

This can be done by induction on s. For $s = 2$ this is trivial:

$$\sum_{n\le N} \frac{1}{n} \le 1 + \ln N.$$

Assume this inequality has been proved for $s-1$. Then

$$\begin{aligned}\sum_{n\le N} \frac{s^{\nu(n)}}{n} &= \sum_{n\le N} \frac{\sum_{d|n} |\mu(d)|(s-1)^{\nu(d)}}{n} \\ &= \sum_{d=1}^{N} \frac{|\mu(d)|(s-1)^{\nu(d)}}{d} \sum_{l\le N/d} \frac{1}{l} \\ &\le \sum_{d=1}^{N} \frac{(s-1)^{\nu(d)}}{d} \sum_{d\le N} \frac{1}{l} \\ &\le (1+\ln N)^{s-1}(1+\ln N) = (1+\ln N)^s.\end{aligned}$$

Thus,

$$\sum_{n\le N} \left(\frac{\varphi(n)}{n}\right)^s = N \sum_{n\le N} \frac{\Phi(n)}{n} + O((1+\ln N)^s).$$

We will prove that

$$\sum_{n>N} \frac{|\Phi(n)|}{n} = O\left(3^s \frac{(1+\ln N)^{s-1}}{N}\right).$$

We have

$$\sum_{n>N} \frac{|\Phi(n)|}{n} \le \sum_{n>N} \frac{s^{\nu(n)}}{n^2}.$$

For $s = 1$ this sum can be estimated by $O(1/N)$. By induction,

$$\begin{aligned}\sum_{n>N} \frac{s^{\nu(n)}}{n^2} &= \sum_{n>N} \frac{\sum_{d|n} |\mu(d)|(s-1)^{\nu(d)}}{n^2} \\ &= \sum_{d=1}^{\infty} \frac{|\mu(d)|(s-1)^{\nu(d)}}{d^2} \sum_{l\ge N/d} \frac{1}{l^2} \\ &\le \sum_{d=1}^{\infty} \frac{(s-1)^{\nu(d)}}{d^2} \sum_{l\ge N/d} \frac{1}{l^2} \\ &\le \frac{\pi^2}{6} \sum_{d\ge N} \frac{(s-1)^{\nu(d)}}{d^2} + \sum_{d<N} \frac{(s-1)^{\nu(d)}}{d^2} \sum_{l\ge N/d} \frac{1}{l^2}.\end{aligned}$$

For $d \le N$ we have

$$\sum_{l\ge N/d} \frac{1}{l^2} \le 2\frac{d}{N}.$$

By the induction assumption,

$$\sum_{d\geq N}\frac{(s-1)^{\nu(d)}}{d^2}\leq 3^{s-1}\frac{(1+\ln N)^{s-2}}{N},$$

and, as we have seen,

$$\sum_{d<N}\frac{(s-1)^{\nu(d)}}{d}\leq (1+\ln N)^{s-1}.$$

Therefore,

$$\begin{aligned}\sum_{n>N}\frac{s^{\nu(n)}}{n^2}&\leq 2\cdot 3^{s-1}\frac{(1+\ln N)^{s-2}}{N}+2\frac{(1+\ln N)^{s-1}}{N}\\&\leq\frac{(1+\ln N)^{s-1}}{N}(2\cdot 3^{s-1}+2)\leq 3^s\frac{(\ln N+1)^s}{N}.\end{aligned}$$

Thus,

$$\sum_{n\leq N}\left(\frac{\varphi(n)}{n}\right)^s=N\sum_{n=1}^{\infty}\frac{\Phi(n)}{n}+O(3^s(\ln N+1)^s).$$

Since $\Phi(d)/d$ is a multiplicative function, we have

$$\begin{aligned}\sum_{d=1}^{\infty}\frac{\Phi(d)}{d}&=\prod_p\left(1+\frac{(1-1/p)^s-1}{p}\right)\\&=\prod_p\left(1-\frac{1}{p}+\frac{1}{p}\left(1-\frac{1}{p}\right)^s\right),\end{aligned}$$

as required.

From the formula for $\sum_{n=1}^{N}(\varphi(n)/n)^s$ it is easy to obtain the following result (Tyan [141]).

THEOREM. *Suppose $P(x)$ is a polynomial of degree s that satisfies the inequality $\max_{0\leq x\leq 1}|P(x)|\leq M$ on $[0,1]$. Then*

$$\frac{1}{N}\sum_{n=1}^{N}P\left(\frac{\varphi(n)}{n}\right)=\int_0^1 P(x)\,dV(x)+O\left(\frac{k^sM\ln^s N}{N}\right),$$

where $k>1$ is a constant and the constant implied by the symbol O is absolute.

Theorems for continuous functions $R(x)$ can be proved by means of polynomial approximations. Naturally, the worse the differential properties of $R(x)$, the worse the remainder term.

THEOREM. *Suppose $R(x)$ is an entire function of order ρ. Let $M = \max_{0\le x\le 1}|R(x)|$. Then for any $\varepsilon > 0$*

$$\frac{1}{N}\sum_{n=1}^{N} R\left(\frac{\varphi(n)}{n}\right) = \int_0^1 R(x)\,dV(x) + O\left(\frac{M+1}{N^{1/(1+\rho)+\varepsilon}}\right).$$

THEOREM. *Suppose $R(x)$ is a function that is analytic inside the ellipse with foci at the points 0 and 1 and with the sum of the semi-axes equal to r ($r > 1/2$). Then*

$$\frac{1}{N}\sum_{n\le N} R\left(\frac{\varphi(n)}{n}\right) = \int_0^1 R(x)\,dV(x)$$
$$+\,O\left(\exp\left[-\ln(2r-\varepsilon)\frac{\ln(N/M)}{\ln\ln N + \ln 12(2r-\varepsilon)}\right]\right),$$

where $\varepsilon > 0$ can be arbitrarily small.

These theorems are proved by the method of moments.

If it is known of $R(x)$ only that some derivative satisfies a Lipschitz condition, then the estimates obtained by the method of moments are unsuitable. As Faĭnleĭb has shown, it is advantageous to use an estimate of the rate at which

$$\frac{1}{N}\sum_{n\le N,\ \varphi(n)\le n\lambda} 1$$

approaches $V(\lambda)$. The following result was established in this way.

THEOREM. *Suppose the function $R(x)$ satisfies a Lipschitz condition on $[0,1]$. Then*

$$\frac{1}{N}\sum_{n\le N} R\left(\frac{\varphi(n)}{n}\right) = A(R) + O(e^{-C\sqrt{\ln N\ln\ln N}}),$$

where $C > 0$ is a constant.

We will mention some other results pertaining to the distribution of values of Euler's function.

W. Sierpiński and A. Schinzel studied the question of the distribution of successive values of $\varphi(n)$. Schinzel [130] proved the following result.

THEOREM. *For any sequence of h positive numbers $a_1,\dots,a_h$ and any $\varepsilon > 0$ there exists a natural number n such that*

$$\left|\frac{\varphi(n+i)}{\varphi(n+i-1)} - a_i\right| < \varepsilon, \qquad i = 1,2,\dots,h.$$

Erdős and Schinzel [40] established, in particular, the following result.

THEOREM. *Suppose $f(n)$ is a positive multiplicative function for which there exists an s such that*

1) *the series*

$$\sum_p \frac{(f(p)-p^s)^2}{p^{2s+1}}$$

converges, and

2) *there exists an interval (a, b), where either $a = 0$ or $b = \infty$, such that for any natural number M the set of numbers $f(n)/n^s$, $(n, M) = 1$, is everywhere dense in (a, b).*

Let there be given a sequence of h nonnegative numbers $a_1, \dots, a_h$ and $\varepsilon > 0$. Then there exist more than $C(a_1, \dots, a_h, \varepsilon)x$ natural numbers $n \le x$ for which

$$\left|\frac{f(n+i)}{f(n+i-1)} - a_i\right| < \varepsilon, \qquad i = 1, 2, \dots, h.$$

Let us verify that the conditions of this theorem (with $s = 1$) are satisfied for the function $\varphi(n)$.

1) $\sum_p (\varphi(p) - p)^2/p^3 = \sum_p 1/p^3 < \infty$.

2) Our verification of the second condition is probably not optimal with regard to complexity. We must prove that for each $\alpha \in (0, 1)$, natural number M, and $\varepsilon > 0$ there exists n such that $(n, M) = 1$ and $|\varphi(n)/n - \alpha| \le \varepsilon$. This will be established once we show that the set of values of $\ln(\varphi(n)/n)$, where n ranges over the numbers relatively prime to prescribed primes $p_1, \dots, p_s$, is everywhere dense in $(-\infty, 0)$. Since both $-\infty$ and 0 are attainable as limits of values of $\ln(\varphi(n)/n)$, $(n, p_1, \dots, p_s) = 1$, it suffices to show that the additive function

$$g(n) = \begin{cases} \ln \dfrac{\varphi(n)}{n}, & \text{if } (n, p_1, \dots, p_s) = 1, \\ 0, & \text{if } (n, p_1, \dots, p_s) > 1, \end{cases}$$

has a continuous distribution function. Since the series

$$\sum_{g(p) \ne 0} \frac{1}{p} = \sum_{\substack{p \ne p_j \\ j=1,2,\dots,s}} \frac{1}{p}$$

diverges, it follows from the Erdős-Wintner theorem that $g(n)$ has a continuous distribution function.

Erdős and Schinzel [40] proved a general theorem on the joint distribution of values of an additive function on successive values of the argument. Let us recall the notation:

$$x^* = \begin{cases} x, & |x| \le 1, \\ 1, & |x| > 1. \end{cases}$$

THEOREM. *Suppose $g(n)$ is an additive function. Assume that*
1) $\sum_p |g(p)|^*/p$ *converges,*
2) $\sum_p (|g(p)|^*)^2/p$ *converges, and*
3) $\sum_{g(p)\neq 0} 1/p$ *diverges.*
Then for any natural number h there exists a continuous distribution function of the h-component vector $(g(m+1), \dots, g(m+h))$.

A similar result was obtained by Kátai ([81], Theorem 3).

§4.9. the characteristic function method

In the analytic scheme, the advantage acquired in studying the distribution of values of positive multiplicative functions $f(n)$ is that the characteristic function connected with the empirical distribution function $\ln f(n)$ is a standardized sum of values of the multiplicative function.

Erdős [37], [38] obtained a result on the distribution of values of nonnegative multiplicative functions; this result was stated in the Introduction.

Broad new possibilities for studying the distribution of values of multiplicative functions of arbitrary sign emerged after Zolotarev [156] suggested a generalization of the characteristic function method, which made possible the reduction of the problem of the distribution of values of a real-valued function of any sign to the summation of multiplicative functions. Zolotarev introduced the concept of the characteristic transform of a distribution function. The *characteristic transform* of a distribution function $F(x)$ is the matrix

$$W(t) = \begin{pmatrix} w_0(t) & 0 \\ 0 & w_1(t) \end{pmatrix},$$

where

$$w_j(t) = \int_{-\infty}^{\infty} |x|^{it} \operatorname{sgn}^j x \, dF(x), \quad j = 0, 1,$$

and it is assumed that $|0|^{it} \operatorname{sgn}^j 0 = 0$.

Zolotarev [156] showed how to recover a distribution function from its characteristic transform.

Applications are based on the following theorem, which expresses a continuous correspondence between characteristic transforms and distribution functions.

THEOREM. *If a sequence of distribution functions $F_n(x)$ is weakly convergent to a distribution function $F(x)$ and if*

$$F_n(0) \to F(0), \qquad F_n(+0) \to F(+0), \tag{1}$$

then the sequence of characteristic transforms $W_n(t)$ of the distribution functions $F_n(x)$ converge uniformly as $n \to \infty$ in each finite interval to the characteristic transform $W(t)$ of $F(x)$. Conversely, if the sequence of characteristic *transforms $W_n(t)$ of the distribution functions $F_n(x)$ converges to a matrix $W(t)$ with continuous entries, then $W(t)$ is the characteristic transform of some distribution function $F(x)$, and the sequence of distribution functions $F_n(x)$ is weakly convergent to $F(x)$.*

We will sketch the proof of this theorem. Let $c_n^+ = 1 - F_n(+0)$ and $c_n^- = F_n(0)$, and let $\varepsilon(x)$ be the singular distribution law

$$\varepsilon(x) = \begin{cases} 0, & x \le 0, \\ 1, & x > 0. \end{cases}$$

It is easy to prove that the entries $w_{j,n}(t)$, $j = 0, 1$, of the characteristic transform $W_n(t)$ can be represented in the form

$$w_{j,n}(t) = c_n^+ f_n^+(t) + (-1)^j c_n^- f_n^-(t),$$

where $f_n^+(t)$ and $f_n^-(t)$ are the characteristic functions of the distribution laws

$$F_n^+(x) = \begin{cases} \dfrac{1}{c_n^+}(F_n(e^x) - F_n(+0)), & \text{if } c_n^+ \ne 0, \\ \varepsilon(x), & \text{if } c_n^+ = 0, \end{cases}$$

and

$$F_n^-(x) = \begin{cases} \dfrac{1}{c_n^-}(F_n(0) - F_n(-e^x + 0)), & \text{if } c_n^- \ne 0, \\ \varepsilon(x), & \text{if } c_n^- = 0. \end{cases}$$

Now suppose the functions $F_n(x)$ are weakly convergent as $n \to \infty$ to some distribution function $F(x)$ and condition (1) holds. Then the limits

$$\lim_{n\to\infty} c_n^+ = c^+, \qquad \lim_{n\to\infty} c_n^- = c^-$$

exist, where $c_+ = 1 - F(+0)$ and $c^- = F(0)$.

1) If $c^+ \ne 0$ and $c^- \ne 0$, then for a sufficiently large n we have

$$F_n^+(x) = \frac{1}{c_n^+}(F_n(e^x) - F_n(+0)),$$
$$F_n^-(x) = \frac{1}{c_n^-}(F_n(0) - F_n(-e^x + 0)).$$

It follows that the sequences of functions $F_n^+(x)$ and $F_n^-(x)$ are weakly convergent to certain distribution functions $F^+(x)$ and $F^-(x)$. Therefore $f_n^+(t)$ and $f_n^-(t)$ approach the characteristic functions $f^+(t)$ and $f^-(t)$ of the distribution functions $F^+(x)$ and $F^-(x)$. Thus,

$$w_{j,n}(t) \to w_j(t) = c^+ f^+(t) + (-1)^j c^- f^-(t), \qquad j = 0, 1.$$

We also see that $w_j(t)$, $j = 0, 1$, are the entries of the characteristic transform of the limit function $F(x)$.

2) Suppose one of c^+ and c^- is equal to zero and the other is nonzero. The sequence of distribution functions $F_n^+(x)$ is weakly convergent to

$$F^+(x) = \frac{1}{c^+}(F(e^x) - F(+0)).$$

Thus the characteristic functions $f_n^+(t)$ converge to $f^+(t)$, and

$$w_{j,n}(t) \to w_j(t) = c^+ f^+(t), \qquad j = 0, 1.$$

Since $F(0) = 0$, it follows that $F^-(x) = \varepsilon(x)$, and the $w_j(t)$ are the entries of the characteristic transform of the limit law $F(x)$.

3) Suppose $c^+ = 0$ and $c^- = 0$. Then as $n \to \infty$

$$w_{j,n}(t) \to w_j(t) = 0, \qquad j = 0, 1.$$

It can be verified directly that the entries of the characteristic transform corresponding to the singular law are identically zero.

We now prove the second part of the theorem. Suppose that as $n \to \infty$ the functions $w_{j,n}(t)$, $j = 0, 1$, approach limit functions $w_j(t)$, $j = 0, 1$. Then the limits

$$\begin{aligned} \lim_{n\to\infty} c_n^+ &= \lim_{n\to\infty} \frac{1}{2}(w_{0,n}(0) + w_{1,n}(0)) = c^+, \\ \lim_{n\to\infty} c_n^- &= \lim_{n\to\infty} \frac{1}{2}(w_{0,n}(0) - w_{1,n}(0)) = c^- \end{aligned}$$

exist. We must again consider various cases.

1) If $c^+ \neq 0$ and $c^- \neq 0$, then for a sufficiently large n the functions

$$f_n^+(t) = \frac{1}{2c_n^+}[w_{0,n}(t) + w_{1,n}(t)]$$

and

$$f_n^-(t) = \frac{1}{2c_n^-}[w_{0,n}(t) - w_{1,n}(t)]$$

are the characteristic functions of distribution functions $F_n^+(x)$ and $F_n^-(x)$ and approach continuous functions $f^+(t)$ and $f^-(t)$. Therefore, $F_n^+(x)$ and $F_n^-(x)$ are weakly convergent to limit distribution functions $F^+(x)$ and $F^-(x)$. It follows that $F_n(e^x)$ is weakly convergent to $1-c^+(1-F^+(x))$, and $F_n(-e^x+0)$ is weakly convergent to $c^-(1 - F^-(x))$. Since the function $F(x)$ defined at points of continuity by the formula

$$F(x) = \begin{cases} c^-(1 - F^-(\ln|x|)), & x < 0, \\ 1 - c^+(1 - F^+(\ln x)), & x > 0, \end{cases}$$

is a distribution function, $F_n(x)$ is weakly convergent to $F(x)$. Since $F(+0) = 1 - c^+$ and $F(0) = c^-$, it follows that $F_n(+0) \to F(+0)$ and $F_n(0) \to F(0)$;

thus, by the first part of our proposition, the limit functions $w_j(t)$, $j = 0, 1$, are the entries of the characteristic transform of $F(x)$.

2) Suppose just one of c^+ and c^- is equal to zero; say $c^+ \neq 0$ and $c^- = 0$. We can prove as in the first case that the functions $F_n(x)$ are weakly convergent to the distribution function $F(x)$ defined by the rule

$$F(x) = \begin{cases} 0 & \text{for } x < 0, \\ 1 - c^+(1 - F^+(\ln x)) & \text{for } x > 0. \end{cases}$$

We then verify the conditions (1), and then it follows that $w_j(t)$, $j = 0, 1$, are the entries of the characteristic transform of $F(x)$.

3) If $c^+ = 0$ and $c^- = 0$, then the inequalities

$$F_n(x) \le F_n(0) = c_n^- \quad \text{for } x \le 0,$$
$$1 - F_n(x) \le 1 - F_n(+0) = c_n^+ \quad \text{for } x > 0$$

imply that $F_n(x)$ is weakly convergent as $n \to \infty$ to the singular law $\varepsilon(x)$. Moreover, $F_n(0) \to 0$ and $F_n(+0) \to 1$. Since the entries of the characteristic transform of $\varepsilon(x)$ are equal to zero, the correspondence of the limit entries of the characteristic transform to the limit distribution law is obvious.

Consider a real-valued multiplicative function $f(n)$ and its corresponding empirical distribution function

$$F_N(x) = \frac{1}{N} \sum_{n \le N,\ f(n) \le x} 1.$$

The entries of the characteristic transform of $F_N(x)$ are equal to

$$w_{j,N}(t) = \frac{1}{N} \sum_{n \le N} |f(n)|^{it} \operatorname{sgn}^j f(n), \qquad j = 0, 1.$$

Thus we see that the question of the limit behavior of the functions $F_N(x)$ has been reduced to the problem of summing multiplicative functions. Here, by analogy with the proof of the Erdős-Wintner theorem, it is natural to use Delange's theorem.

It was by precisely this method that Bakštys [7] obtained an analogue of the Erdős-Wintner theorem and an extension of the theorem of Erdős mentioned in the Introduction (see [37] and [38]) to multiplicative functions of arbitrary sign for which the series

$$\sum_p \frac{(|f(p)| - 1)^*}{p} \quad \text{and} \quad \sum_p \frac{((|f(p)| - 1)^*)^2}{p} \tag{2}$$

converge.

A distribution function $F(x)$ will be called *symmetric* if, for any $x > 0$, $1 - F(x) = F(-x + 0)$.

We will now state the two theorems of Bakštys. Suppose $f(n)$ is a real-valued multiplicative function.

THEOREM. *The distribution function*

$$\frac{1}{N} \sum_{f(n)\le x,\ n\le N} 1 \tag{3}$$

converges as $N \to \infty$ to some nonsymmetric limit distribution function if and only if the two series (2) *and the series*

$$\sum_{f(p)<0} \frac{1}{p} \tag{4}$$

converge and there exists a natural number α such that $f(2^\alpha) \neq -1$.

Note that this theorem is not applicable to the Möbius function, since it follows from the relation

$$\lim_{N\to\infty} \frac{1}{N} \sum_{n\le N} \mu^2(n) = \frac{6}{\pi^2}$$

and the asymptotic law of the distribution of primes

$$\lim_{N\to\infty} \frac{1}{N} \sum_{n\le N} \mu(n) = 0$$

that the Möbius function has the limit distribution function

$$\lim_{N\to\infty} \frac{1}{N} \sum_{\substack{n\le N\\ \mu(n)=\alpha}} 1 = \begin{cases} 3/\pi^2, & \alpha=-1,\\ 1-6/\pi^2, & \alpha=0,\\ 3/\pi^2, & \alpha=+1,\end{cases}$$

which is symmetric.

THEOREM. *If the distribution function* (3) *converges as $N \to \infty$ to some limit distribution function different from*

$$\varepsilon(x) = \begin{cases} 0 & \textit{for } x \le 0,\\ 1 & \textit{for } x > 0,\end{cases}$$

then the limit function is symmetric if and only if at least one of the following conditions holds:

1) *the series* (4) *diverges*;
2) $f(2^\alpha) = -1$ *for all natural numbers* α.

§4.10. Distribution of values of unstable additive functions

In considering the distribution of values of unstable additive functions the main example is the function $\nu(n)$.

In §4.4 we spoke of an analogue of Tchebycheff's inequality for $\nu(n)$. Erdős and Kac [39] proved an analogue of the central limit theorem for $\nu(n)$.

THEOREM. *Suppose $g(n)$ is a strongly additive real arithmetical function such that as $n \to \infty$*

$$B(n) = \Big(\sum_{p \le n} \frac{g^2(p)}{p}\Big)^{1/2} \to \infty$$

and $|g(p)| \le 1$. Then as $N \to \infty$

$$\frac{1}{N} \sum_{\substack{n \le N \\ g(n) \le \sum_{p \le N} g(p)/p + xB(N)}} 1 \to \frac{1}{\sqrt{2\pi}} \int_{-\infty}^{x} e^{-u^2/2}\, du.$$

Clearly the function $\nu(n)$ satisfies the conditions of the Erdős-Kac theorem.

Kubilius introduced the class H of additive functions. By definition, the class H comprises those strongly additive real functions for which

1) $B(n) \to \infty$, and

2) there exists a function $r(n) \to \infty$, $\ln r(n)/\ln n \to 0$, such that

$$B(r(n))/B(n) \to 1.$$

We will show that the additive functions considered in the Erdős-Kac theorem are of class H. The first condition is mutual. Also,

$$1 + \frac{\ln(\ln n/\ln r(n)) + O(1/\ln r(n))}{B^2(r(n))} \ge \frac{B^2(n)}{B^2(r(n))} \ge 1.$$

Since $\sum_{p \le n} g^2(p)/p$ diverges, there exists a nondecreasing function $F(n)$ tending to infinity such that

$$\sum_{p \le n} \frac{g^2(p)}{p} \ge F(n).$$

Thus,

$$1 + o(1) + \frac{\ln(\ln n/\ln r(n))}{F(r(n))} \ge \frac{B^2(n)}{B^2(r(n))} \ge 1.$$

We put $\varphi(x) = e^{\ln x \, \ln F(x)}$, and define $r(n)$ as the function inverse to $\varphi(n)$, $r(n) \to \infty$. We have

$$n = e^{\ln r(n) \, \ln F(r(n))}.$$

Then

$$\frac{\ln n}{\ln r(n)} = \ln F(r(n)) \to \infty.$$

Thus,

$$1 + o(1) + \frac{\ln \ln F(r(n))}{F(r(n))} \ge \frac{B^2(n)}{B^2(r(n))} \ge 1$$

and $B(n)/B(r(n)) \to 1$.

We now state Kubilius's theorem on functions of class H (see [87], Theorem 4.1).

THEOREM. *For a distribution function*

$$\frac{1}{N} \sum_{\substack{n \le N \\ g(n) \le \sum_{p \le N} g(p)/p + uB(N)}} 1,$$

where $g(n)$ is a function of class H, to converge to a limit law with variance 1 it is necessary and sufficient that there exist a nondecreasing function $K(u)$ with variation 1 such that at all points of continuity of $K(u)$

$$\frac{1}{B^2(n)} \sum_{\substack{p \le n \\ g(p) \le uB(n)}} \frac{g^2(p)}{p} \to K(u).$$

The logarithm of the characteristic function of the limit law can be calculated from the formula

$$\ln f(t) = \int_{-\infty}^{\infty} \frac{e^{itu} - 1 - itu}{u^2}\, dK(u),$$

where the value of the integrand at $u = 0$ is taken to be $-t^2/2$.

We will show that the Erdős-Kac theorem is a special case of Kubilius's theorem. Suppose $g(n)$ is bounded in modulus. Then it is easy to see that

$$K(u) = \begin{cases} 1, & u > 0, \\ 0, & u < 0. \end{cases}$$

Therefore, $\ln f(t) = -t^2/2$, which corresponds to the normal law.

In studying questions of the distribution of values of real-valued additive functions $g(n)$, Levin and Faĭnleĭb [96] start from the behavior of these functions on the set of primes. By this is meant the behavior as $N \to \infty$ of the sequence of distribution functions

$$\Phi_N(w) = \frac{1}{\pi(N)} \sum_{p \le N,\ g(p) < w} 1.$$

From the behavior of $\Phi_N(w)$ they determined the behavior of the sequence of distribution functions

$$F_N(x) = \frac{1}{N} \sum_{n \le N,\ g(n) < x} 1$$

as $N \to \infty$.

The question of the remainder terms in integral limit theorems has been the subject of many investigations. Kubilius (see [87], p. 126 (English p. 118)) and Barban [10] mention the following theorem.

THEOREM. *Suppose $g_1(n), \dots, g_s(n)$ are strongly additive real functions such that as $N \to \infty$*

$$B_j(N) = \left(\sum_{p \le N} \frac{g_j^2(p)}{p} \right)^{1/2} \to \infty, \qquad j = 1, 2, \dots, s,$$

and

$$\frac{1}{B_j(N)} \max_{p \le N} |g_j(p)| \le \mu(N), \qquad j = 1, 2, \dots, s,$$

where $\mu(N)$ is nonincreasing and tends to zero. Let there be given distinct nonnegative integers $a_1, \dots, a_s$. Then, uniformly in N as $N \to \infty$ and uniformly in x for $N \ge N_0$,

$$\frac{1}{N} \sum_{n \le N,\ \sum_{j=1}^{s} \frac{g_j(n+a_j) - \sum_{p \le N} g_j(p)/p}{B_j(N)} < x\sqrt{s}} 1$$
$$= \frac{1}{\sqrt{2\pi}} \int_{-\infty}^{x} e^{-u^2/2}\, du + O\left(\mu(N) \left(e^{-x^2/2} \ln \frac{1}{\mu(N)} + 1 \right) \right).$$

W. LeVeque and Kubilius obtained results for $\nu(n)$ more precise than those afforded by the general theorem. Kubilius showed that

$$\frac{1}{N} \sum_{n \le N,\ \nu(n) \le \ln\ln N + x\sqrt{\ln\ln N}} 1 = \frac{1}{\sqrt{2\pi}} \int_{-\infty}^{x} e^{-u^2/2}\, du + O\left(\frac{1}{\sqrt{\ln\ln N}} \right).$$

A more general result in this direction was obtained by Rényi and Turán [125]; namely, they proved the following result.

THEOREM. *Suppose $g(n)$ is a real additive function such that $g(p) = 1$ for each prime p and there exists a constant c such that $g(p^r) = O(r^c)$. Then*

$$\frac{1}{N} \sum_{\substack{n \le N \\ g(n) \le \ln\ln N + x\sqrt{\ln\ln N}}} 1 = \frac{1}{\sqrt{2\pi}} \int_{-\infty}^{x} e^{-u^2/2}\, du + O\left(\frac{1}{\sqrt{\ln\ln N}} \right).$$

The following result is due to Levin and Faĭnleĭb ([94], Theorem 2.3.2).

THEOREM. *Suppose $g(n)$ is a real additive function satisfying the following conditions: first,*

$$\sum_{p \le x} \frac{e^{i\xi g(p)} p}{p} \ln p = \tau(e^{i\xi}) \ln x + B(e^{i\xi}) + O(e^{-c(\ln x)^{\alpha}})$$

uniformly in ξ, where $\tau(e^{i\xi})$ and $B(e^{i\xi})$ are thrice differentiable functions in a neighborhood of $\xi = 0$; second, if $|\xi| \le \pi$, then $\operatorname{Re} \tau(e^{i\xi}) \le 1 - \beta\xi^2$ *with $\beta > 0$; finally,*

$$g(n) = O(e^{(\ln n)^{\alpha - \varepsilon}}), \qquad \varepsilon > 0.$$

Then

$$\frac{1}{N} \sum_{\substack{n \le N \\ g(n) \le \tau'(1) \ln\ln N + x\sqrt{a \ln\ln N}}} 1 = \frac{1}{\sqrt{2\pi}} \int_{-\infty}^{x} e^{-u^2/2}\, du + O\left(\frac{1}{\sqrt{\ln\ln N}}\right)$$

uniformly in x, *where* $a = \tau'(1) + \tau''(1)$.

We will show that this theorem implies the Rényi-Turán theorem. Indeed, under the conditions of the theorem we have, uniformly in ξ,

$$\sum_{p \le x} \frac{e^{i\xi} \ln p}{p} = e^{i\xi} \ln x + Be^{i\xi} + O(e^{-a\sqrt{\ln x}}).$$

Thus $\tau(x) = x$, $B(x) = Bx$, and $\alpha = 1/2$. Also, $\operatorname{Re} e^{i\xi} = \cos\xi \le 1 - \beta\xi^2$, where $\beta > 0$, for $|\xi| \le \pi$. So if $n = p_1^{\alpha_1} \cdots p_s^{\alpha_s}$, then

$$g(n) = O(\alpha_1^c + \cdots + \alpha_s^c) = O(\nu(n)(\ln n/\ln 2)^c) = O(e^{(\ln n)^{1/4}}).$$

There was obtained for the function $\nu(n)$ an asymptotic expansion of the limit distribution function in powers of $1/\sqrt{\ln\ln N}$ (see Delange [30], Kubilius [87], §9). Levin and Faĭnleĭb ([94], Theorem 2.3.3) generalized and developed this result.

One aspect of the theory of distribution of values of additive functions has been the development of problems and results pertaining to the distribution of values of these functions when the argument ranges over a subsequence of the natural sequence. We will not dwell on these questions here, referring the reader instead to Barban's survey article ([11], §7), which discusses the relevant results and provides a guide to the literature.

§4.11. Asymptotic expansions for sums of multiplicative functions

The material pertaining to summation of those multiplicative functions whose summation formulas have asymptotic expansions can be successfully systematized by means of the iterative method of Levin and Faĭnleĭb [94].

For a given multiplicative function $f(n)$ we put $m(x) = \sum_{n \le x} f(n)$.

We introduce for the function $f(n)$ an analogue $\Lambda_f(n)$ of the von Mangoldt function; $\Lambda_f(n)$ is defined by the equality

$$f(n) \ln n = \sum_{d|n} f(d) \Lambda_f\left(\frac{n}{d}\right). \qquad (1)$$

If we put $F(s) = \sum_{n=1}^{\infty} f(n)/n^s$, then

$$F'(s) = -\sum_{n=2}^{\infty} \frac{f(n)\ln n}{n^s} \quad \text{and} \quad -\frac{F'(s)}{F(s)} = \sum_{n=1}^{\infty} \frac{\Lambda_f(n)}{n^s}.$$

We will show that $\Lambda_f(1) = 0$ and that $\Lambda_f(n) = 0$ if n is not a power of a prime. Indeed,

$$F(s) = \prod_p \left(1 + \frac{f(p)}{p^s} + \frac{f(p^2)}{p^{2s}} + \cdots\right),$$

and hence

$$\begin{aligned}
-\frac{F'(s)}{F(s)} &= (-\ln F(s))' \\
&= \sum_p \left(-\ln\left(1 + \frac{f(p)}{p^s} + \frac{f(p^2)}{p^{2s}} + \cdots\right)\right)' \\
&= \sum_p \frac{f(p)/p^s + 2f(p^2)/p^{2s} + 3f(p^3)/p^{3s} + \cdots}{1 + f(p)/p^s + f(p^2)/p^{2s} + \cdots} \\
&= \sum_p \sum_j \frac{\Lambda_f(p^j)}{p^{js}}.
\end{aligned}$$

It remains only to equate the coefficients with those of the expression

$$-\frac{F'(s)}{F(s)} = \sum_{n=1}^{\infty} \frac{\Lambda_f(n)}{n^s}.$$

REMARK. In §4.5, in the proof of Delange's first theorem, the multiplicative function f was connected with certain numbers $c_r^{(p)}(f)$ ($r = 1, 2, \ldots$ and p ranges over the primes). These numbers are uniquely determined by the equality

$$f(n) \ln n = \sum_{p^j | n} c_j^{(p)}(f) f\left(\frac{n}{p^j}\right) \ln p. \tag{2}$$

Comparing (1) and (2) for an n equal to a power of a prime, we see that

$$\Lambda_f(n) = \begin{cases} 0, & \text{if } n = 1 \text{ or } n \text{ is not a power of a prime}, \\ c_j^{(p)}(f) \ln p, & \text{if } n = p^j,\ j \neq 0. \end{cases}$$

In the iterative method, the asymptotic behavior of the sum $m(x)$ is deduced from that of the sum

$$\sum_{n \leq x} \Lambda_f(n). \tag{3}$$

In specific problems the data about the sum (3) are consequences of the laws of distribution of prime numbers (or prime ideals). Usually these are equalities of the form

$$\sum_{n \leq x} \Lambda_f(n) = \tau \ln x + B + h(x), \tag{4}$$

where $h(x)$ is nonincreasing and

$$h(x) = O\left(\min\left(1, \frac{1}{\ln^N x}\right)\right).$$

We impose on $f(n)$ the requirement that

$$\sum_{n\le x} |f(n)| = O(\ln^A x). \tag{5}$$

We now describe the iterative process. The starting estimate for $m(x)$ is one that follows from (5) in the obvious way:

$$m(x) = \sum_{n\le x} f(n) = O\Big(\sum_{n\le x} |f(n)|\Big) = O(\ln^A x). \tag{6}$$

For simplicity, let us assume that τ is a real number. Summing (1) over all $n \le x$, we obtain

$$\begin{aligned}\sum_{n\le x} f(n)\ln n &= \sum_{n\le x}\sum_{d|n} f(d)\Lambda_f\left(\frac{n}{d}\right) = \sum_{d\le x} f(d) \sum_{k\le x/d} \Lambda_f(k)\\ &= \tau \sum_{n\le x} f(n)\ln\frac{x}{n} + B\sum_{n\le x} f(n) + \sum_{n\le x} f(n)h\left(\frac{x}{n}\right).\end{aligned}$$

We write this equality in the form

$$\begin{aligned}&-\sum_{n\le x} f(n)\ln\frac{x}{n} + \ln x\sum_{n\le x} f(n)\\ &\qquad = \tau \sum_{n\le x} f(n)\ln\frac{x}{n} + B\sum_{n\le x} f(n) + \sum_{n\le x} f(n)h\left(\frac{x}{n}\right).\end{aligned}$$

Recalling that $m(x) = \sum_{n\le x} f(n)$, we write this relation as follows:

$$m(x)\ln x - (\tau+1)\int_1^x m(v)\frac{dv}{v} = Bm(x) + \sum_{n\le x} f(n)h\left(\frac{x}{n}\right). \tag{7}$$

Since $|h(x)| \le c$, we obtain from (6) that

$$m(x)\ln x - (\tau+1)\int_1^x m(v)\frac{dv}{v} = O(\ln^A x).$$

Assume $A - \tau - 2 \ne -1$. Then

$$\frac{m(x)}{x(\ln x)^{\tau+1}} - \frac{\tau+1}{x(\ln x)^{\tau+2}}\int_1^x \frac{m(v)}{v}\,dv = O\left(\frac{\ln^{A-\tau-2} x}{x}\right).$$

Replacing x by u in this relation and integrating from 2 to x, we obtain

$$\int_2^x \frac{m(u)\,du}{u(\ln u)^{\tau+1}} - (\tau+1)\int_2^x \frac{du}{u(\ln u)^{\tau+2}}\int_1^u \frac{m(v)}{v}\,dv = c_2 + O((\ln x)^{A-\tau-1}).$$

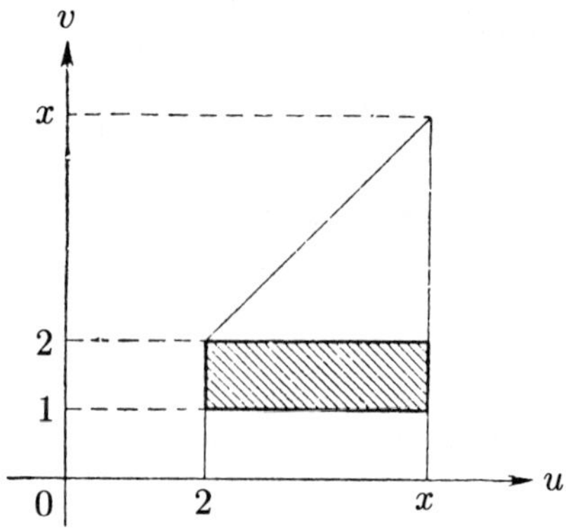

FIGURE 7

Interchanging the order of integration in the second integral (see Figure 7), we obtain

$$\int_2^x \frac{m(u)\,du}{u(\ln u)^{\tau+1}} - (\tau+1)\int_1^x \frac{m(v)}{v}\,dv \int_{\max(2,v)}^x \frac{du}{u(\ln u)^{\tau+2}} = c_2 + O((\ln x)^{A-\tau-1}),$$

or

$$\int_2^x \frac{m(u)\,du}{u(\ln u)^{\tau+1}} - (\tau+1)\int_2^x \frac{m(v)}{v}\left(\int_v^x \frac{du}{u(\ln u)^{\tau+2}}\right)dv - (\tau+1)\int_1^2\int_2^x \frac{m(v)}{v}\frac{1}{u(\ln u)^{\tau+2}}\,du\,dv = c_2 + O((\ln x)^{A-\tau-1}),$$

$$\int_2^x \frac{m(u)}{u(\ln u)^{\tau+1}}\,du + \int_2^x \frac{m(v)}{v}\left(\frac{1}{(\ln x)^{\tau+1}} - \frac{1}{(\ln v)^{\tau+1}}\right)dv + \int_1^2 \frac{m(v)}{v}\left(\frac{1}{(\ln x)^{\tau+1}} - \frac{1}{(\ln 2)^{\tau+1}}\right)dv = c_2 + O((\ln x)^{A-\tau-1}).$$

After cancellation we have

$$\frac{1}{(\ln x)^{\tau+1}}\int_1^x \frac{m(v)}{v}\,dv = c_2' + O((\ln x)^{A-\tau-1}),$$

i.e.,

$$\int_1^x \frac{m(v)}{v}\,dv = c_2'(\ln x)^{\tau+1} + O((\ln x)^A).$$

So along with the equality

$$m(x)\ln x - (\tau+1)\int_1^x m(v)\frac{dv}{v} = O((\ln x)^A)$$

we obtain

$$m(x)\ln x = c_2''(\ln x)^{\tau+1} + O((\ln x)^A),$$

or

$$m(x) = c_2''(\ln x)^\tau + O((\ln x)^{A-1}).$$

If $\tau > A - 1$, we obtain one term of the asymptotic formula. If $\tau \le A - 1$, we have simply improved the original estimate (6). In either case we have the start of the iterative process.

Here is an example to illustrate the method. Consider the function defined by

$$b(n) = \begin{cases} 0, & \text{if } n \text{ is not a sum of two squares,} \\ 1, & \text{if } n \text{ is a sum of two squares.} \end{cases}$$

The function $b(n)$ is multiplicative.

We will prove the following

LEMMA. *For any natural number F, as $x \to \infty$*

$$\sum_{n\le x} \frac{b(n)}{n} = a_0\sqrt{\ln x} + \sum_{r=1}^{F} \frac{a_r}{(\ln x)^{r-1/2}} + O\left(\frac{1}{(\ln x)^{F+1/2}}\right),$$

where a_0 is a nonzero constant and $a_1, \dots, a_F$ are also constants.

PROOF. Let $f(n) = b(n)/n$. In view of the theorem of number theory on the structure of the canonical decomposition of integers representable as a sum of two squares and the fact that the function $f(n) = b(n)/n$ is multiplicative,

$$\sum_{n=1}^{\infty} \frac{f(n)}{n^s} = \frac{1}{1-(1/2)/2^s} \prod_{p\equiv 1 \pmod 4} \frac{1}{1-(1/p)/p^s} \prod_{p\equiv 3 \pmod 4} \frac{1}{1-(1/p)/p^{2s}}.$$

Therefore,

$$\ln \sum_{n=1}^{\infty} \frac{f(n)}{n^s} = -\ln\left(1 - \frac{1/2}{2s}\right) - \sum_{p\equiv 1 \pmod 4} \ln\left(1 - \frac{1/p}{p^s}\right) - \sum_{p\equiv 3 \pmod 4} \ln\left(1 - \frac{1/p}{p^{2s}}\right)$$

and

$$-\left(\ln\sum_{n=1}^{\infty}\frac{f(n)}{n^s}\right)' = -\frac{(\sum_{n=1}^{\infty} f(n)/n^s)'}{\sum_{n=1}^{\infty} f(n)/n^s} = \frac{\sum_{n=1}^{\infty} f(n)\ln n/n^s}{\sum_{n=1}^{\infty} f(n)/n^s}$$

$$= \frac{(1/2)(e^{-s\ln 2})'}{1-(1/2)/2^s} + \sum_{p\equiv 1 (\mathrm{mod}\, 4)} \frac{1}{p}\frac{(e^{-s\ln p})'}{1-(1/p)/p^s}$$

$$+ \sum_{p\equiv 3(\mathrm{mod}\, 4)} \frac{1}{p^2}\frac{(e^{-2s\ln p})'}{1-(1/p^2)/p^{2s}}$$

$$= \frac{\ln 2/2^{s+1}}{1-1/2^{s+1}} + \sum_{p\equiv 1(\mathrm{mod}\, 4)} \frac{\ln p/p^{s+1}}{1-1/p^{s+1}}$$

$$+ \sum_{p\equiv 3(\mathrm{mod}\, 4)} \frac{2\ln p/p^{2(s+1)}}{1-1/p^{2(s+1)}}$$

$$= \sum_{k=1}^{\infty}\frac{\ln 2}{2^{k(s+1)}} + \sum_{p\equiv 1(\mathrm{mod}\, 4)} \ln p\sum_{k=1}^{\infty}\frac{1}{p^{k(s+1)}}$$

$$+ \sum_{p\equiv 3(\mathrm{mod}\, 4)} 2\ln p\sum_{k=1}^{\infty}\frac{1}{p^{2k(s+1)}}$$

$$= \sum_{n=1}^{\infty}\frac{\Lambda_f(n)}{n^s},$$

where

$$\Lambda_f(n) = \begin{cases} \ln p/p^k, & n = p^k,\ p = 2,\ p\equiv 1 \pmod 4, \\ 2\ln p/p^{2k}, & n = p^{2k},\ p\equiv 3 \pmod 4. \end{cases}$$

Thus,

$$f(n)\ln n = \sum_{d|n} f(d)\Lambda_f\left(\frac{n}{d}\right).$$

From the relation

$$\sum_{p\equiv 1 \ (\mathrm{mod}\, 4)} \ln p = \frac{x}{2} + R(x), \qquad R(x) = O(xe^{-c\sqrt{\ln x}}),$$

we obtain, by means of Abel's transformation,

$$\begin{aligned}\sum_{p\equiv 1 \pmod 4} \frac{\ln p}{p} &= \int_2^x \frac{\sum_{p\equiv 1 \pmod 4,\ p\le u} \ln p}{u^2}\,du + \frac{\sum_{p\equiv 1 \pmod 4,\ p\le x} \ln p}{x} \\ &= \frac{1}{2}\int_2^x \frac{du}{u} + \int_2^x \frac{R(u)}{u^2}\,du + \frac{1}{2} + O(e^{-c\sqrt{\ln x}}) \\ &= \frac{\ln x}{2} - \frac{\ln 2}{2} + \frac{1}{2} + \int_2^\infty \frac{R(u)}{u^2}\,du + O\left(\int_x^\infty \frac{e^{-c\sqrt{\ln u}}}{u}\,du\right) \\ &\quad + O(e^{-c\sqrt{\ln x}}) \\ &= \frac{\ln x}{2} + c_1 + O\left(\frac{1}{(\ln x)^N}\right).\end{aligned}$$

Given any natural number N, we have

$$\sum_{n\le x} \Lambda_f(n) = \frac{\ln x}{2} + c_2 + O\left(\min\left(1, \frac{1}{(\ln x)^N}\right)\right).$$

Also,

$$\sum_{n\le x} \frac{b(n)}{n} = O\Bigl(\sum_{n\le x} \frac{1}{n}\Bigr) = O(\ln x).$$

Thus the conditions for applying the iterative method are satisfied (with $A = 1$ and $\tau = 1/2$). Carrying out the calculation by the above scheme once, we obtain

$$\sum_{n\le x} \frac{b(n)}{n} = a_0\sqrt{\ln x} + O(1).$$

We can now take $A = 1/2$. Carrying out the calculations by the scheme of the method of integral equations a second time, we obtain

$$\sum_{n\le x} \frac{b(n)}{n} = a_0\sqrt{\ln x} + O\left(\frac{1}{(\ln x)^{1/2}}\right). \tag{8}$$

We now proceed by induction. Assume that

$$m(x) = \sum_{\nu=0}^{r-1} a_\nu (\ln x)^{1/2-\nu} + O\left((\ln x)^{1/2-\sum_{\rho=0}^{r-1}(N/(N+1))^\rho} (\ln\ln x)^{r-1}\right). \tag{9}$$

For $r = 1$ assertion (8) is stronger than required. If we can show that under these conditions

$$m(x) = \sum_{\nu=0}^{r} a_\nu (\ln x)^{1/2-\nu} + O\left((\ln x)^{1/2-\sum_{\rho=0}^{r}(N/(N+1))^\rho} (\ln\ln x)^{r}\right),$$

then the lemma will be proved. Under the induction assumption (9) consider the sum

$$\sum_{n\le x}\frac{b(n)}{n}h\left(\frac{x}{n}\right)$$

appearing in (7). We take $0<\theta<1-\varepsilon_0$, where ε_0 is fixed, and make the following estimates:

$$\begin{aligned}\sum_{n\le x}\frac{b(n)}{n}h\left(\frac{x}{n}\right) &= \sum_{x^{1-\theta}<n\le x}\frac{b(n)}{n}h\left(\frac{x}{n}\right)+\sum_{n\le x^{1-\theta}}\frac{b(n)}{n}h\left(\frac{x}{n}\right)\\ &= \sum_{x^{1-\theta}<n\le x}\frac{b(n)}{n}h\left(\frac{x}{n}\right)+h(x^{\theta})\sum_{n\le x^{1-\theta}}\frac{b(n)}{n}\\ &= \sum_{x^{1-\theta}<n\le x}\frac{b(n)}{n}h\left(\frac{x}{n}\right)+O\left(\frac{1}{\theta^N(\ln x)^{N-1/2}}\right)\\ &= \int_{u=1-\theta}^{1}h(x^{1-u})\,dm(x^u)+O\left(\frac{1}{\theta^N(\ln x)^{N-1/2}}\right).\end{aligned}$$

Put

$$R(x)=m(x)-\sum_{\nu=0}^{r-1}a_\nu(\ln x)^{1/2-\nu}.$$

We then obtain

$$\begin{aligned}\sum_{n\le x}\frac{b(n)}{n}h\left(\frac{x}{n}\right) &= \sum_{\nu=0}^{r-1}a_\nu\left(\frac{1}{2}-\nu\right)(\ln x)^{1/2-\nu}\int_{1-\theta}^{1}\frac{h(x^{1-u})}{u^{1/2+\nu}}\,du\\ &\quad+\int_{u=1-\theta}^{1}h(x^{1-u})\,dR(x^u)+O\left(\frac{1}{\theta^N(\ln x)^{N-1/2}}\right).\end{aligned}$$

Integration by parts yields

$$\begin{aligned}&\left|\int_{u=1-\theta}^{1}h(x^{1-u})\,dR(x^u)\right|\\ &\quad=\left|h(1)R(x)-h(x^{\theta})R(x^{1-\theta})+\int_{1-\theta}^{1}R(x^u)\,dh(x^{1-u})\right|\\ &\quad\le\max_{x^{1-\theta}\le z\le x}|R(z)|\operatorname*{var}_{1\le z\le x^{\theta}}h(z)\\ &\quad\quad+O\left((\ln x)^{1/2-\sum_{\rho=0}^{r-1}(N/(N+1))^{\rho}}(\ln\ln x)^{r-1}\right),\end{aligned}$$

since $h(x^\theta) = O(1)$ and the order of the estimate for $R(x^{1-\theta})$ is the same as for $R(x)$. But

$$\begin{aligned}\operatorname*{var}_{1\le u\le x^\theta} h(u) &\le \operatorname*{var}_{1\le u\le x^\theta} \sum_{n\le u} \Lambda_f(n) + \operatorname*{var}_{1\le u\le x^\theta} \frac{\ln u}{2} \\ &\le \sum_{n\le x^\theta} |\Lambda_f(n)| + O(\theta \ln x) \\ &= \sum_{n\le x^\theta} \Lambda_f(n) + O(\theta\ln x) = O(\theta \ln x).\end{aligned}$$

Thus if we make the change of variable $z = 1-u$ in the integrals, we obtain

$$\begin{aligned}\sum_{n\le x} \frac{b(n)}{n} h\left(\frac{x}{n}\right) &= \sum_{\nu=0}^{r-1} a_\nu \left(\frac{1}{2}-\nu\right)(\ln x)^{1/2-\nu} \int_0^\theta h(x^z)(1-z)^{-1/2-\nu}\,dz \\ &\quad + O\left(\theta (\ln x)^{3/2-\sum_{\rho=0}^{r-1}(N/(N+1))^\rho} (\ln\ln x)^{r-1}\right) \\ &\quad + O\left(\frac{1}{\theta^N (\ln x)^{N-1/2}}\right) \\ &= \sum_{\nu=0}^{r-2} a_\nu \left(\frac{1}{2}-\nu\right)(\ln x)^{1/2-\nu} \int_0^\theta h(x^z)(1-z)^{-1/2-\nu}\,dz \\ &\quad + a_{r-1}\left(\frac{3}{2}-r\right)(\ln x)^{3/2-r} \int_0^\theta h(x^z)(1-z)^{-1/2-r}\,dz \\ &\quad + O\left(\theta (\ln x)^{3/2-\sum_{\rho=0}^{r-1}(N/(N+1))^\rho} (\ln\ln x)^{r-1}\right) \\ &\quad + O\left(\frac{1}{\theta^N (\ln x)^{N-1/2}}\right); \qquad (10)\end{aligned}$$

if $r = 1$, the first sum (from $\nu = 0$ to $r-2$) is empty. In view of the estimate for $h(x)$,

$$(\ln x)^{3/2-r} \int_0^\theta h(x^z)(1-z)^{-1/2-r}\,dz = O\left((\ln x)^{3/2-r-N}\theta^{1-N}\right) + O\left((\ln x)^{1/2-r}\right).$$

We transform the expression involving $h(x^z)(1-z)^{-1/2-\nu}$ by the binomial formula:

$$\int_0^\theta h(x^z)(1-z)^{-1/2-\nu}\,dz = \sum_{\mu=0}^{r-2-\nu} (-1)^\mu \binom{-\nu-\frac{1}{2}}{\mu} \int_0^\theta h(x^z) z^\mu\,dz + O\left(\int_0^\theta |h(x^z)| z^{r-\nu-1}\,dz\right).$$

In (7) we express the quantity $m(x)$ appearing in the right-hand side by (9) (the induction assumption), and in place of $\sum_{n\le x}(b(n)/n)h(x/n)$ we substitute (10). This gives us

$$\begin{aligned}
&m(x)\ln x-\frac{3}{2}\int_1^x m(u)\frac{du}{u}\\
&=B\sum_{\nu=0}^{r-1}a_\nu(\ln x)^{1/2-\nu}+\sum_{\nu=0}^{r-2}\left(\frac{1}{2}-\nu\right)a_\nu(\ln x)^{1/2-\nu}\\
&\times\sum_{\mu=0}^{r-2-\nu}(-1)^\mu\binom{-\frac{1}{2}-\nu}{\mu}\int_0^\theta h(x^z)z^\mu\,dz\\
&+O\left(\sum_{\nu=0}^{r-2}(\ln x)^{1/2-\nu}\int_0^\theta |h(x^z)|z^{r-1-\nu}\,dz\right)\\
&+O(N^{1-\theta}(\ln x)^{3/2-r-N})+O\left(\theta(\ln x)^{3/2-\sum_{\rho=0}^{r-1}(N/(N+1))^\rho}(\ln\ln x)^r\right)\\
&+O\left(\frac{1}{\theta^N}\frac{1}{(\ln x)^{N-1/2}}\right)+O((\ln x)^{1/2-r}).
\end{aligned}$$

But

$$\int_0^\theta h(x^z)z^\mu\,dz=\frac{1}{(\ln x)^{\mu+1}}\int_1^{x^\theta}\frac{h(z)(\ln z)^\mu}{z}\,dz.$$

In view of the estimate $h(z)=O(1/(\ln z)^N)$, the integral

$$\int_1^\infty\frac{h(z)(\ln z)^\mu}{z}\,dz$$

converges. Let us denote it by b_μ. Then

$$\begin{aligned}
\int_0^\theta h(z)z^\mu\,dz&=\frac{b_\mu}{(\ln x)^{\mu+1}}+O\left(\frac{1}{(\ln x)^{\mu+1}}\int_{x^\theta}^\infty\frac{h(z)(\ln z)^\mu}{z}\,dz\right)\\
&=\frac{b_\mu}{(\ln x)^{\mu+1}}+O\left(\frac{1}{(\ln x)^{\mu+1}}\int_{x^\theta}^\infty\frac{dz}{(\ln z)^{N-\mu}z}\right)\\
&=\frac{b_\mu}{(\ln x)^{\mu+1}}+O\left(\frac{1}{\theta^{N-\mu-1}(\ln x)^N}\right).
\end{aligned}$$

Therefore,

$$\begin{aligned}
&m(x)\ln x - \frac{3}{2}\int_1^x m(u)\frac{du}{u} \\
&= \sum_{\nu=0}^{r-1} c_\nu (\ln x)^{1/2-\nu} + O\left(\frac{1}{\theta^N}\sum_{\nu=0}^{r-2}(\ln x)^{1/2-\nu-N}\sum_{\mu=0}^{r-\nu-2}\theta^{\mu+1}\right) \\
&\quad + O\left(\sum_{\nu=0}^{r-2}(\ln x)^{1/2-\nu-N}\theta^{r-\nu-N}\right) + O(\theta^{1-N}(\ln x)^{3/2-r-N}) \\
&\quad + O\left(\theta(\ln x)^{3/2-\sum_{\rho=0}^{r-1}(N/(N+1))^\rho}(\ln\ln x)^r\right) \\
&\quad + O\left(\frac{1}{\theta^N}\frac{1}{(\ln x)^{N-1/2}}\right) + O((\ln x)^{1/2-r}) \\
&= \sum_{\nu=0}^{r-1} c_\nu (\ln x)^{1/2-\nu} + O\left(\frac{\theta}{\theta^N(\ln x)^{N-1/2}}\right) \\
&\quad + O\left(\frac{\theta^2}{\theta^N}\sum_{\nu=0}^{r-2}(\ln x)^{1/2-\nu-N}\right) + O(\theta^{1-N}(\ln x)^{3/2-r-N}) \\
&\quad + O\left(\theta(\ln x)^{3/2-\sum_{\rho=0}^{r-1}(N/(N+1))^\rho}(\ln\ln x)^r\right) + O\left(\frac{1}{\theta^N(\ln x)^{N-1/2}}\right) \\
&= \sum_{\nu=0}^{r-1} c_\nu (\ln x)^{1/2-\nu} + O\left(\frac{1}{\theta^N}\frac{1}{(\ln x)^{N-3/2}}\right) \\
&\quad + O\left(\theta(\ln x)^{3/2-\sum_{\rho=0}^{r-1}(N/(N+1))^\rho}(\ln\ln x)^r\right) + O((\ln x)^{1/2-r}).
\end{aligned}$$

Put

$$\theta = (\ln x)^{-(N/(N+1))^r};$$

then, since N is large and therefore $1/(\ln x)^{N-3/2}$ is small, we obtain

$$\begin{aligned}
&m(x)\ln x - \frac{3}{2}\int_1^x m(u)\frac{du}{u} \\
&= \sum_{\nu=0}^{r-1} c_\nu (\ln x)^{1/2-\nu} + O\left((\ln x)^{3/2-\sum_{\rho=0}^{r-1}(N/(N+1))^\rho}(\ln\ln x)^r\right). \qquad (11)
\end{aligned}$$

Now according to the original scheme

$$\begin{aligned}
&\frac{m(x)}{x(\ln x)^{3/2}} - \frac{3}{2}\frac{1}{(\ln x)^{5/2}}\int_1^x m(u)\frac{du}{u} \\
&\qquad = \sum_{\nu=0}^{r-1}\frac{c_\nu}{(\ln x)^{r+\nu}} + O\left(\frac{(\ln x)^{-1-\sum_{\rho=0}^{r}(N/(N+1))^\rho}}{x}(\ln\ln x)^r\right).
\end{aligned}$$

Therefore,

$$\frac{1}{(\ln x)^{3/2}}\int_1^x \frac{m(u)\,du}{u} = c' + \sum_{\nu=0}^{r-1}\frac{c'_\nu}{(\ln x)^{\nu+1}} + O\left((\ln x)^{-\sum_{\rho=0}^{r}(N/(N+1))^\rho}(\ln\ln x)^r\right).$$

It follows that

$$\int_1^x \frac{m(u)}{u}\,du = c'(\ln x)^{3/2} + \sum_{\nu=0}^{r-1}\frac{c'_\nu(\ln x)^{1/2}}{(\ln x)^\nu} + O\left((\ln x)^{3/2-\sum_{\rho=0}^{r}(N/(N+1))^\rho}(\ln\ln x)^r\right).$$

This last relation, together with (11), yields

$$m(x) = \sum_{\nu=0}^{r} a_\nu(\ln x)^{1/2-\nu} + O\left((\ln x)^{1/2-\sum_{\rho=0}^{r}(N/(N+1))^\rho}(\ln\ln x)^r\right).$$

Of course, the length of our induction on r is restricted by the choice of N, but we can take N to be arbitrarily large.

The lemma is proved.

We will prove a refinement of a previously mentioned theorem of Landau [90].

THEOREM. *For any natural number F*

$$\sum_{n\le x} b(n) = \frac{a_0}{2}\frac{x}{\sqrt{\ln x}} + x\sum_{\nu=1}^{F}\frac{l_\nu}{(\ln x)^{\nu-1/2}} + O\left(\frac{x}{(\ln x)^{F+1/2}}\right),$$

where $a_0 \ne 0$ and $l_1, \dots, l_F$ are constants.

PROOF. By Abel's transformation,

$$\sum_{n\le x} b(n) = \sum_{n\le x}\frac{b(n)}{n}n = -\int_2^x\left(\sum_{n\le u}\frac{b(n)}{n}\right)du + x\sum_{n\le x}\frac{b(n)}{n} + O(1).$$

By the lemma, with $F \ge 1$,

$$\sum_{n\le x} b(n) = -\int_2^x\left(a_0\sqrt{\ln x} + \sum_{r=1}^{F}\frac{a_r}{(\ln x)^{r-1/2}} + O\left(\frac{1}{(\ln x)^{F+1/2}}\right)\right)dx + x\left(a_0\sqrt{\ln x} + \sum_{r=1}^{F}\frac{a_r}{(\ln x)^{r-1/2}} + O\left(\frac{1}{(\ln x)^{F+1/2}}\right)\right).$$

Integrating by parts, we obtain

$$\sum_{n\le x} b(n) = \int_2^x \left(\frac{a_0}{2}\frac{1}{\sqrt{\ln x}} - \sum_{r=1}^{F}\frac{r}{2}\frac{a_r}{(\ln x)^{r+1/2}}\right)dx + O\left(\frac{x}{(\ln x)^{F+1/2}}\right).$$

The theorem now follows by successive integration by parts.

§4.12. Summation involving power reduction

As an example of an application of the method of Dirichlet generating series we consider the problem of the asymptotic behavior of the sum

$$S(N) = \sum_{n\le N} \tau(n^2). \tag{1}$$

The solution of this problem is based on the work of Stronina [137].

THEOREM. *As $N \to \infty$*

$$S(N) = NP(\ln N) + O\left(N^{1/2}\exp\left(-c\frac{(\ln N)^{3/5}}{(\ln\ln N)^{1/5}}\right)\right), \tag{2}$$

where $P(x)$ is a second-degree polynomial and $c > 0$ is a constant.

PROOF. Since the function $\tau(n)$ is multiplicative, so is $\tau(n^2)$. We introduce the function of a complex variables s

$$F(s) = \sum_{n=1}^{\infty}\frac{\tau(n^2)}{n^s}.$$

The function $F(s)$ can be written as an Euler product:

$$\begin{aligned} F(s) &= \prod_p \left(1 + \frac{3}{p^s} + \frac{5}{p^{2s}} + \frac{7}{p^{3s}} + \cdots\right) \\ &= \prod_p \frac{1+1/p^s}{(1-1/p^s)^2} = \prod_p \frac{1-1/p^{2s}}{(1-1/p^s)^3} = \frac{\varsigma^3(s)}{\varsigma(2s)}. \end{aligned}$$

Therefore,

$$\sum_{n=1}^{\infty}\frac{\tau(n^2)}{n^s} = \sum_{d=1}^{\infty}\frac{\mu(d)}{(d^2)^s}\sum_{n=1}^{\infty}\frac{\tau_3(n)}{n^s},$$

where $\tau_3(n)$ is the number of representations of the natural number n as a product of three natural factors. Equating coefficients, we obtain

$$\tau(n^2) = \sum_{m=d^2n}\mu(d)\tau_3(n). \tag{3}$$

This formula can also be obtained in an elementary way.

In view of (3),

$$S(N) = \sum_{d^2 n \le N}\sum \mu(d)\tau_3(n), \tag{4}$$

where the summation extends over all lattice points with $d > 0$, $n > 0$, and $d^2 n \le N$.

Take an ε such that $0 < \varepsilon < 1$, to be defined precisely later, and split the sum $S(N)$ in (4) as follows:

$$S(N) = \sum_{d \le \varepsilon\sqrt{N}} \mu(d) \sum_{\varepsilon^{-2} \le n \le N/d^2} \tau_3(n) + \sum_{n \le \varepsilon^{-2}} \tau_3(n) \sum_{d \le \sqrt{N/n}} \mu(d).$$

Let $M(x) = \sum_{n \le x} \mu(n)$. We have

$$S(N) = \sum_{d \le \varepsilon\sqrt{N}} \mu(d) \sum_{n \le N/d^2} \tau_3(n) - M(\varepsilon\sqrt{N}) \sum_{n \le \varepsilon^{-2}} \tau_3(n)$$
$$+ O\Big(\sum_{n \le \varepsilon^{-2}} \tau_3(n) \Big| M\left(\sqrt{N/n}\right)\Big| \Big).$$

For the subsequent calculation we will need, first of all, an estimate of $M(x)$:

$$M(x) = O(x\delta(x)), \quad \text{where } \delta(x) = \exp\left(-c\frac{(\ln x)^{3/5}}{(\ln\ln x)^{1/5}} \right),$$

$c > 0$ being a constant (see Walfisz [151], p. 191, formula (15)). Second, we will need the following formula of Atkinson (see Titchmarsh [139], p. 270, and, for more details, Atkinson [2]):

$$\sum_{n \le N} \tau_3(n) = N\overline{P}(\ln N) + O(N^{37/75} \ln^K N),$$

where $\overline{P}(x)$ is some second-degree polynomial and K is a sufficiently large natural number. Suppose $\overline{P}(x) = \lambda_2 x^2 + \lambda_1 x + \lambda_0$. Let $\theta' = \frac{37}{75} + \frac{1}{200}$; note that $\theta' < \frac{1}{2}$. We have

$$\sum_{d \le \varepsilon\sqrt{N}} \mu(d) \sum_{n \le N/d^2} \tau_3(n)$$
$$= \lambda_2 N \sum_{d \le \varepsilon\sqrt{N}} \frac{\mu(d)}{d^2} \ln^2 \frac{N}{d^2} + \lambda_1 N \sum_{d \le \varepsilon\sqrt{N}} \frac{\mu(d)}{d^2} \ln \frac{N}{d^2}$$
$$+ \lambda_0 N \sum_{d \le \varepsilon\sqrt{N}} \frac{\mu(d)}{d^2} + O\Big(\sum_{d \le \varepsilon\sqrt{N}} \left(\frac{N}{d^2}\right)^{\theta'} \Big).$$

It is obvious that

$$\sum_{d\le\varepsilon\sqrt{N}}\left(\frac{N}{d^2}\right)^{\theta'}=N^{\theta'}\sum_{d\le\varepsilon\sqrt{N}}\frac{1}{d^{2\theta'}}=O(\varepsilon^{1-2\theta'}N^{1/2}).$$

Also,

$$\begin{aligned}
&\lambda_2N\sum_{d\le\varepsilon\sqrt{N}}\frac{\mu(d)}{d^2}\ln^2\frac{N}{d^2}+\lambda_1N\sum_{d\le\varepsilon\sqrt{N}}\frac{\mu(d)}{d^2}\ln\frac{N}{d^2}+\lambda_0N\sum_{d\le\varepsilon\sqrt{N}}\frac{\mu(d)}{d^2}\\
&=\lambda_2N\ln^2N\sum_{d\le\varepsilon\sqrt{N}}\frac{\mu(d)}{d^2}-4\lambda_2N\ln N\sum_{d\le\varepsilon\sqrt{N}}\frac{\mu(d)\ln d}{d^2}\\
&\quad+4\lambda_2N\sum_{d\le\varepsilon\sqrt{N}}\frac{\mu(d)}{d^2}\ln^2d+\lambda_1N\ln N\sum_{d\le\varepsilon\sqrt{N}}\frac{\mu(d)}{d^2}\\
&\quad-2\lambda_1N\sum_{d\le\varepsilon\sqrt{N}}\frac{\mu(d)\ln d}{d^2}+\lambda_0N\sum_{d\le\varepsilon\sqrt{N}}\frac{\mu(d)}{d^2}\\
&=\overline{\lambda}_2N\ln^2N+\overline{\lambda}_1N\ln N+\overline{\lambda}_0N\\
&\quad+O\biggl(N\ln^2N\biggl|\sum_{d>\varepsilon\sqrt{N}}\frac{\mu(d)}{d^2}\biggr|+N\ln N\biggl|\sum_{d>\varepsilon\sqrt{N}}\frac{\mu(d)\ln d}{d^2}\biggr|\\
&\qquad\qquad+N\biggl|\sum_{d>\varepsilon\sqrt{N}}\frac{\mu(d)\ln^2d}{d^2}\biggr|\biggr),
\end{aligned}\tag{5}$$

since all of the series involved converge. Abel's transformation yields

$$\begin{aligned}
\biggl|\sum_{d>\varepsilon\sqrt{N}}\frac{\mu(d)}{d^2}\biggr|&=\biggl|\sum_{d\ge[\varepsilon\sqrt{N}]+1}\frac{M(d)-M(d-1)}{d^2}\biggr|\\
&\le\frac{|M([\varepsilon\sqrt{N}])|}{(\varepsilon\sqrt{N})^2}+\sum_{d\ge[\varepsilon\sqrt{N}]+1}|M(d)|\left(\frac{1}{d^2}-\frac{1}{(d+1)^2}\right)\\
&=O\left(\frac{M([\varepsilon\sqrt{N}])}{(\varepsilon\sqrt{N})^2}\right)+O\biggl(\sum_{d>\varepsilon\sqrt{N}}\frac{|M(d)|}{d^3}\biggr).
\end{aligned}$$

Using the estimate for $M(x)$, we obtain

$$\begin{aligned}\left|\sum_{d>\varepsilon\sqrt{N}} \frac{\mu(d)}{d^2}\right| &= O\left(\frac{\delta(\varepsilon\sqrt{N})}{\varepsilon N^{1/2}}\right) + O\Big(\sum_{d>\varepsilon\sqrt{N}} \frac{\delta(d)}{d^2}\Big) \\ &= O\left(\frac{\delta(\varepsilon\sqrt{N})}{\varepsilon N^{1/2}}\right) + O\Big(\delta(\varepsilon\sqrt{N}) \sum_{d>\varepsilon\sqrt{N}} \frac{1}{d^2}\Big) \\ &= O\left(\frac{\delta(\varepsilon\sqrt{N})}{\varepsilon\sqrt{N}}\right). \qquad (6)\end{aligned}$$

Similarly,

$$\begin{aligned}&\sum_{d>\varepsilon\sqrt{N}} \frac{\mu(d)\ln d}{d^2} \\ &\quad= O\left(\frac{M(\varepsilon\sqrt{N})\ln\varepsilon\sqrt{N}}{(\varepsilon\sqrt{N})^2}\right) + O\Big(\sum_{d\geq[\varepsilon\sqrt{N}]+1} \frac{|M(d)|\ln d}{d^2}\Big) \\ &\quad= O\left(\frac{\delta(\varepsilon\sqrt{N})\ln\varepsilon\sqrt{N}}{\varepsilon\sqrt{N}}\right) + O\Big(\sum_{d\geq\varepsilon\sqrt{N}} \frac{\delta(d)\ln d}{d^2}\Big).\end{aligned}$$

Since for any positive k the function $\delta(x)\ln^k x$ is decreasing for sufficiently large x, it follows that

$$\begin{aligned}\sum_{d>\varepsilon\sqrt{N}} \frac{\mu(d)\ln d}{d^2} &= O\left(\frac{\delta(\varepsilon\sqrt{N})\ln\varepsilon\sqrt{N}}{\varepsilon\sqrt{N}}\right) \\ &\quad + O\Big(\delta(\varepsilon\sqrt{N})\ln\varepsilon\sqrt{N} \sum_{d>\varepsilon\sqrt{N}} \frac{1}{d^2}\Big) \\ &= O\left(\frac{\delta(\varepsilon\sqrt{N})\ln\varepsilon\sqrt{N}}{\varepsilon\sqrt{N}}\right). \qquad (7)\end{aligned}$$

Analogously,

$$\sum_{d>\varepsilon\sqrt{N}} \frac{\mu(d)\ln^2 d}{d^2} = O\left(\frac{\delta(\varepsilon\sqrt{N})\ln^2\varepsilon\sqrt{N}}{\varepsilon\sqrt{N}}\right). \qquad (8)$$

Let $P(x) = \overline{\lambda}_2 x^2 + \overline{\lambda}_1 x + \overline{\lambda}_0$. Then we obtain

$$\begin{aligned}
&\sum_{d\le\varepsilon\sqrt{N}} \mu(d) \sum_{n\le N/d^2} \tau_3(n) \\
&\quad = NP(\ln N) + O\left(N^{1/2}\ln^2 N\frac{\delta(\varepsilon\sqrt{N})}{\varepsilon}\right. \\
&\qquad\qquad \left. + N^{1/2}\ln N\frac{\delta(\varepsilon\sqrt{N})}{\varepsilon}\ln\varepsilon\sqrt{N} + N^{1/2}\frac{\delta(\varepsilon\sqrt{N})\ln^2\varepsilon\sqrt{N}}{\varepsilon}\right) \\
&\qquad + O(\varepsilon^{1-2\theta'}N^{1/2}) \\
&\quad = NP(\ln N) + O\left(N^{1/2}\ln^2 N\frac{\delta(\varepsilon\sqrt{N})}{\varepsilon}\right) + O(N^{1/2}\varepsilon^{1-2\theta'}).
\end{aligned}$$

Obviously

$$\begin{aligned}
M(\varepsilon\sqrt{N}) \sum_{n\le\varepsilon^{-2}} \tau_3(n) &= O(\varepsilon\sqrt{N}\delta(\varepsilon\sqrt{N})\frac{1}{\varepsilon^2}\ln^2\frac{1}{\varepsilon}) \\
&= O\left(N^{1/2}\ln^2\frac{1}{\varepsilon}\frac{\delta(\varepsilon\sqrt{N})}{\varepsilon}\right).
\end{aligned}$$

Also,

$$\begin{aligned}
\sum_{n\le\varepsilon^{-2}} \tau_3(n) M\left(\sqrt{N/n}\right) &= O\Bigg(\sqrt{N}\sum_{n\le\varepsilon^{-2}}\frac{\tau_3(n)}{n^{1/2}}\delta\left(\sqrt{N/n}\right)\Bigg) \\
&= O\Bigg(N^{1/2}\delta(\varepsilon\sqrt{N})\sum_{n\le\varepsilon^{-2}}\frac{\tau_3(n)}{n^{1/2}}\Bigg)
\end{aligned}$$

and, by Abel's transformation,

$$\begin{aligned}
\sum_{n\le\varepsilon^{-2}}\frac{\tau_3(n)}{n^{1/2}} &= O\left(\int_1^{\varepsilon^{-2}}\frac{\sum_{n\le u}\tau_3(n)}{u^{3/2}}\,du\right) + O\Bigg(\varepsilon\sum_{n\le\varepsilon^{-2}}\tau_3(n)\Bigg) \\
&= O\left(\int_1^{\varepsilon^{-2}}\frac{\ln^2 u}{u^{1/2}}\,du\right) + O\left(\frac{1}{\varepsilon}\ln^2\frac{1}{\varepsilon}\right) = O\left(\frac{1}{\varepsilon}\ln^2\frac{1}{\varepsilon}\right).
\end{aligned}$$

Thus,

$$\begin{aligned}
S(N) = NP(\ln N) &+ O\left(N^{1/2}\ln^2 N\frac{\delta(\varepsilon\sqrt{N})}{\varepsilon}\right) \\
&+ O(N^{1/2}\varepsilon^{1-2\theta'}) + O\left(N^{1/2}\ln^2\frac{1}{\varepsilon}\frac{\delta(\varepsilon\sqrt{N})}{\varepsilon}\right).
\end{aligned}$$

Coarsening this relation somewhat, we obtain

$$S(N) = NP(\ln N) + O\left(\varepsilon^{1-2\theta'} \ln^2 \frac{1}{\varepsilon} \ln^2 N \cdot N^{1/2} \left(1 + \frac{\delta(\varepsilon\sqrt{N})}{\varepsilon^{2(1-\theta')}}\right)\right).$$

Take

$$\varepsilon = \delta\left(\sqrt[4]{N}\right)^{1/2(1-\theta')}.$$

Then

$$1 + \frac{\delta(\varepsilon\sqrt{N})}{\varepsilon^{2(1-\theta')}} = 1 + \frac{\delta(\varepsilon\sqrt{N})}{\delta\left(\sqrt[4]{N}\right)}.$$

Since ε gives a reduction weaker than any power of N, it follows that, for sufficiently large N, $\sqrt[4]{N} \leq \varepsilon\sqrt{N}$, and, since $\delta(x)$ decreases,

$$1 + \frac{\delta(\varepsilon\sqrt{N})}{\varepsilon^{2(1-\theta')}} \leq 2.$$

Taking into account everything said above, we obtain

$$\begin{aligned} S(N) &= NP(\ln N) + O\left(N^{1/2} \exp\left(-c' \frac{(\ln N)^{3/5}}{(\ln\ln N)^{1/5}}\right) \ln^k N\right) \\ &= NP(\ln N) + O\left(N^{1/2} \exp\left(-c'' \frac{(\ln N)^{3/5}}{(\ln\ln N)^{1/5}}\right)\right). \end{aligned}$$

The theorem is proved.

Note that the value of the sum $S(N)$ is the number of lattice points lying on the surface of the cone $z^2 = xy$ in the region $1 \leq z \leq N$, $x, y \geq 1$.

An example is instructive only if it clarifies some general idea. It is therefore appropriate to analyze the above problem concerning the sum

$$S(N) = \sum_{n \leq N} \tau(n^2). \tag{1}$$

In studying this sum we relied on the asymptotic formula for the sum of values of the function $\tau_3(n)$ (the number of representations of n as a product of three natural factors), namely the formula

$$\sum_{n \leq N} \tau_3(n) = N\overline{P}(\ln N) + O(N^{\lambda+\varepsilon}), \tag{9}$$

where $\overline{P}(x)$ is a second-degree polynomial, λ is a constant such that $0 < \lambda < 1$, and $\varepsilon > 0$ can be arbitrarily small.

We have seen that the Dirichlet generating series of $\tau_3(n)$ has the form

$$Z(s) = \sum_{n=1}^{\infty} \frac{\tau_3(n)}{n^s} = \varsigma^3(s). \tag{10}$$

An effective investigative tool is the functional equation method. Let us explain what we mean by this.

Assuming that N is one-half of an odd number, we can write the inversion formula for the coefficients of the Dirichlet generating series (10) in the form

$$\sum_{n\le N} \tau_3(n) = \frac{1}{2\pi i}\int_{1+\delta-i\infty}^{1+\delta+i\infty} \frac{N^s}{s}\varsigma^3(s)\,ds,$$

where $\delta > 0$ is a constant. In the integral appearing in the right-hand side of this last equality we can replace the straight line over which the integral is taken by a contour C consisting of five lines joining the points $1+\delta-i\infty$, $1+\delta-iN^\alpha$, $-\delta-iN^\alpha$, $-\delta+iN^\alpha$, $1+\delta+iN^\alpha$, and $1+\delta+i\infty$, where α is a constant such that $1/2 \le \alpha \le 2/3$. We denote these parts of the contour C by C_1, C_2, C_3, C_4, and C_5, respectively. In replacing the path of integration we must consider the residues arising from the poles of the integrand at $s=1$ and $s=0$. Thus we obtain

$$\begin{aligned}\sum_{n\le N} \tau_3(n) &= N\overline{P}(\ln N) - \frac{1}{8} + \frac{1}{2\pi i}\int_C \frac{N^s}{s}\varsigma^3(s)\,ds \\ &= N\overline{P}(\ln N) + \sum_{j=1}^{5}\frac{1}{2\pi i}\int_{C_j}\frac{N^s}{s}\varsigma^3(s)\,ds + O(1). \qquad (11)\end{aligned}$$

Here we give only the idea behind the investigation of the sum (9) by the functional equation method; for more details see Atkinson [2]. We can show that

$$\int_{C_1}\frac{N^s}{s}\varsigma^3(s)\,ds = O(N^{1+\delta-\alpha}\ln N),$$

and an analogous estimate holds for the integral along C_5. Also, on the basis of estimates for the zeta-function in the critical strip it can be shown that

$$\int_{C_2}\frac{N^s}{s}\varsigma^3(s)\,ds = O(N^{1+\delta-\alpha}\ln^2 N),$$

and an analogous estimate holds for the integral along the straight line C_4. Thus,

$$\begin{aligned}\sum_{n\le N} \tau_3(n) &= N\overline{P}(\ln N) + \frac{1}{2\pi i}\int_{-\delta-iN^\alpha}^{-\delta+iN^\alpha}\frac{N^s}{s}\varsigma^3(s)\,ds \\ &\quad + O(N^{1+\delta-\alpha}\ln^2 N) \\ &= N\overline{P}(\ln N) + \frac{1}{2\pi i}\int_{1+\delta-iN^\alpha}^{1+\delta+iN^\alpha}\frac{N^{1-s}}{1-s}\varsigma^3(1-s)\,ds \\ &\quad + O(N^{1+\delta-\alpha}\ln^2 N). \qquad (12)\end{aligned}$$

The Riemann zeta-function satisfies the functional equation

$$\pi^{-s/2}\Gamma\left(\frac{s}{2}\right)\varsigma(s)=\pi^{-(1-s)/2}\Gamma\left(\frac{1-s}{2}\right)\varsigma(1-s). \tag{13}$$

The functional equation for the Dirichlet generating series of $\tau_3(n)$ can be obtained from (13) by cubing:

$$\pi^{-3s/2}\Gamma^3\left(\frac{s}{2}\right)Z(s)=\pi^{-3(1-s)/2}\Gamma^3\left(\frac{1-s}{2}\right)Z(1-s). \tag{14}$$

The main feature of the argument is the use of the functional equation (14). Transforming (12) by means of the functional equation, we obtain

$$\begin{aligned}\sum_{n\le N}\tau_3(n)&=N\overline{P}(\ln N)\\&\quad+\frac{\pi^{3/2}}{2\pi i}\sum_{n=1}^{\infty}\tau_3(n)\int_{1+\delta-iN^\alpha}^{1+\delta+iN^\alpha}\frac{\Gamma^3(s/2)N^{1-s}}{(\pi^3 n)^s\Gamma^3((1-s)/2)}\frac{ds}{1-s}\\&\quad+O(N^{1+\delta-\alpha}\ln^2 N).\end{aligned} \tag{15}$$

Transformations of (15), which we will not reproduce here, lead to the formula

$$\begin{aligned}\sum_{n\le N}\tau_3(n)&=N\overline{P}(\ln N)+\frac{N^{1/3}}{\pi\sqrt{3}}\sum_{n\le X}\frac{\tau_3(n)}{n^{2/3}}\cos 6\pi(nN)^{1/3}\\&\quad+O(N^{1+\delta-\alpha}\ln^2 N),\end{aligned} \tag{16}$$

where $X=N^{3\alpha-1}/8\pi^3$, $1/2\le\alpha<2/3$, and $\delta>0$.

We take $\alpha=1/2$ and trivially estimate the sum in the right-hand side of (16):

$$\begin{aligned}\sum_{n\le N}\tau_3(n)&=N\overline{P}(\ln N)+O\left(N^{1/3}\sum_{n\le c\sqrt{N}}\frac{1}{n^{2/3-\varepsilon}}\right)+O(N^{1/2+\varepsilon})\\&=N\overline{P}(\ln N)+O(N^{1/2+\varepsilon}).\end{aligned} \tag{9'}$$

Thus the calculations based on the functional equation for the Dirichlet series

$$Z(s)=\sum_{n=1}^{\infty}\frac{\tau_3(n)}{n^s}=\varsigma^3(s) \tag{10}$$

reduce to (9′).

The function (10) is only one of an extensive set of functions for which there are functional equations. In order to avoid "getting lost" in details we must have theorems of a general nature on the summation of coefficients of Dirichlet series that satisfty functional equations of a specific form. Such theorems were proved by Landau [90a] and also by Chandrasekharan and Narasimhan [19]. We state a theorem of Landau.

THEOREM. *Assume that c_ν and l_ν are complex numbers, $\alpha \geq 0$, α_i and γ_i are real numbers, δ_i and β_i are positive numbers, μ and ν are natural numbers, and $\lambda_1 < \lambda_2 < \cdots$. Assume moreover that the following conditions hold:*

a) $c_n = O(n^{\alpha+\varepsilon})$ *for any* $\varepsilon > 0$.

b) *The function* $Z(s) = \sum_{n=1}^{\infty} c_n/n^s$, *defined for* $\sigma > 1+\alpha$, *is meromorphic in the whole plane and has a finite number of poles in the strip* $\sigma_1 \leq \sigma \leq \sigma_2$.

c) *The series* $\sum_{n=1}^{\infty} l_n e^{\lambda_n s}$ *is absolutely convergent for* $\sigma < 0$.

d) *For* $\sigma < 0$ *we have the functional equation*

$$\Gamma(\alpha_1 + \beta_1 s) \cdots \Gamma(\alpha_\mu + \beta_\mu s) Z(s) = \Gamma(\gamma_1 - \delta_1 s) \cdots \Gamma(\gamma_\nu - \delta_\nu s) \sum_{n=1}^{\infty} l_n e^{\lambda_n s}.$$

e) $\beta_1 + \cdots + \beta_\mu = \delta_1 + \cdots + \delta_\nu$.

f) *If* $\eta = \gamma_1 + \cdots + \gamma_\nu - (\alpha_1 + \cdots + \alpha_\mu) + \frac{1}{2}(\mu - \nu)$, *then* $\eta \geq 1/2$ *and* $\eta \geq \alpha + 1/2$.

g) *For a fixed strip* $\sigma_1 \leq \sigma \leq \sigma_2$ *there exists a constant* $\gamma = \gamma(\sigma_1, \sigma_2)$ *such that for* $\sigma_1 \leq \sigma \leq \sigma_2$ *and large* $|t|$ *we have* $Z(s) = O(e^{\gamma|t|})$.

Then for any $\varepsilon > 0$

$$\sum_{n \leq N} c_n = R(N) + O(N^{(\alpha+1)(2\eta-1)/(2\eta+1)+\varepsilon}),$$

where $R(N)$ *is the sum of the residues of the function* $N^s Z(s)/s$ *for all poles of* $Z(s)/s$ *in the strip*

$$(\alpha+1)\frac{2\eta-1}{2\eta+1} < \sigma \leq \alpha + 1.$$

Formula (9′) is a direct consequence of Landau's theorem. We can write (14) in the form

$$\begin{aligned} &\Gamma\left(\frac{s}{2}\right)\Gamma\left(\frac{s}{2}\right)\Gamma\left(\frac{s}{2}\right)\zeta^3(s) \\ &\quad = \Gamma\left(\frac{1-s}{2}\right)\Gamma\left(\frac{1-s}{2}\right)\Gamma\left(\frac{1-s}{2}\right)\sum_{n=1}^{\infty} \frac{\pi^{-3/2}\tau_3(n)}{n} e^{s \ln \pi^3 n}. \end{aligned}$$

Here $\mu = \nu = 3$, $\alpha_1 = \alpha_2 = \alpha_3 = 0$, $\beta_1 = \beta_2 = \beta_3 = 1/2$, and $\gamma_1 = \gamma_2 = \gamma_3 = \delta_1 = \delta_2 = \delta_3 = 1/2$. Since $\tau_3(n) = O(n^\varepsilon)$, it follows that $\alpha = 0$. Also, $\eta = 3/2$. Verification of the conditions on the growth of $Z(s)$ will offer no difficulties to the reader familiar with the theory of the zeta-function. Thus,

$$\sum_{n \leq N} \tau_3(n) = N\overline{P}(\ln N) + O(N^{1/2+\varepsilon}). \tag{9′}$$

If we return to the study of the sum (1), we note that (9′) is insufficient for the proof of the asymptotic formula (2). Here we need (9) with $\lambda < 1/2$.

Fundamental for the study of the sum of values of the function $\tau_3(n)$ is formula (16). In the proof of Landau's theorem there is a step analogous to (16). Estimate (9′) can be obtained in a crude way from (16), by means of an estimate of the modulus of the terms of the series on the right-hand side. The reserve by which we can improve the precision of the estimates comes from calculating the interference of the terms of the series. This is a problem on estimating trigonometric sums, in this case the triple trigonometric sum

$$\sum_{\substack{xyz\le X \\ x\ge 1,\ y\ge 1,\ z\ge 1}}\sum\sum \exp[6\pi i(xyzN)^{1/3}]. \tag{17}$$

To investigate (17) we use the methods of estimating trigonometric sums due to Weyl and van der Corput (an account of these methods can be found in [139]). Furthermore, to improve the estimates we use various other methods. It was in just this way that estimate (9) with $\lambda = \frac{37}{75}$ was obtained (see Atkinson [2]), which we used in deducing the asymptotic formula (2); the strongest result was obtained by Chen: $\lambda = \frac{5}{11}$ (see [19a]).

The method described above can be extended to some cases where there is no functional equation of the desired form for the Dirichlet generating series

$$Z(s) = \sum_{n=1}^{\infty} \frac{f(n)}{n^s}. \tag{18}$$

The idea is to reduce the problem of summing the coefficients of series (18) to the same problem for the coefficients of a Dirichlet series for which there is a functional equation. We did just this when we reduced the estimate of the sum (1) to (9).

A tool for such a reduction is a theorem of Faĭnleĭb [46].

THEOREM. *Suppose $f(n)$ and $g(n)$ are multiplicative functions and $0 < \lambda < 1$. Suppose the function $\tilde{g}(n)$ is defined by*

$$\sum_{d|n} \tilde{g}(d) g\left(\frac{n}{d}\right) = \begin{cases} 1 & \text{if } n = 1, \\ 0 & \text{if } n \ne 1. \end{cases} \tag{19}$$

Assume also that $g(n)$ is such that

$$\sum_{r=1}^{\infty} \frac{|\tilde{g}(p^r)|}{p^{r\lambda}} < \infty \tag{20}$$

for all primes p, and that as $N \to \infty$

$$\sum_{n\le N} g(n) = NQ(\ln N) + O(N^{\lambda}), \tag{21}$$

where $Q(u)$ is a polynomial of degree k. Then, if

$$\sum_p \sum_r \frac{1}{p^{r\lambda}} |f(p^r) - g(p^r)| < \infty, \tag{22}$$

it follows that

$$\sum_{n \leq N} f(n) = NP(\ln N) + O(N^{\lambda+\varepsilon}), \tag{23}$$

where $P(u)$ is a polynomial of degree k.

Note that condition (19) can be written in the form

$$\sum_{n=1}^{\infty} \frac{g(n)}{n^s} \sum_{m=1}^{\infty} \frac{\tilde{g}(m)}{m^s} = 1. \tag{19'}$$

We will apply Faĭnleĭb's theorem to the problem of summing the values of the function $\tau(n^2)$. Put $g(n) = \tau_3(n)$. Then it follows from (19′) that $\tilde{g}(n) = \mu(n)\tau_3(n)$.

We have mentioned the fact that condition (21) holds with $\lambda = 1/2 + \varepsilon$. Obviously,

$$\sum_p \frac{|\tilde{g}(p^r)|}{p^{r(1/2+\varepsilon)}} = 1 + \frac{3}{p^{1/2+\varepsilon}} < \infty.$$

Put $f(n) = \tau(n^2)$. Since $\tau(p^2) = \tau_3(p)$ and

$$\tau(p^{2r}) = O(p^{r\varepsilon/2}), \qquad \tau_3(p^r) = O(p^{r\varepsilon/2}),$$

it follows that

$$\sum_p \sum_r \frac{1}{p^{r(1/2+\varepsilon)}} |\tau(p^{2r}) - \tau_3(p^r)| = O\left(\sum_p \sum_{r=2}^{\infty} p^{-r(1/2+\varepsilon/2)}\right)$$
$$= O\left(\sum_p \frac{1}{p^{1+\varepsilon}}\right) = O(1).$$

Equality (23) yields

$$\sum_{n \leq N} \tau(n^2) = N\overline{P}(\ln N) + O(N^{1/2+\varepsilon}).$$

We see that Faĭnleĭb's theorem is insufficient to give the asymptotic formula (2). The problem, as D. Ismoilov has kindly pointed out, is that the "nearness" of the two functions $g(n)$ and $f(n)$ expressed by (22) is insufficient. In order to obtain (2) from general considerations we must strengthen (22), namely connect $g(n)$ and $f(n)$ by the relation

$$f(n) = \sum_{d^2|n} \mu(d) g\left(\frac{n}{d^2}\right). \tag{24}$$

We have the following result.

THEOREM. *Suppose the functions $f(n)$ and $g(n)$ are connected by condition (24) and $g(n) \geq 0$. Suppose that as $N \to \infty$*

$$\sum_{n \leq N} g(n) = NQ(\ln N) + O(N^{\lambda}),$$

where $Q(x)$ is a kth degree polynomial and λ is a constant such that $0 < \lambda < 1/2$. Then

$$\sum_{n \leq N} f(n) = NP(\ln N) + O(N^{1/2} \exp[-c(\ln N)^{3/5}(\ln \ln N)^{-1/5}]),$$

where $P(x)$ is a kth degree polynomial and $c > 0$ is a constant depending on λ.

The asymptotic formula (2) is a direct consequence of this result.

We have had to be very brief. A systematic exposition of the various questions touched upon here would require a separate monograph.

Bibliography

1. P. S. Aleksandrov, *Introduction to the general theory of sets and functions*, GITTL, Moscow, 1948; German transl., VEB Deutscher Verlag Wiss., Berlin, 1956; 2nd German ed., 1964.

2. F. V. Atkinson, *A divisor problem*, Quart. J. Math. Oxford Ser. **12** (1941), 193–200.

3. F. V. Atkinson and Lord Cherwell, *The mean-values of arithmetical functions*, Quart. J. Math. Oxford Ser. **20** (1949), 65–79.

4. Raymond Ayoub, *An introduction to the analytic theory of numbers*, Amer. Math. Soc., Providence, R. I., 1963.

5. G. Babaev, *Distribution of lattice points on algebraic surfaces*, Tadzhik. Gos. Univ., Dushanbe, 1966. (Russian)

6. Paul Bachmann, *Die analytische Zahlentheorie*, Teubner, Leipzig, 1894.

7. A. Bakshtis [Bakštys], *On limit distribution laws for multiplicative arithmetical functions*. I, Litovsk. Mat. Sb. **8** (1968), 5–20. (Russian)

8. ——, *On limit distribution laws for multiplicative arithmetical functions*. II, Litovsk. Mat. Sb. **8** (1968), 201–219. (Russian)

9. ——, *On limit distribution laws for multiplicative arithmetical functions*. III, Litovsk. Mat. Sb. **8** (1968), 643–680. (Russian)

10. M. B. Barban, *On a theorem of J. P. Kubilius*, Izv. Akad. Nauk UzSSR Ser. Fiz.-Mat. Nauk **1961**, no. 5, 3–9. (Russian)

11. ——, *The "large sieve" method and its application to number theory*, Uspekhi Mat. Nauk **21** (1966), no. 1(127), 51–102; English transl. in Russian Math. Surveys **21** (1966).

12. Arne Beurling, *Analyse de la loi asymptotique de la distribution des nombres premiers généralisés*. I, Acta Math. **68** (1937), 255–291.

13. Z. I. Borevich and I. R. Shafarevich, *Number theory*, "Nauka", Moscow, 1964; English transl., Academic Press, 1966.

14. B. M. Bredikhin, *Free numerical semigroups with power densities*, Mat. Sb. **46(88)** (1958), 143–158. (Russian)

15. ——, *The remainder term in the asymptotic formula for* $\nu_G(x)$, Izv. Vyssh. Uchebn. Zaved. Mat. **1960**, no. 6(19), 40–49. (Russian)

16. A. A. Bukhshtab, *Number theory*, 2nd ed., "Prosveshchenie", Moscow, 1966. (Russian)

17. J. W. S. Cassels, *An introduction to Diophantine approximation*, Cambridge Univ. Press, 1957.

18. G. Castelnuovo, *Sur quelques problèmes se rattachant au calcul des probabilités*, Ann. Inst. H. Poincaré **3** (1932/33), 465–490.

19. K. Chandrasekharan and Raghavan Narasimhan, *Functional equations with multiple gamma factors and the average order of arithmetical functions*, Ann. of Math. (2) **76** (1962), 93–136.

19a. Chen Sing-run [Ch'en Ching-jun], *On the divisor problem for* $d_3(n)$; Acta Math. Sinica **14** (1964), 549–558; English transls., Chinese Math. Acta **5** (1964), 591–601, and Sci. Sinica **14** (1965), 19–29.

20. N. G. Chudakov, *Introduction to the theory of Dirichlet's L-functions*, OGIZ, Moscow, 1947. (Russian)

21. K. A. Corrádi and I. Kátai, *On the theory of multiplicative functions*, Ann. Univ. Sci. Budapest. Eötvös Sect. Math. **9** (1966), 147–155.

22. Harald Cramér, *Random variables and probability distributions*, Cambridge Univ. Press, 1937.

23. Harold Davenport, *Über numeri abundantes*, S.-B. Preuss. Akad. Wiss. Phys.-Math. Kl. **1933**, 830–837.

24. ——, *On some infinite series involving arithmetical functions*. II, Quart. J. Math. Oxford Ser. **8** (1937), 313–320.

25. ——, *The higher arithmetic*, Hutchinson's Univ. Library, London, and Longmans, Green, New York, 1952.

26. Hubert Delange, *Quelques théorèmes taubériens relatifs à l'intégrale de Laplace et leurs applications arithmétiques*, Rend. Sem. Mat. Univ. e Politec. Torino **14** (1954/55), 87–103.

27. ——, *Un théorème sur les fonctions arithmétiques multiplicatives et ses applications*, Ann. Sci. École Norm. Sup. (3) **78** (1961), 1–29.

28. ——, *Sur les fonctions arithmétiques multiplicatives*, Ann. Sci. École Norm. Sup. (3) **78** (1961), 273–304.

29. ——, *Application de la méthode du crible à l'étude des valeurs moyennes de certaines fonctions arithmétiques*, Sém. Delange-Pilot, 3e année 1961/62, Théorie des Nombres, Secrétariat Math., Paris, 1963, exposé 16.

30. ——, *Sur le nombre des diviseurs premiers de* n, Acta Arithmetica **7** (1961/62), 191–215.

31. ——, *On a class of multiplicative arithmetical functions*, Scripta Math. **26** (1963), 121–141.

32. ____, *Sur un théorème de Rényi*, Acta Arith. **11** (1965), 241–252.

33. A. A. Drozdova and G. A. Freĭman, *Estimates of certain arithmetical functions*, Elabuzh. Gos. Ped. Inst. Uchen. Zap. **3** (1958), 160–165. (Russian)

34. Paul Erdős, *On the density of some sequences of numbers*. III, J. London Math. Soc. **13** (1938), 119–127.

35. ____, *On the smoothness of the asymptotic distribution of additive arithmetical functions*, Amer. J. Math. **61** (1939), 722–725.

36. ____, *On an elementary proof of some asymptotic formulas in the theory of partitions*, Ann. of Math. (2) **43** (1942), 437–450.

37. ____, *Some remarks about additive and multiplicative functions*, Bull. Amer. Math. Soc. **52** (1946), 527–537.

38. ____, Remarks and corrections to [37], Bull. Amer. Math. Soc. **53** (1947), 761–763.

39. P. Erdős and M. Kac, *The Gaussian law of errors in the theory of additive number theoretic functions*, Amer. J. Math. **62** (1940), 738–742.

40. P. Erdős and A. Schinzel, *Distributions of the values of some arithmetical functions*, Acta Arith. **6** (1960/61), 473–485.

41. Paul Erdős and Aurel Wintner, *Additive arithmetical functions and statistical independence*, Amer. J. Math. **61** (1939), 713–721.

42. ____, *Additive functions and almost periodicity* (B^2), Amer. J. Math. **62** (1940), 635–645.

43. Carl-Gustav Esseen, *Fourier analysis of distribution functions. A mathematical study of the Laplace-Gaussian law*, Acta Math. **77** (1945), 1–125.

44. A. S. Faĭnleĭb, *On the distribution of values of Euler's function*, Mat. Zametki **1** (1967), 645–652; English transl. in Math. Notes **1** (1967).

45. ____, *A generalization of Esseen's inequality and its application in probabilistic number theory*, Izv. Akad. Nauk SSSR Ser. Mat. **32** (1968), 859–879; English transl. in Math. USSR Izv. **2** (1968).

46. ____, *Some asymptotic formulas for sums of multiplicative functions and their applications*, Litovsk. Mat. Sb. **7** (1967), 535–546. (Russian)

47. P. Fatou, *Séries trigonométriques et séries de Taylor*, Acta Math. **30** (1906), 335–400.

48. G. M. Fikhtengol′ts, *A course in differential and integral calculus*. Vol. I, 2nd ed., ONTI, Moscow, 1948; German transls. of later eds., VEB Deutscher Verlag Wiss., Berlin, 1968, 1972.

49. ____, *A course in differential and integral calculus*. Vol. II, 2nd ed., ONTI, Moscow, 1948; German transls. of later eds., VEB Deutscher Verlag Wiss., Berlin, 1966, 1972.

50. ———, *A course in differential and integral calculus*. Vol. III, 1st ed., ONTI, Moscow, 1949; German transls. of later eds., VEB Deutscher Verlag Wiss., Berlin, 1967, 1972.

51. G. A. Freĭman, *An elementary method for solving problems on the partitioning of numbers into an unbounded number of summands*, Trudy Moskov. Mat. Obshch. **4** (1955), 113–124. (Russian)

52. ———, *Inverse problems of additive number theory*, Kazan. Gos. Univ. Uchen. Zap. **115** (1955), no. 14, 109–115. (Russian)

53. ———, *Inverse problems of additive number theory*, Izv. Akad. Nauk SSSR Ser. Mat. **19** (1955), 275–284. (Russian)

54. ———, *An elementary method for proving limit theorems in probability theory*, Vestnik Leningrad. Univ. **1956**, no. 1 (Ser. Mat. Mekh. Astr. vyp. 1), 57–73.

55. ———, *Dense sequences in the theory of partitions*, Elabuzh. Gos. Ped. Inst. Uchen. Zap. **3** (1958), 120–137. (Russian)

56. ———, *Waring's problem with an increasing number of terms*, Elabuzh. Gos. Ped. Inst. Uchen. Zap. **3** (1958), 105–119. (Russian)

57. ———, *Foundations of a structural theory of set addition*, Kazan. Gos. Ped. Inst. and Elabuzh. Gos. Ped. Inst., Kazan, 1966; English transl., Amer. Math. Soc., Providence, R. I., 1973.

58. Géza Freud, *Restglied eines Tauberschen Satzes*. I, Acta Math. Acad. Sci. Hungar. **2** (1951), 299–308.

59. A. O. Gel′fond, *Transcendental and algebraic numbers*, GITTL, Moscow, 1952; English transl., Dover, New York, 1960.

60. A. O. Gel′fond and Yu. V. Linnik, *Elementary methods in analytic number theory*, Fizmatgiz, Moscow, 1962; English transls., Rand McNally, Chicago, Ill., 1965; Pergamon Press, 1966.

61. B. V. Gnedenko, *Course in the theory of probability*, 5th ed., "Nauka", Moscow, 1969; English transl. of 4th ed., Chelsea, New York, 1967.

62. B. V. Gnedenko and A. N. Kolmogorov, *Limit distributions for sums of independent random variables*, GITTL, Moscow, 1949; English transl., Addison-Wesley, 1954; rev. ed., 1968.

63. G. Kalász, *Über die Mittelwerte multiplikativer zahlentheoretischer Funktionen*, Acta Math. Acad. Sci. Hungar. **19** (1968), 365–403.

64. Paul R. Halmos, *Measure theory*, Van Nostrand, Princeton, N. J., 1950.

65. G. H. Hardy, *Ramanujan: twelve lectures on subjects suggested by his life and work*, Cambridge Univ. Press, 1940.

66. ———, *Divergent series*, Clarendon Press, Oxford, 1949.

67. G. H. Hardy and J. E. Littlewood, *Tauberian theorems concerning power series and Dirichlet's series whose coefficients are positive*, Proc. London Math. Soc. (2) **13** (1913/14), 174–191.

68. G. H. Hardy and S. Ramanujan, *The normal number of prime factors of a number n*, Quart. J. Pure Appl. Math. **48** (1917), 76–92.

69. G. H. Hardy, J. E. Littlewood, and G. Pólya, *Inequalities*, Cambridge Univ. Press, 1934.

70. C. B. Haselgrove and H. N. V. Temperley, *Asymptotic formulae in the theory of partitions*, Proc. Cambridge Philos. Soc. **50** (1954), 225–241.

71. Edwin Hewitt and Kenneth A. Ross, *Abstract harmonic analysis*. Vol. I, Springer-Verlag, 1963.

72. I. A. Ibragimov and Yu. V. Linnik, *Independent and stationary sequences of random variables*, "Nauka", Moscow, 1965; English transl., Noordhoff, 1971.

73. I. Il′yasov, *Summation of composite functions of the Euler function*, Dokl. Akad. Nauk SSSR **178** (1968), 529–532; English transl. in Soviet Math. Dokl. **9** (1968).

74. A. E. Ingham, *The distribution of prime numbers*, Cambridge Univ. Press, 1932; reprint Stechert-Hafner, New York, 1964.

75. ——, *A Tauberian theorem for partitions*, Ann. of Math. (2) **42** (1941), 1075–1090.

76. Børge Jessen and Aurel Wintner, *Distribution functions and the Riemann zeta function*, Trans. Amer. Math. Soc. **38** (1935), 48–88.

77. Mark Kac, *Statistical independence in probability theory, analysis and number theory*, Carus Math. Monographs, no. 12, Wiley, 1959; Russian transl., IL, Moscow, 1963.

78. ——, *Probability and related topics in physical sciences*, Lectures in Appl. Math., vol. 1A, Interscience, New York, 1960; reprint, Amer. Math. Soc., Providence, R. I., 1976.

79. E. R. van Kampen, *Infinite product measures and infinite convolutions*, Amer. J. Math. **62** (1940), 417–448.

80. Imre Kátai, Remark on [32], Magyar Tud. Akad. Mat. Fiz. Oszt. Közl. **16** (1966), 269–273. (Hungarian; English summary)

81. ——, *On the distribution of arithmetical functions*, Acta Math. Acad. Sci. Hungar. **20** (1969), 69–87.

82. A. Ya. Khinchin [Khintchine], *Mathematical foundations of quantum statistics*, GITTL, Moscow, 1951; English transl., Graylock Press, Albany, N. Y., 1960.

83. Jacob Korevaar, *An estimate of the error in Tauberian theorems for power series*, Duke Math. J. **18** (1951), 723–734.

84. ——, *A very general form of Littlewood's theorem*, Nederl. Akad. Wetensch. Proc. Ser. A **57**=Indag. Math. **16** (1954), 36–45.

85. ——, *Another numerical Tauberian theorem for power series*, Nederl. Akad. Wetensch. Proc. Ser. A **57**=Indag. Math. **16** (1954), 46–56.

86. N. M. Korobov, *Estimates of trigonometric sums and their applications*, Uspekhi Mat. Nauk **13** (1958), no. 4(82), 185–192. (Russian)

87. I. P. Kubilyus [J. Kubilius], *Probabilistic methods in the theory of numbers*, rev. ed., Gos. Izdat. Politich. i Nauchn. Lit. Litovsk. SSR, Vilnius, 1962; English transl., Amer. Math. Soc., Providence, R. I., 1964.

88. ——, *On asymptotic distribution laws for additive arithmetical functions*, Litovsk. Mat. Sb. **5** (1965), 261–273. (Russian)

89. R. O. Kuzmin [Kuz′min], *Sur la distribution des valeurs de quelques fonctions arithmétiques*, C. R. (Dokl.) Acad. Sci. URSS **15** (1937), 117–118.

90. Edmund Landau, *Über die Einteilung der positiven ganzen Zahlen in vier Klassen nach der Mindestzahl der zu ihrer additiven Zusammensetzung erforderlichen Quadrate*, Arch. Math. und Phys. (3) **13** (1908), 305–312.

90a. ——, *Über die Anzahl der Gitterpunkte in gewissen Bereichen*, Nachr. Ges. Wiss. Gottingen Math.-Phys. Kl. **1912**, 687–771.

91. ——, *Handbuch der Lehre von der Verteilung der Primzahlen.* Vol. II, Teubner, Leipzig, 1909; 2nd ed., Chelsea, New York, 1953.

92. ——, *Einführung in die elementare und analytische Theorie der algebraischen Zahlen und der Ideale*, 2nd ed., Teubner, Leipzig, 1927; reprint, Chelsea, New York, 1949.

93. Serge Lang, *Algebraic numbers*, Addison-Wesley, 1964.

94. B. V. Levin and A. S. Faĭnleĭb, *Application of some integral equations to problems of number theory*, Uspekhi Mat. Nauk **22** (1967), no. 3(135), 119–198; English transl. in Russian Math. Surveys **22** (1967).

95. ——, *A summation method for multiplicative functions*, Izv. Akad. Nauk SSSR Ser. Mat. **31** (1967), 697–710; English transl. in Math. USSR Izv. **1** (1967).

96. ——, *Integral limit theorems for certain classes of additive arithmetical functions*, Trudy Moskov. Mat. Obshch. **18** (1968), 19–54; English transl. in Trans. Moscow Math. Soc. **18** (1968).

97. B. M. Levitan, *Almost-periodic functions*, GITTL, Moscow, 1953. (Russian)

98. Paul Lévy, *Sur les séries dont les termes sont des variables éventuelles indépendantes*, Studia Math. **3** (1931), 119–155.

99. Indar S. Luthar, *A generalization of a theorem of Landau*, Acta Arith. **12** (1966/67), 223–228.

100. F. Mertens, *Über einige asymptotische Gesetze der Zahlentheorie*, J. Reine Angew. Math. **77** (1873), 289–338.

101. A. Mitalauskas and V. Statulyavichyus [Statulevičius], *Local limit theorems and asymptotic expansions for sums of independent lattice random variables*, Litovsk. Mat. Sb. **6** (1966), 569–583. (Russian)

102. D. A. Moskvin, L. P. Postnikova, and A. A. Yudin, *An arithmetical method of obtaining local limit theorems for lattice random variables*, Teor. Veroyatnost. i Primenen. **15** (1970), 86–96; English transl. in Theor. Probab. Appl. **15** (1970).

103. D. S. Mustafin, *On the classification of arithmetical functions*, Izv. Akad. Nauk SSSR Ser. Mat. **29** (1965), 877–886. (Russian)

104. S. V. Nagaev, *Some limit theorems for large deviations*, Teor. Veroyatnost. i Primenen. **10** (1965), 231–254; English transl. in Theor. Probab. Appl. **10** (1965).

105. I. P. Natanson, *Constructive function theory*, GITTL, Moscow, 1949; English transl., Vols. I, II, III, Ungar, New York, 1964, 1965.

106. J. von Neumann, *Zur Prüferschen Theorie der idealen Zahlen*, Acta Litt. ac Sci. Univ. Szeged. **2** (1926), 193–227; reprinted in his *Collected works*. Vol. I, Pergamon Press, 1961.

107. E. V. Novoselov, *Topological theory of divisibility of integers*, Elabuzh. Gos. Ped. Inst. Uchen. Zap. **8** (1960), 3–23. (Russian)

108. ——, *Integration on a bicompact ring and its applications to number theory*, Izv. Vyssh. Uchebn. Zaved. Mat. **1961**, no. 3(22), 66–79. (Russian)

109. ——, *Some formulas connected with a reduced residue system*, Dokl. Akad. Nauk SSSR **143** (1962), 1274–1277; English transl. in Soviet Math. Dokl. **3** (1962).

110. ——, *Foundations of classical analysis and of the theory of analytic functions in a polyadic domain*, Izv. Vyssh. Uchebn. Zaved. Mat. **1963**, no. 5(36), 71–88. (Russian)

111. ——, *A new method in probabilistic number theory*, Izv. Akad. Nauk SSSR Ser. Mat. **28** (1964), 307–364; English transl. in Amer. Math. Soc. Transl. (2) **52** (1966).

112. Bertil Nyman, *A general prime number theorem*, Acta Math. **81** (1949), 299–307.

113. S. S. Pillai and S. D. Chowla, *On the error terms in some asymptotic formulae in the theory of numbers*. I, J. London Math. Soc. **5** (1930), 95–101.

114. L. S. Pontryagin, *Continuous groups*, 2nd ed., GITTL, Moscow, 1954; English transl., *Topological groups*, Gordon and Breach, New York, 1966.

115. A. G. Postnikov, *A Tauberian theorem for Dirichlet series*, Dokl. Akad. Nauk SSSR **92** (1953), 487–490. (Russian)

116. ———, *On an application of the central limit theorem of probability theory*, Uspekhi Mat. Nauk **10** (1955), no. 1(63), 147–149. (Russian)

117. ———, *Additive problems with an increasing number of summands*, Izv. Akad. Nauk SSSR Ser. Mat. **20** (1956), 751–764. (Russian)

118. Karl Prachar, *Primzahlverteilung*, Springer-Verlag, 1957.

119. Heinz Prüfer, *Neue Begrundung der algebraischen Zahlentheorie*, Math. Ann. **94** (1925), 198–243.

120. D. A. Raĭkov, *Generalization of the Ikehara-Landau theorem*, Mat. Sb. **3**(**45**) (1938), 559–568. (Russian; English summary)

121. Srinivasa Ramanujan, *Collected papers*, Cambridge Univ. Press, 1927.

122. R. A. Rankin, *Representations of a number as the sum of a large number of squares*, Proc. Roy. Soc. Edinburgh Sect. A **65** (1960/61), 318–331.

123. Alfréd Rényi, *On the density of certain sequences of integers*, Acad. Serbe Sci. Publ. Inst. Math. **8** (1955), 157–162.

124. ———, *A new proof of a theorem of Delange*, Publ. Math. Debrecen **12** (1965), 323–329.

125. A. Rényi and P. Turán, *On a theorem of Erdős-Kac*, Acta Arith. **4** (1958), 71–84.

126. Marcel Riesz, *Ein Konvergenzsatz für Dirichletsche Reihen*, Acta Math. **40** (1915/16), 349–361.

127. N. P. Romanov, *Concerning the distribution of prime numbers*, Mat. Sb. **23**(**65**) (1948), 259–278. (Russian)

128. K. F. Roth and G. Szekeres, *Some asymptotic formulae in the theory of partitions*, Quart. J. Math. Oxford Ser. (2) **5** (1954), 241–259.

129. A. E. Saltykov, *On Euler's function*, Vestnik Moskov. Univ. Ser. I Mat. Mekh. **1960**, no. 6, 34–50. (Russian)

130. A. Schinzel, *On functions $\varphi(n)$ and $\sigma(n)$*, Bull. Acad. Polon. Sci. Cl. III **3** (1955), 415–419.

131. Theodor Schneider, *Über die Approximation algebraischen Zahlen*, J. Reine Angew. Math. **175** (1936), 182–192.

132. L. G. Shnirel′man [Schnirelmann], *Prime numbers*, ONTI, Moscow, 1940. (Russian)

133. Isac Schoenberg, *Über die asymptotische Verteilung reallen Zahlen* mod 1, Math. Z. **28** (1928), 171–199.

134. Wolfgang Schwarz, *Über die Summe $\sum_{n\le x}\varphi(f(n))$ und verwandte Probleme*, Monatsh. Math. **66** (1962), 43–54.

135. S. Kh. Sirazhdinov and T. A. Azlarov, *On a uniform local theorem*, Izv. Akad. Nauk UzSSR Ser. Fiz.-Mat. Nauk **1963**, no. 2, 32–37. (Russian)

136. V. G. Sprindzhuk, *Mahler's problem in metric number theory*, "Nauka i Tekhnika", Minsk, 1967; English transl., Amer. Math. Soc., Providence, R. I., 1969.

137. M. I. Stronina, *Lattice points on circular cones*, Izv. Vyssh. Uchebn. Zaved. Mat. **1969**, no. 8(87), 112–116. (Russian)

138. V. Kh. Tashbaev, *An inverse additive problem*, Mat. Sb. **52**(**94**) (1960), 947–952. (Russian)

139. E. C. Titchmarsh, *The theory of the Riemann zeta-function*, 2nd ed., Clarendon Press, Oxford, 1951.

140. Paul Turán, *On a theorem of Hardy and Ramanujan*, J. London Math. Soc. **9** (1934), 274–276.

141. M. M. Tyan, *On the question of the distribution of values of the Euler function* $\varphi(n)$, Litovsk. Mat. Sb. **6** (1966), 105–119. (Russian)

142. L. P. Usol′tsev, *An additive problem with an increasing number of summands involving the exponential function*, Izv. Vyssh. Uchebn. Zaved. Mat. **1967**, no. 3(58), 96–104. (Russian)

143. Ya. V. Uspenskiĭ [J. V. Uspensky], *Asymptotic expressions for numerical functions occurring in the problem of partitioning numbers into summands*, Izv. Rossiĭsk. Akad. Nauk (6) **14** (1920), 199–218. (Russian)

144. R. Uzhdavinis [Uždavinys], *An analogue of the Erdős-Wintner theorem for the sequence of values of an integral polynomial*, Litovsk. Mat. Sb. **7** (1967), 329–338. (Russian)

145. B. A. Venkov, *Elementary number theory*, ONTI, Moscow, 1937; English transl, Noordhoff, 1970.

146. ——, *On a monotone function*, Leningrad. Gos. Univ. Uchen. Zap. Ser. Mat. No. 16 (1949), 3–19. (Russian)

147. I. M. Vinogradov, *A new estimate of the function* $\zeta(1+it)$, Izv. Akad. Nauk SSSR Ser. Mat. **22** (1958), 161–164. (Russian)

148. ——, *Foundations of the theory of numbers*, 7th ed., "Nauka", Moscow, 1965; English transl. of 6th ed., Pergamon Press, 1955.

149. ——, *Trigonometrical sums in number theory*, rev. ed., "Nauka", Moscow, 1971; English transl., Statist. Publ. Soc., Calcutta, 1978.*

150. Georges [G. F.] Voronoï, *Sur un problème du calcul des fonctions asymptotiques*, J. Reine Angew. Math. **126** (1903), 241–282.

151. Arnold Walfisz, *Weylsche Exponentialsummen in der neuren Zahlentheorie*, VEB Deutscher Verlag Wiss., Berlin, 1963.

152. S. Wigert, *Sur l'ordre de grandeur du nombre des diviseurs d'un entier*, Ark. Mat. Astr. och Fys. **3** (1906/07), no. 18.

---

* *Editor's note.* There is also a translation of a later edition in his *Selected works*, Springer-Verlag, 1985.

153. Aurel Wintner, *The theory of measure in arithmetical semi-groups*, Waverly Press, Baltimore, Md., 1944.

154. Eduard Wirsing, *Das asymptotische Verhalten von Summen über multiplikative Funktionen*, Math. Ann. **143** (1961), 75–102.

155. ——, *Das asymptotische Verhalten von Summen über multiplikative Funktionen.* II, Acta Math. Acad. Sci. Hungar. **18** (1967), 411–467.

156. V. M. Zolotarev, *On a general theory of multiplication of independent random variables*, Dokl. Akad. Nauk SSSR **142** (1962), 788–791; English transl. in Soviet Math. Dokl. **3** (1962).

157. T. M. Zuparov, *Uniform local theorems for a class of random variables*, Izv. Akad. Nauk UzSSR Ser. Fiz.-Mat. Nauk **12** (1968), no. 5, 13–19. (Russian)

P. D. T. A. Elliott

Appendix

Professor Postnikov's book represents an interesting personal selection of a number of topics from analytic number theory. At the editor's request I append here some further references, including a discussion of more recent advances in areas close to those which he considers.

Chapter 2. §2.4. Linnik [*42*], [*43*]* and A. I. Vinogradov [*55*] applied estimates for trigonometric sums involving powers g^x of an integer g to show, in particular, that every sufficiently large even integer may be expressed as the sum of two primes and a bounded number of powers of 2. For a refined treatment of these and related results, see Gallagher [*31*].

§2.6. This section is modelled on a Selberg version of the first elementary proof of the Prime Number Theorem. For an account of the latter see Elliott [*17*], Chapter 19. For a discussion of appropriate analogues of Tchebyshev's classical bounds for $\pi(x)$, within the format of generalized integers, see Diamond [*8*].

§2.7. Hardy and Ramanujan gave an asymptotic expansion for $p(n)$ in terms of familiar (hyperbolic) functions. By slightly modifying their choice of representing function, Rademacher replaced their asymptotic series by one that converged. An account of this may be found in Rademacher [*46*].

Chapter 3. §3.3. The theory of almost periodic functions in analysis was invented by H. Bohr. An account of this theory, together with various generalisations in analysis, may be found in Besicovitch [*3*]. For discussions of generalisations of the notion of almost periodic functions appropriate to the study of arithmetic functions in number theory, see Schwarz [*50*] and Mauclaire [*44*].

---

*References cited in italic will be found in the bibliography at the end of this appendix.

For $\alpha \geq 1$ we define a *seminorm* $\| \|_\alpha$ on complex-valued arithmetic functions f by

$$\|f\|_\alpha = \limsup_{x\to\infty}\left(x^{-1}\sum_{n\leq x}|f(n)|^\alpha\right)^{1/\alpha}.$$

Those functions for which this seminorm is finite form a linear space L^α over the complex numbers. Factoring out the functions with a seminorm zero then gives a Banach space.

We say that f is *almost periodic* B^α if for each $\varepsilon > 0$ there is a trigonometric polynomial

$$P_\varepsilon(t) = \sum_{j=1}^{k} a_j e^{2\pi i\beta_j t}$$

with real β_j, so that $\|f - P_\varepsilon\| < \varepsilon$. If the polynomial P_ε may be chosen periodic in t, then we say that f is *limit-periodic* B^α.

For any arithmetic function f, the Fourier-Bohr spectrum is the set of real β (mod 1) such that

$$\limsup_{x\to\infty} x^{-1}\left|\sum_{n\leq x} f(n)e^{-2\pi i n\beta}\right| > 0.$$

A complete characterisation of multiplicative functions g which, for a given $\alpha \geq 1$, are almost periodic B^α with a nonempty spectrum was given by Daboussi [*4*]:

It is necessary and sufficient that there be a Dirichlet character χ so that the series

$$\sum p^{-1}(\chi(p)g(p) - 1), \qquad \sum_{|g(p)|\leq 2} \frac{|\chi(p)g(p) - 1|^2}{p},$$

$$\sum_{|g(p)|>2} \frac{|g(p)|^\alpha}{p}, \qquad \sum_{p,k\geq 2} \frac{|g(p^k)|^\alpha}{p^k},$$

taken over the prime numbers, converge. When this condition is satisfied the function g is actually limit-periodic B^α.

For those multiplicative functions g satisfying $|g(n)| \leq 1$, Daboussi showed that the Fourier-Bohr spectrum cannot contain irrationals (mod 1). The restriction on the size of g was later weakened by Daboussi and Delange [*7*] to membership of the class L^α for some $\alpha > 1$.

Various results of this and related types may be found in the survey of Schwarz [*50*]. See also Indlekofer [*35*].

§§3.4 and 3.5. For an application of Novoselov's construction, see Babu [*2*]. On the whole, the approach of Novoselov has not been widely adopted.

The axiom of choice readily guarantees the existence of a finitely additive measure on the natural integers, which assigns to every infinite sequence with an asymptotic density that density as measure. However, it remains a problem to relate such a measure to a specific arithmetic function.

It is natural to employ the limit as $N \to \infty$ of the frequencies

$$\nu_N(n; n \in A) = \frac{1}{N} \sum_{n \le N, n \in A} 1$$

which appear in §3.3. This affords an immediate realisation of the idea of Kac that, suitably interpreted, divisibility of an integer by differing primes represents independent events. A similar philosophy underlies the first theorem of §3.4. Unfortunately, when considering multiplicative functions, for example, this approach is fruitful only if the function becomes rapidly smaller on the larger primes. Replacing the asymptotic frequencies by the use of a particular ν_N allows only the approximate independence of divisibility by primes. The study of arithmetic functions within the aesthetic of the classical theory of probability, with its emphasis on sums and products of independent random variables, therefore involves a careful balance between the convenience of a measure, with respect to which appropriate events are independent, and the loss of generality for the class of functions which may be considered. This dilemma is succinctly summed up by Galambos [*30*], in his recent review of a paper by Nanopoulos [*45*].

For works related to §3.4 see Schwarz and Spilker [*51*] and Mauclaire [*44*], where the introduction of Riesz representations forms an attractive feature. For works which allow a larger class of functions to be considered, but involve a less obvious construction of an appropriate measure, see Kubilius [*37*] (which is an improved version of reference [87] in Postnikov's bibliography) and Elliott [*17*].

Chapter 4. §4.2. It is interesting to note that Erdős and Shapiro [*28*] proved that the frequencies

$$\nu_N\left(n; \sum_{m \le n} \frac{\phi(m)}{m} - \frac{6n}{\pi^2} \le u\right)$$

possess a continuous limiting distribution as $N \to \infty$.

§4.3. This result of Wirsing prompted much research. An important feature is the consideration of $f(n) \log n$, rather than $f(n)$ directly.

§§4.5 and 4.6. For an extensive discussion of the theorems of Delange, Wirsing and Halász, see Elliott [*17*], Chapter 6. The mean-value theorem of Delange has the following generalisation.

Let $a > 1$. A multiplicative function g belongs to the space L^α and has a nonzero mean value

$$\lim_{x\to\infty} x^{-1} \sum_{n\le x} g(n)$$

if and only if the series

$$\sum_{|g(p)|\le 3/2} p^{-1}|g(p)-1|^2, \qquad \sum_{|g(p)|>3/2} p^{-1}|g(p)|^\alpha,$$

$$\sum_{p,k\ge 2}\sum p^{-k}|g(p^k)|^\alpha, \qquad \sum p^{-1}(g(p)-1),$$

taken over the prime numbers, converge, and if for each prime p,

$$\sum_{k=1}^{\infty} p^{-k} g(p^k) \neq -1.$$

When these conditions are satisfied, the limit

$$\lim_{x\to\infty} x^{-1} \sum_{n\le x} |g(n)|^\alpha$$

actually exists.

For $\alpha = 2$ this was first proved by Elliott [*15*]. Alternative proofs of the necessity of the conditions on the primes were given by Daboussi and Delange [*6*], and of their sufficiency by Schwarz and Spilker [*52*]. Generalisations to $\alpha > 1$ were given independently by Elliott [*18*] and Daboussi [*5*], using quite different methods.

Not surprisingly, the consideration of multiplicative functions g, with limiting mean-value zero, causes more difficulty.

Suppose that g belongs to L^α for some $\alpha > 1$. Then in order that

$$\lim_{x\to\infty} x^{-1} \sum_{n\le x} g(n) = 0$$

it is necessary and sufficient that one of the following four conditions be satisfied:

(i) *One of the series*

$$\sum_{||g(p)|-1|\le 1/2} p^{-1}|1-|g(p)||^2, \qquad \sum_{||g(p)|-1|>1/2} p^{-1}|1-|g(p)||^\alpha$$

diverges.

(ii) *The condition* (i) *fails, but for each real value of t the series*

$$\sum p^{-1}(|g(p)| - \operatorname{Re} g(p)p^{it})$$

diverges.

(iii) *The conditions* (i) *and* (ii) *fail, but there is a real t so that the series in condition* (ii) *converges and*

$$\sum_{k=1}^{\infty} g(p^k)p^{-k(1+it)} = -1$$

for some prime p.

(iv) *The conditions* (i), (ii), *and* (iii) *fail, but*

$$\sum_{p\le x} \frac{1-|g(p)|}{p} \to \infty$$

as $x \to \infty$.

This theorem was proved by Elliott [*18*].

For multiplicative functions g which assume only nonnegative real values, the following result of Elliott [*20*] may be compared with that of Wirsing in §4.3.

Suppose that a finite limiting mean-value

$$\lim_{x\to\infty} x^{-1} \sum_{n\le x} g(n)$$

exists. Then for each δ, $0 < \delta < 1$, *so does*

$$\lim_{x\to\infty} x^{-1} \sum_{n\le x} g(n)^{\delta}.$$

Moreover, these new limits are all zero, unless the series $\sum p^{-1}(g(p)^{1/2}-1)^2$, *taken over the prime numbers, converges.*

As shown in that same reference, this result has an interesting application to Ramanujan's modular coefficient function $\tau(n)$.

In his work on multiplicative functions Wirsing used approximate integral equations. For real-valued functions satisfying $|g(n)| \le 1$ he was able to establish the existence of a limiting mean-value, so settling an old conjecture of Erdős and Wintner. This same result can be obtained by the method of Halász (ref. [63] of Postnikov), using Dirichlet series. For a proof of that result along lines quite different from those of Wirsing or Halász, see Hildebrand [*34*].

§4.4. Other proofs of the Turán-Kubilius inequality may be found in Elliott [*12*], [*16*]. In the notation of Postnikov, let λ_N denote the supremum of

$$\frac{1}{ND^2(N)} \sum_{n=1}^{N} |f(n) - A(N)|^2,$$

taken over all complex-valued additive functions not identically zero on the interval $1 \le n \le N$. The Turán-Kubilius inequality guarantees a uniform bound for λ_N.

The asymptotic best value $3/2 + o(1)$ for the constant was obtained by Kubilius (announced in a 1981 Budapest conference on number theory), [*38*], [*39*], and (with a different method) by Hildebrand [*32*]. In fact Kubilius' method led to the sharper estimate $\lambda_N = 3/2 + O((\log N)^{-1/2})$. This he has recently improved [*40*], by replacing the exponent $-1/2$ in the error term with -1.

The Turán-Kubilius inequality may be viewed within the format of functional analysis, so that it has a dual:

$$\sum_{p\le N} p\left|\sum_{\substack{n=1\\ n\cong 0(\operatorname{mod} p)}}^{N} a_n - \frac{1}{p}\sum_{n=1}^{N} a_n\right|^2 + \sum_{\substack{p^k\le N\\ k\ge 2}} p^k\left|\sum_{\substack{n=1\\ n\cong 0(\operatorname{mod} p^k)}}^{N} a_n\right|^2 < C_1\sum_{n=1}^{N}|a_n|^2$$

valid for all complex a_n. Here $n \cong 0 \pmod{p^k}$ denotes that p^k divides n, but p^{k+1} does not. The use of a dual Turán-Kubilius inequality to study arithmetic functions seems to have begun with Elliott [*14*]. Explicitly or implicity such a notion underlies many of the recent developments mentioned in the commentary on Chapter 3 and on the present chapter. An extensive overview of these matters, including the relation of the Turán-Kubilius inequality to a certain Hermitian operator, and various generalisations of it involving differing norms, is given in Elliott [*22*]. See also Elliott [*13*] for a connection with the large sieve.

Generalisations of the Turán-Kubilius inequality on the integers and involving powers other than squares may be found in Elliott [*19*], [*22*], [*23*]. See also Ruzsa [*47*], [*48*] and Hildebrand [*33*]. Besides this, one may give generalisations in which the argument n in the additive function $f(n)$ is replaced by other sequences of arithmetic interest, such as the shifted primes $p+1$, or polynomials in n, as in Elliott [*17*], Chapter 4, and Alladi [*1*].

§**4.7.** Another derivation of Lévy's criterion for an infinite convolution to be continuous may be found in Elliott [*17*], Lemma 1.22. It goes by way of a concentration function estimate, from the theory of probability, due to Kolmogorov and Rogozin.

A direct proof that the divergence of the series $\sum p^{-1}$ $(f(p) \ne 0)$ guarantees the continuity of the limit law for $\nu_N(n; f(n) \le z)$, when that exists, was indicated by Szüsz in [*53*] and a letter to the present author. It employs Wiener's criterion for characteristic functions, and Tchebyshev's inequality in the theory of probability. A detailed account may be found in Elliott [*17*], [*21*]. At present there does not seem to be any truly short proof of this result.

There is a generalisation of this last result, due to Erdős [*26*], Elliott and Ryavec [*24*], as in Elliott [*17*], which shows that suitably interpreted it remains true even if the frequencies $\nu_N(n; f(n) \le z)$ do not possess a limiting distribution as $N \to \infty$.

§**4.8.** Erdős [*27*] proved that there is an absolute constant c so that the inequality

$$\nu_N\left(n; a < \frac{\phi(n)}{n} \le a\left(1 + \frac{1}{t}\right)\right) \le \frac{c}{\log t}$$

holds uniformly for all real $a > 0$, $2 \le t \le N$. This concentration function estimate is better than the one employed in the text by Postnikov, and leads (see Elliott [*17*], Chapter 5) to the sharper estimate

$$\Phi_N(x) = \Phi(x) + O\left(\frac{\log\log N}{\log N \log\log\log N}\right).$$

It was conjectured by Erdős that this error term should be $O((\log N)^{-1})$.

The elementary but ingenious method of Erdős yields an upper bound of the form $c_1(\log(x/\varepsilon))^{-1}$ for $\Phi(x + \varepsilon x) - \Phi(x)$, provided $0 < \varepsilon < x < 1$. Diamond and Rhoads [*11*] improved this to an estimate for the modulus of continuity of $\Phi(x)$, with a bound $c_2(\log(1/\varepsilon))^{-1}$ to hold uniformly in $x > 0$, $0 < \varepsilon < 1$. For this they employ characteristic functions.

§**4.9.** For other applications to number theory of the theory of products of random variables, see Elliott [*17*], Chapter 7. This includes results of Galambos [*29*] concerning the convergence of frequencies $\nu_N(n; g(n) \le z)$ to a symmetric limit law, and of Levin, Timofeev, and Tuliaganov [*41*] characterizing those multiplicative functions g for which the frequencies $\nu_N(n; g(n) - \alpha(N) \le z\beta(N))$ converge when the renormalising constants $\alpha(N), \beta(N) > 0$ are suitably chosen.

The value distribution of positive-valued arithmetic functions h may be studied in terms of frequencies $\nu_N(n; \log h(n) - \alpha(N) \le z\beta(N))$. For those which grow rapidly there is another perspective. Let $N(y)$ denote the number of integers n for which $h(n) \le y$, and characterize those h for which $\delta = \lim y^{-1}N(y)$, $y \to \infty$, exists and is positive. For example, when h is Euler's totient function, $\delta = \varsigma(2)\varsigma(3)/\varsigma(6)$, in terms of the Riemann zeta function. This is already asserted in Erdős [*25*]. For a sharpening of this result see Diamond [*9*]. A detailed account of a general theory for multiplicative functions h, employing the method of Halász (ref. [63] of Postnikov), is given by Diamond and Erdős [*10*].

§**4.10.** The important pioneering results stated in this section have been improved or generalised by many authors, and form part of an extensive and rapidly growing area known as Probabilistic Number Theory. For an account of such matters up until 1980 see Elliott [*17*]; see also Elliott [*21*] (especially the Supplement), Babu [*2*], and Alladi [*1*]. There are, moreover, many interesting recent developments.

§**4.11.** It is interesting to compare the treatment of the integral equation (7) in Postnikov's account of the method of Levin and Faĭnleĭb, with the proof of Lemma 4 which he gives in §4.3 when establishing Wirsing's theorem on mean values.

§**4.12.** For further results concerning the Riemann zeta function and divisor problems see Ivić [*36*], and the remarks of Heath-Brown in the new edition of the well-known book of Titchmarsh [*54*].

Bibliography to the Appendix

[**1**] K. Alladi, *A study of the moments of additive functions using Laplace transforms and sieve methods*, Number Theory (Proc. Fourth Matscience Conf., Ootacamund, 1984), Lecture Notes in Math., vol. 1122, Springer-Verlag, 1985, pp. 1–37.

[**2**] G. J. Babu, *Probabilistic methods in the theory of arithmetic functions*, Lectures in Math., vol. 2, Macmillan of India, Delhi, 1978.

[**3**] A. S. Besicovitch, *Almost periodic functions*, Cambridge Univ. Press, 1932.

[**4**] H. Daboussi, *Caractérisation des fonctions multiplicatives p. p. B^λ à spectre non vide*, Ann. Inst. Fourier (Grenoble) **30** (1980), fasc. 3, 141–166.

[**5**] ——, *Sur les fonctions multiplicatives ayant une valeur moyenne non nulle*, Bull. Soc. Math. France **109** (1981), 183–205.

[**6**] H. Daboussi and H. Delange, *On a theorem of P. D. T. A. Elliott on multiplicative functions*, J. London Math. Soc. (2) **14** (1976), 345–356.

[**7**] ——, *On multiplicative arithmetical functions whose modulus does not exceed one*, J. London Math. Soc. (2) **26** (1982), 245–264.

[**8**] H. Diamond, *Chebyshev estimates for Beurling generalized prime numbers*, Proc. Amer. Math. Soc. **39** (1973), 503–508.

[**9**] ——, *The distribution of values of Euler's phi function*, Analytic Number Theory, Proc. Sympos. Pure Math., vol. 24, Amer. Math. Soc., Providence, R. I., 1973, pp. 63–75.

[**10**] H. Diamond and P. Erdős, *Multiplicative functions whose values are uniformly distributed in* $(0,\infty)$, Proc. Queen's Number Theory Conf. 1979 (P. Ribenboim, editor), Queens Papers in Pure and Appl. Math., Queen's Univ., Kingston, Ont., 1980, pp. 329–378.

[**11**] H. Diamond and D. Rhoads, *The modulus of continuity of the distribution function of* $\Phi(n)/n$, Topics in Classical Number Theory (Proc. Internat. Conf., Budapest, 1981), Vol. I, Colloq. Math. Soc. János Bolyai, vol. 34, North-Holland, 1984, pp. 335–353.

[**12**] P. D. T. A. Elliott, *The Turán-Kubilius inequality, and a limitation theorem for the large sieve*, Amer. J. Math. **92** (1970), 293–300.

[**13**] ——, *On inequalities of large sieve type*, Acta Arith **18** (1971), 405–422.

[**14**] ——, *On connections between the Turán-Kubilius inequality and the large sieve: some applications*, Analytic Number Theory, Proc. Sympos. Pure Math., vol. 24, Amer. Math. Soc., Providence, R. I., 1973, pp. 77–82.

[**15**] ——, *A mean value theorem for multiplicative functions*, Proc. London Math. Soc. (3) **31** (1975), 418–438.

[**16**] ——, *The Turán-Kubilius inequality*, Proc. Amer. Math. Soc. **65** (1977), 8–10.

[**17**] ——, *Probabilistic number theory.* I: *Mean value theorems.* II: *Central limit theorems*, Springer-Verlag, 1979, 1980.

[**18**] ——, *Mean value theorems for multiplicative functions bounded in mean α-power, $\alpha >$ 1*, J. Austral. Math. Soc. Ser. A **29** (1980), 177–205.

[**19**] ——, *High power analogues of the Turán-Kubilius inequality and an application to number theory*, Canad. J. Math. **32** (1980), 893–907.

[**20**] ——, *Multiplicative functions and Ramanujan's τ-function*, J. Austral. Math. Soc. Ser. A **30** (1981), 461–468.

[**21**] ——, *Arithmetic functions and integer products*, Springer-Verlag, 1985.

[**22**] ——, *Functional analysis and additive arithmetic functions*, Bull. (N.S.) Amer. Math. Soc. **16** (1987), 179–223.

[**23**] ——, *The norms of the composition of arithmetic operators*, Bull. London Math. Soc. **19** (1987) (to appear).

[**24**] P. D. T. A. Elliott and C. Ryavec, *The distribution of the values of additive arithmetical functions*, Acta Math. **126** (1971), 143–164.

[**25**] P. Erdős, *Some remarks on Euler's ϕ function and some related problems*, Bull. Amer. Math. Soc. **51** (1945), 540–544.

[**26**] ——, *On the distribution function of additive functions*, Bull. Amer. Math. Soc. **52** (1946), 527–537.

[**27**] ——, *On the distribution of numbers of the form $\sigma(n)/n$ and some related questions*, Pacific J. Math. **52** (1974), 59–65.

[**28**] P. Erdős and H. N. Shapiro, *The existence of a distribution function for an error term related to the Euler function*, Canad. J. Math. **7** (1955), 63–75.

[**29**] J. Galambos, *On the distribution of strongly multiplicative functions*, Bull. London Math. Soc. **3** (1971), 307–312.

[**30**] ——, Review of [A45], Zbl. Math. **597** (1987), 10053.

[**31**] P. X. Gallagher, *Primes and powers of* 2, Invent. Math. **29** (1975), 125–142.

[**32**] A. Hildebrand, *An asymptotic formula for the variance of an additive function*, Math. Z. **183** (1983), 145–170.

[**33**] ——, *Sur les moments d'une fonction additive*, Ann. Inst. Fourier (Grenoble) **33** (1983), fasc. 3, 1–22.

[**34**] ——, *On Wirsing's mean-value theorem for multiplicative functions*, Bull. London Math. Soc. **18** (1986), 147–152.

[**35**] K.-H. Indlekofer, *Some remarks on almost-even and almost-periodic functions*, Arch. Math. (Basel) **37** (1981), 353–358.

[**36**] A. Ivić, *The theory of the Riemann zeta-function with applications*, Wiley, 1985.

[**37**] J. Kubilius, *Probabilistic methods in the theory of numbers*, Transl. Math. Monographs, vol. 11, Amer. Math. Soc., Providence, Rhode Island, 1964.

[**38**] ——, *On the estimation of the second central moment for strongly additive arithmetic functions*. Litovsk. Mat. Sb. **23** (1983), no. 1, 122–133; English transl. in Lithuanian Math. J. **23** (1983).

[**39**] ——, *On the estimate of the second central moment for arbitrary additive arithmetic functions*, Litovsk. Mat. Sb. **23** (1983), no. 2, 110–117; English transl. in Lithuanian Math. J. **23** (1983).

[**40**] ——, *Sharpening of the estimate of the second central moment for additive arithmetical functions*, Litovsk. Mat. Sb. **25** (1985), no. 3, 104–110; English transl. in Lithuanian Math. J. **25** (1985).

[**41**] B. V. Levin, N. M. Timofeev, and S. T. Tuliaganov, *The distribution of the values of multiplicative functions*, Litovsk. Mat. Sb. **13** (1973), no. 1, 87–100; English transl. in Math. Trans. Acad. Sci. Lithuanian SSR **13** (1973).

[**42**] Yu. V. Linnik, *Prime numbers and powers of* 2, Trudy Mat. Inst. Steklov. **38** (1951), 152–169. (Russian)

[**43**] ——, *Addition of prime numbers and powers of one and the same number*, Mat. Sb. **32(74)** (1953), 3–60. (Russian)

[44] J.-L. Mauclaire, *Intégration et théorie des nombres.* Hermann, Paris, 1986.

[45] P. Nanopoulos, *Weak convergence of probability nets on the integers and the Erdős-Wintner theorem as a three-series theorem*, Bull. Soc. Math. Grèce (N.S.) **25** (1984), 83–94.

[46] H. Rademacher, *Collected papers.* Vol. 2, MIT Press, Cambridge, Mass., 1974, pp. 100–122. See also *On the partition function* $p(n)$, Proc. London Math. Soc. (2) **43** (1937), 241–254.

[47] I. Z. Ruzsa, *On the variance of additive functions*, Studies in Pure Mathematics (To the Memory of Paul Turán), Birkhäuser, 1983, pp. 577–586.

[48] ——, *Generalized moments of additive functions*, J. Number Theory **18** (1984), 27–33.

[49] W. Schwarz, *Fourier-Ramanujan-Entwicklungen zahlentheoretischer Funktionen und Anwendungen*, Festschrift zum 75-jährigen Jubiläum der Wissenschaftlichen Gesellschaft an der J. W. Goethe-Universität Frankfurt am Main, Franz Steiner Verlag GmbH., Wiesbaden, 1981, pp. 394–415.

[50] ——, *Remarks on the theorem of Elliott and Daboussi, and applications*, Elementary and Analytic Theory of Numbers (H. Iwaniec, editor), Banach Center Publ., vol. 17, PWN, Warsaw, 1985, pp. 463–498.

[51] W. Schwarz and J. Spilker, *Eine Anwendung des Approximationssatzes von Weierstrass-Stone auf Ramanujan-Summen*, Nieuw Arch. Wisk. (3) **19** (1971), 198–209.

[52] ——, *Remarks on Elliott's theorem on mean-values of multiplicative functions*, Recent Progress in Analytic Number Theory (Durham, 1979). Vol. 1 (H. Halberstram and C. Hooley, editors), Academic Press, 1981, pp. 325–339. See also the remarks at the end of [A50].

[53] P. Szüsz, *Remark to a theorem of P. Erdős*, Acta Arith. **26** (1974), 97–100.

[54] E. C. Titchmarsh, *The theory of the Riemann zeta-function*, Clarendon Press, Oxford, 1951; new ed. (edited by D. R. Heath-Brown), 1986.

[55] A. I. Vinogradov, *On an "almost binary" problem*, Izv. Akad. Nauk SSSR Ser. Mat. **20** (1956), 713–750. (Russian)

ABCDEFGHIJ — 898